Jonas Bürgler

Bidentate and Tridentate P-Stereogenic Ferrocenyl Phosphines

Jonas Bürgler

Bidentate and Tridentate P-Stereogenic Ferrocenyl Phosphines

Synthesis, Coordination Properties and Applications in Asymmetric Catalysis

Südwestdeutscher Verlag für Hochschulschriften

Impressum/Imprint (nur für Deutschland/only for Germany)
Bibliografische Information der Deutschen Nationalbibliothek: Die Deutsche Nationalbibliothek verzeichnet diese Publikation in der Deutschen Nationalbibliografie; detaillierte bibliografische Daten sind im Internet über http://dnb.d-nb.de abrufbar.
Alle in diesem Buch genannten Marken und Produktnamen unterliegen warenzeichen-, marken- oder patentrechtlichem Schutz bzw. sind Warenzeichen oder eingetragene Warenzeichen der jeweiligen Inhaber. Die Wiedergabe von Marken, Produktnamen, Gebrauchsnamen, Handelsnamen, Warenbezeichnungen u.s.w. in diesem Werk berechtigt auch ohne besondere Kennzeichnung nicht zu der Annahme, dass solche Namen im Sinne der Warenzeichen- und Markenschutzgesetzgebung als frei zu betrachten wären und daher von jedermann benutzt werden dürften.

Verlag: Südwestdeutscher Verlag für Hochschulschriften GmbH & Co. KG
Dudweiler Landstr. 99, 66123 Saarbrücken, Deutschland
Telefon +49 681 37 20 271-1, Telefax +49 681 37 20 271-0
Email: info@svh-verlag.de

Zugl.: Zürich, ETH, Diss., 2011

Herstellung in Deutschland:
Schaltungsdienst Lange o.H.G., Berlin
Books on Demand GmbH, Norderstedt
Reha GmbH, Saarbrücken
Amazon Distribution GmbH, Leipzig
ISBN: 978-3-8381-2830-6

Imprint (only for USA, GB)
Bibliographic information published by the Deutsche Nationalbibliothek: The Deutsche Nationalbibliothek lists this publication in the Deutsche Nationalbibliografie; detailed bibliographic data are available in the Internet at http://dnb.d-nb.de.
Any brand names and product names mentioned in this book are subject to trademark, brand or patent protection and are trademarks or registered trademarks of their respective holders. The use of brand names, product names, common names, trade names, product descriptions etc. even without a particular marking in this works is in no way to be construed to mean that such names may be regarded as unrestricted in respect of trademark and brand protection legislation and could thus be used by anyone.

Publisher: Südwestdeutscher Verlag für Hochschulschriften GmbH & Co. KG
Dudweiler Landstr. 99, 66123 Saarbrücken, Germany
Phone +49 681 37 20 271-1, Fax +49 681 37 20 271-0
Email: info@svh-verlag.de

Printed in the U.S.A.
Printed in the U.K. by (see last page)
ISBN: 978-3-8381-2830-6

Copyright © 2011 by the author and Südwestdeutscher Verlag für Hochschulschriften GmbH & Co. KG and licensors
All rights reserved. Saarbrücken 2011

For my parents Armin and Helen and my sister Simone

"A scientist in his laboratory is not only a technician: he is also a child placed before natural phenomena which impress him like a fairy tale."

Marie Curie

"Men love to wonder, and that is the seed of science."

Ralph Waldo Emerson

Table of Contents

Abstract	ix
Zusammenfassung	xi

1　P-Stereogenic Phosphines　　　　1

   1.1　Introduction　　　　1

   1.2　Some General Aspects of P-Stereogenic Phosphines　　　　2

      1.2.1　Pyramidal Inversion/Inversion Barrier　　　　3

   1.3　Preparation of P-Stereogenic Phosphines　　　　4

      1.3.1　Protection and Deprotection Methods　　　　5

      1.3.2　Resolution of Racemates　　　　6

      1.3.3　Stereoselective Synthesis of P-Stereogenic Phosphines Using Chiral Auxiliaries　　　　8

      1.3.4　Enantioselective Synthesis of P-Stereogenic Phosphines by Asymmetric Catalysis　　　　18

      1.3.5　P-Stereogenic Phosphines with a Ferrocene Backbone　　　　20

   1.4　Applications of P-Stereogenic Phosphines in Asymmetric Catalysis　　　　25

      1.4.1　Asymmetric Hydrogenation Reactions　　　　25

      1.4.2　Palladium-Catalyzed Asymmetric Allylic Alkylation　　　　29

      1.4.3　Miscellanous Examples　　　　30

2　Secondary Phosphines and Secondary Phosphine Oxides　　　　33

   2.1　General Aspects of Planar Chiral Ferrocene Systems　　　　33

      2.1.1　Diastereoselective ortho-Lithiation　　　　33

      2.1.2　Substitution of the Dimethylamino Group with Retention of Configuration　　　　34

      2.1.3　Stereochemistry of Chiral Ferrocenes　　　　34

   2.2　General Aspects of Secondary Phosphine Oxides (SPO)　　　　35

	2.2.1	Introduction	35
	2.2.2	Synthesis of P-Stereogenic Secondary Phosphine Oxides	36
	2.2.3	Application of Secondary Phosphine Oxides in Transition Metal-Catalysis	37
	2.2.4	Secondary Phosphine Oxides with a Ferrocene Backbone	39
2.3		Aim of this Work	41
2.4		Synthesis of Ferrocene-Based Secondary Phosphines	42
2.5		Synthesis of Secondary Phosphine Oxides	46
2.6		Synthesis of Transition Metal-Complexes	52
2.7		Ferrocenyl Secondary Phosphines and SPOs in Asymmetric Catalysis	62
2.8		Conclusion and Outlook	64

3 P-Stereogenic Trifluoromethyl Phosphine Ligands — 65

3.1		Introduction	65
	3.1.1	Nucleophilic Trifluoromethylation of Phosphines	66
	3.1.2	Electrophilic Trifluoromethylation of Phosphines	68
3.2		Intramolecular Substitution of Trifluoromethyl Groups	71
	3.2.1	Stereoselective Synthesis of 1,2-Diphospholes	71
	3.2.2	Ring Opening of 1,2-Diphospholes	75
	3.2.3	Transition-Metal Complexes	78
3.3		Intermolecular Substitution of Trifluoromethyl Groups	80
	3.3.1	Synthesis of Trifluoromethylphosphines via Nucleophilic Substitution	80
	3.3.2	Synthesis of P-Stereogenic Trifluoromethyl Analogues of PPFA	81
	3.3.3	Synthesis of P-Stereogenic Trifluoromethylphosphines of the Josiphos Type	86
	3.3.4	Transition-Metal Complexes of Bidentate Trifluoromethylphosphine Ligands	87
3.4		Some Mechanistic Aspects of the Formation of 1,2-Diphospholes	94
	3.4.1	Cyclization with Bis(trifluoromethyl)phosphines as Substrates	94
	3.4.2	Cyclization with Trifluoromethylphosphines as Substrates	95

		3.4.3 Cyclization in the Presence of Electrophiles	100

    3.4.3   Cyclization in the Presence of Electrophiles   100

3.5   Trifluoromethyl Phosphines in Asymmetric Catalysis   102

    3.5.1   Rhodium-Catalyzed Hydrogenation of Olefins   102

    3.5.2   Pd-Catalyzed Asymmetric Allylic Alkylation   106

3.6   Conclusion and Outlook   107

4   P-Stereogenic Tridentate Ligands of the Pigiphos-Type   109

4.1   Introduction   109

    4.1.1   Ferrocene-based Tridentate Phosphine Ligands   110

    4.1.2   Transition-Metal Complexes of Tridentate Phosphine Ligands   111

    4.1.3   Transition-Metal Asymmetric Catalysis with Pigiphos Ligands   112

4.2   Aim of the Project   116

4.3   Pigiphos Ligands with a P-Stereogenic Phosphine on the Backbone   117

    4.3.1   Ligand Synthesis   117

    4.3.2   Transition-Metal Complexes of P-Stereogenic Pigiphos   121

4.4   Pigiphos Ligands with P-Stereogenic Phosphines on the Cp-Ring   125

    4.4.1   Ligand Synthesis   125

    4.4.2   Nickel(II) Complexes of P-Stereogenic Pigiphos   128

4.5   Conclusion and Outlook   129

5   General Conclusions and Outlook   131

6   Experimental Part   135

6.1   General Remarks   135

    6.1.1   Techniques   135

    6.1.2   Chemicals   135

    6.1.3   Analytical Techniques and Instruments   135

6.2   Secondary Phosphines and SPO   137

6.3   Trifluoromethylphosphines   156

6.4   Transition-Metal Complexes   176

	6.5	Tridentate Phosphine Ligands of the Pigiphos-Type	185
	6.6	Transition-Metal Complexes of Pigiphos Ligands	192
7	**Literature**		**197**
8	**Appendix**		**207**
	8.1	Abbreviations	207
	8.2	Crystallographic Data and Tables	209
	8.3	Fit Results of 113	231
	8.4	Danksagung	235

Abstract

This thesis is concerned with the synthesis of P-stereogenic ferrocenyl phosphines and their application as ligands in asymmetric catalysis. The synthesis of all compounds described herein starts from the commercially available precursor Ugi's amine. The secondary phosphines of type **85** are synthesized in a few steps and in high yields (Chapter 2).

The secondary phosphine **85** was shown to be a highly versatile precursor for the synthesis of P-stereogenic ferrocenyl phosphines. Diphenylphosphine-substituted compound **89** was used for the synthesis of stable bidentate palladium (**112**) and rhodium (**116**) complexes. However, the secondary phosphino group in **89** showed special reactivity under certain conditions. Deprotonation of **112** yielded phosphido-bridged complex **113** and intramolecular allylation was observed when **89** was reacted with the dinuclear allylpalladium(II) chloride to afford the tertiary phosphine complex **115**. The bis(trifluoromethyl)phosphine **110** was used for the selective synthesis of secondary phosphine oxide (SPO) **111** in high yields and good diastereomeric ratio. The P-SPO bidentate ligand **111** was applied for the synthesis of transition-metal complexes and showed different coordination behavior for palladium (**117**, P-coordination) and rhodium (**118**, O-coordination).

When the secondary phosphines **110** and **131** were treated with base, intramolecular substitution of a trifluoromethyl group occurred, affording the 1,2-diphospholes **129** and **132** stereoselectively and in moderate yields (Chapter 3). These diphospholes were then

Abstract

transformed into a variety of bidentate ligands by applying an electrophilic methylation-nucleophilic ring-opening sequence. Thus, the twofold P-stereogenic Josiphos-like diphosphines **133-137** were obtained in moderate yields and as single regio- and stereoisomers.

133: R = Ad, R' = Me
134: R = Ad, R' = Ph
135: R = Ad, R' = i-Pr
136: R = Cy, R' = Ph
137: R = Cy, R' = i-Pr

110: R = Ad
131: R = Cy

129: R = Ad
132: R = Cy

The use of a trifluoromethyl substituent as a leaving group on phosphines was shown to be rather general, since lithiated Ugi's amine reacted with several bis(trifluormethyl) substituted alkyl and aryl phosphines to yield the corresponding ferrocenyl trifluoromethyl phosphines. Separation of the stereoisomers by flash column chromatography and substitution of the amine functionality afforded the P-stereogenic diphosphine ligands **142**, **151** and **152**.

(S_P)-**142**: R = Ph, R' = CF_3
(R_P)-**142**: R = CF_3, R' = Ph
(S_P)-**151**: R = Np, R' = CF_3
(S_P)-**152**: R = Ad, R' = CF_3
(R_P)-**152**: R = CF_3, R' = Ad

The catalytic potential of the trifluoromethyl phosphine ligands was demonstrated in the rhodium-catalyzed hydrogenation of functionalized olefins where, in general, high activities and moderate to excellent selectivities were observed. The best of these special ligands was diphosphine **137**, which provided turnover frequencies up to 6000 per hour and >99% *ee* for the hydrogenation of dimethyl itaconate and methyl α-acetamido acrylate.

Furthermore, the secondary phosphines **91** and **110** were applied as nucleophiles for the synthesis of P-stereogenic versions of the tridentate ligand Pigiphos (Chapter 4). The major stereoisomer of triphosphine **179** was isolated in good yields and used as ligand for cationic palladium(II) (**183**) and nickel(II) complexes (**184**). The coordination behavior of these tridentate ligands for transition metals were examined by X-ray crystal structural and spectroscopical analysis of the palladium and nickel complexes **183-185**.

100

91: R = Cy
110: R = CF_3

179: R = CF_3
181: R = Cy

183: R = CF_3, M = Pd
184: R = CF_3, M = Ni
185: R = Cy, M = Ni

Zusammenfassung

Die vorliegende Dissertation befasst sich mit der Synthese von P-stereogenen Ferrocenylphosphinen und deren Anwendung in der asymmetrischen Katalyse. Das kommerziell erhältliche Ugi-Amin dient als Ausgangsverbindung für die Synthese aller Ferrocen-Derivate, die in dieser Arbeit beschrieben werden. Sekundäre Phosphin-Verbindungen vom Typ **85** konnten in wenigen Schritten hergestellt und in guten Ausbeuten erhalten werden (Kapitel 2).

	81: R = Ph, R' = Cy
	87: R = Ph, R' = Ph
	88: R = Ph, R' = Np
	89: R = Ph, R' = Ad
	110: R = CF$_3$, R' = Ad
	131: R = CF$_3$, R' = Cy

Es konnte gezeigt werden, dass sekundäre Phosphine des Typs **85** vielseitige Ausgangsverbindungen für die Synthese von P-stereogenen Ferrocenylphosphinen darstellen. Einerseits konnten aus der Diphenylphosphin-Verbindung **89** die typischen bidentaten Palladium-(**112**) und Rhodium-Komplexe (**116**) hergestellt werden, andererseits weist die sekundäre Phosphin-Funktion in **89** eine spezielle Reaktivität unter gewissen Reaktionsbedingungen auf. Durch Deprotonierung des Palladium-Komplexes **112** konnte der überbrückte Phosphido-Komplex **113** hergestellt werden und intramolekulare Allylierung von **89** mit dem dinuklearen Allylpalladiumchlorid Komplex ergab **115** mit zwei koordinierenden, tertiären Phosphinen. Die sekundäre Phosphin-Einheit in **110** konnte selektiv oxidiert werden, wobei der sekundäre Phosphinoxid-Ligand **111** in guter Ausbeute erhalten wurde. Dieser wurde mit Palladium- und Rhodiumprekursoren zu den entsprechenden Komplexen (**117** bzw. **118**) umgesetzt wobei ein unerwartetes metallspezifisches Koordinationsverhalten (P-Koordination für Pd und O-Koordination für Rh) beobachtet werden konnte.

Zusammenfassung

Die Umsetzung der sekundären Phosphin-Verbindungen **110** und **131** mit einer Base ergab, nach intramolekularer Substitution einer CF_3-Gruppe, stereoselektiv die 1,2-Diphosphole **129** and **132** (Kapitel 3). Diese wurden durch elektrophile Methylierung, gefolgt von einer nukleophilen Ringöffnung, in die Josiphos-verwandten Diphosphine **133-137** überführt. Es konnte gezeigt werden, dass diese Reaktionssequenz sowohl regio- als auch stereoselektiv verläuft, so dass jeweils nur ein Isomer des Produkts erhalten wurde.

Die Verwendung von Trifluoromethylsubstituenten als Abgangsgruppe konnte auf andere Alkyl- und Arylphosphine ausgeweitet werden und mit lithiiertem Ugi-Amin als Nukleophil wurden verschiedene P-stereogene Trifluoromethylphosphine vom PPFA-Typ hergestellt. Die beiden erhaltenen Diastereoisomere konnten durch Säulenchromatographie getrennt und zu den Josiphos-ähnlichen Diphosphinen **142**, **151** und **152** umgesetzt werden.

Das Potential der obengenannten Trifluoromethylphosphin-Liganden in der asymmetrischen Katalyse wurde anhand der Rhodium-katalysierten Hydrierung funktionalisierter Olefine demonstriert; alle eingesetzten Liganden zeigten hohe Aktivitäten und mässige bis exzellente Selektivitäten. Besonders hervorzuheben ist der Ligand **137**, welcher Umsatzfrequenzen von bis zu 6000 pro Stunde und Enantiomerenüberschüsse von >99% ee sowohl für Dimethylitaconat als auch für Methyl α-Acetamidoacrylat erreichte.

Desweiteren wurden die sekundären Phosphine **91** und **110** auch als Nukleophile in der Synthese von P-stereogenen tridentaten Liganden vom Pigiphos-Typ verwendet (Kapitel 4). Das Koordinationsverhalten dieser tridentaten Liganden wurde anhand von Kristallstrukturen sowie spektroskopischen Analysen der Palladium- und Nickelkomplexe **183-185** untersucht.

1 P-Stereogenic Phosphines

1.1 Introduction

Since it was first found that certain P-stereogenic phosphines are configurationally stable at ambient temperature and can therefore be resolved in their enantiomers,[1] interest within the chemical community in the preparation as well as the application of these compounds has grown. The idea to substitute the triphenylphosphine ligands of the very successful Wilkinson's Catalyst[2] by a P-stereogenic phosphine was discussed by several researchers in the 1960ies, but it was William S. Knowles,[3] who first presented the asymmetric hydrogenation of α-phenylacrylic acid giving an enantiomeric excess of 15%. This was one of the first examples of a non-enzymatic asymmetric catalytic reaction. Four years and several modifications later, he presented the rhodium-catalyzed reduction of α-acylamidoacrylic acid with 95% *ee* using a bidentate P-stereogenic ligand called DIPAMP (Figure 1).[4-5]

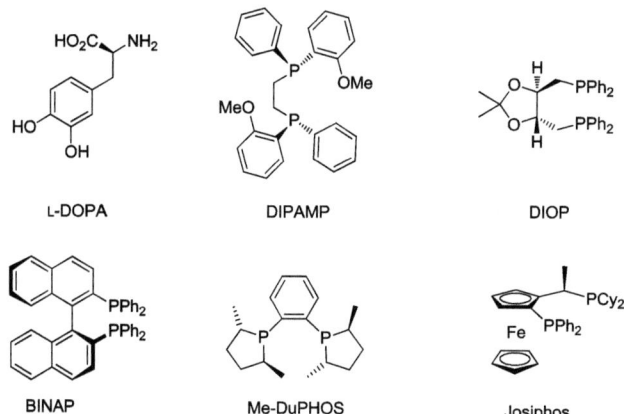

Figure 1: Various chiral bidentate phosphine ligands, which showed high activities and selectivites. Among them the P-steregeonic diphosphine DIPAMP that was used in the synthesis of L-DOPA.

His developments not only had a deep impact on the academic community but also on industry, since he was working for Monsanto and applied DIPAMP in the large scale process for the synthesis of L-DOPA, a pharmaceutical for the treatment of Parkinson's disease. Knowles once stated:[6]

> "We felt strongly that, if one wanted to get high ee values, the asymmetry would have to be directly on the phosphorus. That is where the action is."

Yet, during the development of DIPAMP, this statement was demonstrated not to be wholly correct when Kagan invented DIOP, a bidentate phosphine ligand with chirality on the carbon backbone, and which gave good enantioselectivity in the hydrogenation of several enamides.[7-8] Kagan's discovery was the starting point of a period in which a vast number of chiral bidentate phosphine ligands were invented, among others such successful ligands as BINAP[9], DuPHOS[10-11] and Josiphos.[12] The interest in P-stereogenic phosphines decreased somewhat due to the success of the "backbone-chiral" phosphines and the synthetic difficulties suffered by P-stereogenic phosphine ligands. Nevertheless, several groups have been working in this field, creating various new methods for the synthesis of enantiomerically pure P-stereogenic phosphines as well as applications in asymmetric catalysis.

In this first chapter, the general aspects of P-stereogenic phosphorus compounds will be discussed and some of the various methods for their diastereo- or enantioselective synthesis will be presented. The application of some of these ligands in asymmetric catalysis will be shown as well as the importance of the stereogenic phosphorus atoms as demonstrated by several examples.

1.2 Some General Aspects of P-Stereogenic Phosphines[13]

Phosphines are *pyramidal* structures consisting of a central phosphorus atom, three substituents and a lone pair. If the three substituents are all different this tetrahedral structure lacks a plane of symmetry and the phosphine is therefore a chiral molecule with a stereogenic phosphorus atom (Figure 2). The interconversion of enantiomers is called in this case *pyramidal inversion* and is a process, which occurs without bond formation and/or breaking. In the literature, these phosphines are called "P-stereogenic", "P-chirogenic" or simply "P-chiral", although chirality is used for molecules or objects thus the latter denomination is not strictly correct. The term P-stereogenic will be used throughout this thesis. The exact denotation of the two different enantiomers follows the normal CIP rules[14-15] for tetrahedral structures introduced in 1966 by Robert, S. Cahn, Christopher, K. Ingold and Vladimir Prelog as it is shown in Figure 2.

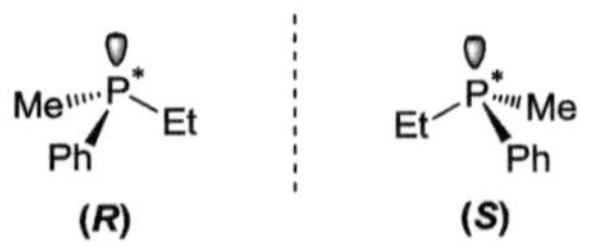

Figure 2: Phosphines bearing three different substituents are chiral molecules with the phosphorus atom as stereogenic center.

1.2.1 Pyramidal Inversion/Inversion Barrier

Whereas in amines the energy barrier for the inversion of the stereocenter is too low (6 kcal/mol for ammonia) to isolate single enantiomers, the pyramidal inversion in phosphines is in general much higher in energy. Phosphorus is less electronegative than nitrogen, therefore its lone pair has more s character than that of nitrogen.[16] This is directly reflected in the bond angles of 94° for PH_3 versus 107° for NH_3. As in the transition state for pyramidal inversion the lone pair has pure p-character, the inversion barrier of phosphines is thus much higher than that of amines.

Mislow and co-workers[1] examined values for the activation energy E_{inv} for the thermal racemization of phosphines of the type $R_1R_2(Me)P$ as shown in Table 1.

Table 1: Activation energy for thermal racemization of tertiary phosphines.[1]

Entry	R^1	R^2	E_{inv} [kcal/mol]
1	C_6H_{11}	n-C_3H_7	35.6
2	C_6H_5	n-C_3H_7	32.1
3	C_6H_5	t-C_4H_9	32.7
4	C_6H_5	p-$CH_3OC_6H_4$	30.8
5	C_6H_5	p-$CH_3C_6H_4$	30.3
6	C_6H_5	p-$CF_3C_6H_4$	29.1

Steric bulk on the phosphorus destabilizes the sp^3 configuration and therefore decreases the activation energy E_{inv}. However, the influence of the size of the substituents on the phosphine is rather small, as can be seen by comparing entry 2 and entry 3. A more pronounced effect is observed if the alkyl group is replaced by an aryl group; the rate of inversion of methyl(p-methoxyphenyl)phenylphosphine (entry 2) is 78 times higher than the one of cyclohexyl-methylphenylphosphine (entry 1). Mislow explains this effect with a stabilization of the transition state of the inversion, which effects a lower energy barrier. In the transition state the phosphine adopts a planar structure with an sp^2 hybridization of the bonding orbitals and a p-hybridization of the lone pair electrons. Aromatic substituents allow delocalization of the lone pair electrons into π-orbitals of the aromatic system, which leads to stabilization of the transition state and therefore a lowering of the activation energy. With electron-withdrawing substituents on the aromatic groups this delocalization is more pronounced than with electron-donating substituents, which results in a lower barrier of inversion for the phosphine with R_2 = p-$CF_3C_6H_4$ compared to the one with R_2 = p-$CH_3OC_6H_4$ or R_2 = p-$CH_3C_6H_4$, respectively (entry 4-6).

Mislow and co-workers also worked out that the electronic nature of the substituents bound to the phosphine has the most important impact on the activation energy of the inversion[17]. In inversion experiments, phosphines bearing σ-acceptors (-F,-Cl, -NH$_2$,-OMe) showed to have a higher energy barrier than phosphines with electron-rich substituents (-PR$_2$, -SiR$_3$, -SnR$_3$). Contrary to the case with aromatic substituents this observation is not explained by a destabilization of the transition state but by ground state stabilization caused by electron-withdrawing substituents on the phosphine. The geometry and thus the hybridization around the phosphorus is dependent on the electronegativity of the substituents as described by Bent's rule:[18]

"Atomic p character concentrates in orbitals directed toward electronegative substituents."

Thus, introduction of electron-withdrawing substituents on the phosphine results in a higher p character of the atomic orbitals on phosphorus that are involved in bonding to these substituents and consequently in a higher s character of the lone pair orbital. Since in the transition state the lone pair orbital has pure p character, the energy required for the change of hybridization is increased resulting in a higher barrier of inversion in the case of phosphines bearing electron-withdrawing substituents. When comparing the electronegativity of substituents and the energy of inversion, Mislow and co-workers could show a linear correlation between the Allred-Rochow electronegativity[19] values of atoms attached to phosphorus and E_{inv} of the corresponding phosphines.

Thus, knowing that P-stereogenic phosphorus compounds are configurationally stable and provide high activity and selectivity in certain asymmetric reactions the question arises how to synthesize such compounds. Are there separation methods for racemates or selective synthetic strategies for the preparation of P-stereogenic phosphines? The next subchapters try to deliver some answers to these questions.

1.3 Preparation of P-Stereogenic Phosphines[13, 20-22]

Since the first isolation of a P-stereogenic organophosphorus compound almost a century ago, the interest in these molecules has continuously increased in academic research as well as in view of industrial applications. A large variety of methods for their synthesis have been published since then and many applications of these chiral compounds are known. In the first years after their discovery, resolution methods from racemates were dominant, whereas later the use of chiral auxiliaries rendered the stereoselective synthesis of these chiral phosphorus compounds possible. In recent years asymmetric syntheses using transition metals and chiral ligands gained in greater importance.

1.3.1 Protection and Deprotection Methods

Tertiary phosphines have the tendency to oxidize when exposed to air, making them difficult to handle and store under normal conditions. In addition, racemization of P-stereogenic phosphines is usually observed at higher temperature. To overcome these problems the corresponding phosphine oxides or phosphine boranes are usually used.

1.3.1.1 Phosphine Oxides

Secondary and tertiary phosphine oxides were, especially before the 1990ies, the key intermediates in the synthesis of P-stereogenic phosphines. They are normally air and moisture stable and therefore easy to handle and purify. Besides this, due to the stable tetrahedral structure, enantioenriched phosphine oxides do not racemize, even at elevated temperature. In order to convert these phosphine oxides into the corresponding phosphines, reducing agents such as silanes or LiAlH$_4$ are normally applied. Methods that provide the products with inversion of configuration[23-25] as well as methods that proceed with retention at the phosphorus center[23-24] are known in literature. The drawback of these methods is that in most cases some loss of enantiopurity[26] or even full racemization is observed,[27] although some optimized applications have been developed.[25]

1.3.1.2 Phosphine Boranes

The protection of phosphines with boranes has become the most important method to overcome synthetic problems in recent years.[28-29] Phosphine boranes are air and moisture stable, do not show racemization even at elevated temperature and are compatible with a wide range of reaction conditions. Additionally, the borane group attached to the phosphorus can activate the P-H bond or an adjacent methyl group and facilitates therefore their deprotonation using a strong base (vide infra). The boranation of P-stereogenic phosphines, typically performed with the BH$_3$·THF complex is in almost all cases quantitative and occurs with total retention of configuration at the phosphorus center. Removal of the borane group is usually accomplished by using an excess of a tertiary amine such as diethylamine or morpholine as it first has been reported by Imamoto.[30] Complete preservation of the stereochemical integrity at phosphorus and excellent conversions are usually observed. The P-B bond of phosphines

1 P-Stereogenic Phosphines

bearing electron-rich substituents is sometimes too strong to be effectively broken with amines. Therefore, Livinghouse and co-workers developed a different method for these special cases. They used strong acids, such as methanesulfonic or tetrafluoroboric acids followed by neutralization with $NaHCO_3$ or K_2CO_3 for the deprotection of phosphine boranes.[31]

1.3.2 Resolution of Racemates

Due to the lack of enantioselective methods or appropriate auxiliaries for the stereoselective synthesis of P-stereogenic phosphines the method of choice in earlier days was the racemic synthesis of the desired phosphine followed by resolution of the racemate using a chiral auxiliary. The major drawbacks of this method are the difficulties of finding the appropriate auxiliary or metal complex for the successful resolution of the enantiomers, as well as the tedious recrystallization or chromatographic separation methods that are often required. Nevertheless, several examples can be found, in which high yields and high optical purity were achieved showing the relevance of these old and elaborate separation methods. From the large variety of resolution methods presented in the literature, only two will be outlined in detail, namely the first reported example of a resolved P-stereogenic phosphorus compound and one of the resolution methods that found the broadest applications and best results in terms of yield and enantiopurity.

1.3.2.1 Chiral Brønstedt Acids

Meisenheimer and Lichtenstadt[32] in 1911 were the first to describe the isolation of P-stereogenic organophosphorus compound **1** in enantiomerically enriched form by using (+)-bromocamphorsulfonic acid for the formation of separable diastereomeric salts (Scheme 1). In spite of this early and important discovery of a P-stereogenic phosphorus compound, another 15 years passed before both enantiomers of **1** could be isolated in good enantiomeric excess.[33]

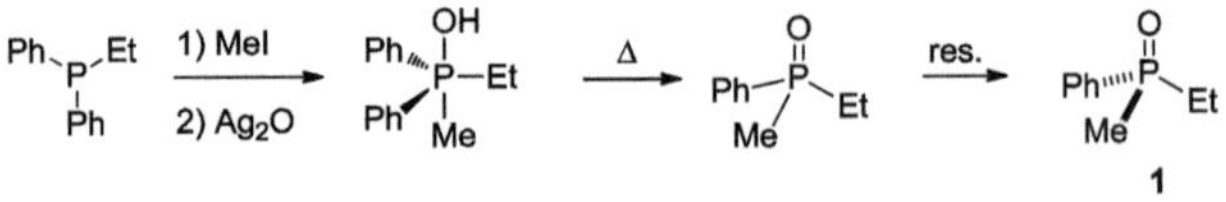

Scheme 1: First isolation of a P-stereogenic phosphorus compound by Meisenheimer and Lichtenstadt.[32]

1.3.2.2 Chiral Palladium Complexes

In the early 1970s two groups introduced the use of chiral palladium complexes as auxiliaries for the resolution of P-stereogenic tertiary phosphines. Otsuka and co-workers[34-35] used the synthetic strategy depicted in Scheme 2 to resolve a variety of monodentate phosphines.

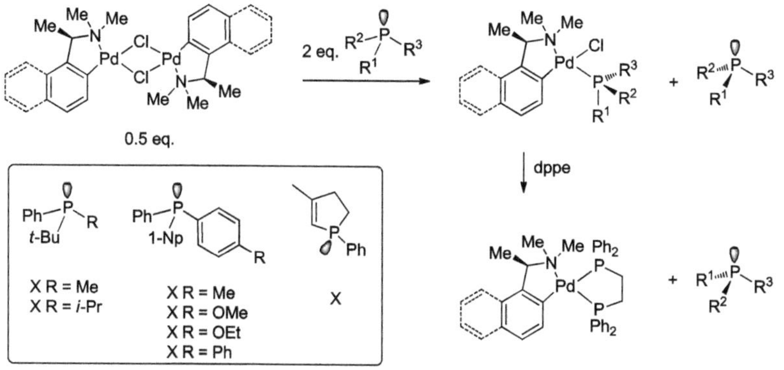

Scheme 2: Chiral palladium complexes for the resolution of monodentate phosphines.[34-35]

This method proved to be efficient and somewhat general and afforded both enantiomers of the desired phosphine in high enantiomeric purity although the choice of the appropriate palladium complex appeared to be crucial in certain cases. Wild and co-workers[36-37] used the same complexes for the resolution of a variety of bidentate and polydentate phosphines and arsines. A typical synthetic procedure using this method is depicted in Scheme 3. Complexation of a racemic mixture of bidentate phosphine **2** to palladium complex **3** led, after counter ion exchange from chloride to hexafluorophosphate, to the precipitation of (*RRR*)-**4**, which was enantiomerically pure after one crystallization step. The bidentate phosphine was then liberated from the complex in a two-step procedure to yield (*RR*)-**2** in over 90% overall yield and high enantiomeric excess. The other enantiomer (*SS*)-**2** was isolated after concentration of the mother liquor left from step 1 and applying the same liberation method as for (*RR*)-**2**. With this example, Wild clearly shows the usefulness of this method; both enantiomers were obtained in high yield and enantiomeric excess and the chiral amine ligand as well as the palladium can, in principle, be recovered.

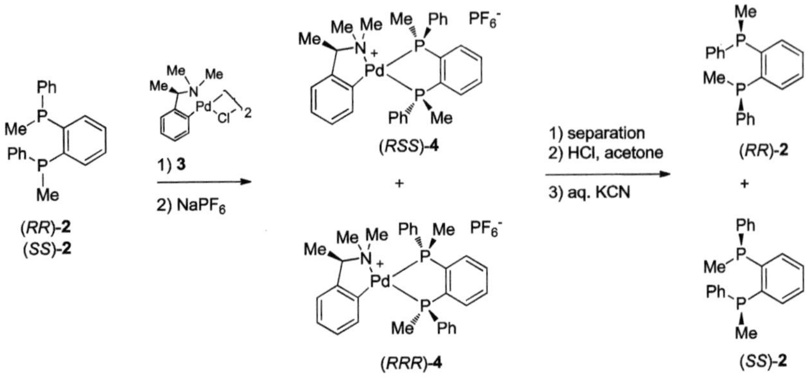

1 P-Stereogenic Phosphines

Scheme 3: Strategy for the resolution of bidentate P-stereogenic phosphines as reported by Wild and co-workers.[36-37]

In addition to the resolution methods involving chiral auxiliaries, more and more separation of P-stereogenic phosphines are achieved by preparative HPLC using chiral columns.[38-39] Even though it is expensive, large scale application, where more than 20 g of product were separated are known in the literature.[40]

1.3.3 Stereoselective Synthesis of P-Stereogenic Phosphines Using Chiral Auxiliaries

The drawbacks of the above discussed resolution methods include long and tedious crystallization or chromatography steps, low overall yields and high dependency on the chiral auxiliary and the substituents on phosphorus, leading to the search for further possibilities. One of these is the use of stoichiometric amounts of chiral molecules from the chiral pool as auxiliaries in stereoselective synthesis of P-stereogenic organophosphorus compounds. The best of these methods in terms of yields, selectivity and applicability are discussed in following sections.

1.3.3.1 (-)-Menthol

In the late sixties, Nudelman[41] and Mislow[42-44] independently introduced the use of (-)-menthol as a chiral auxiliary for the stereoselective synthesis of P-stereogenic phosphine oxides. Starting from a dichlorophosphine, phosphinic acid chloride **5** was obtained in a 3 step procedure as illustrated in Scheme 4. Reaction of **5** with (-)-menthol afforded the desired menthyl phosphinates **6**, which could be separated by fractional crystallization. Both authors mentioned above showed that the menthyl group could be substituted by a variety of alkyl and aryl Grignard reagents with inversion of the stereocenter to yield the tertiary phosphine oxide, which itself could be reduced to the P-stereogenic phosphine without loss of enantiopurity.

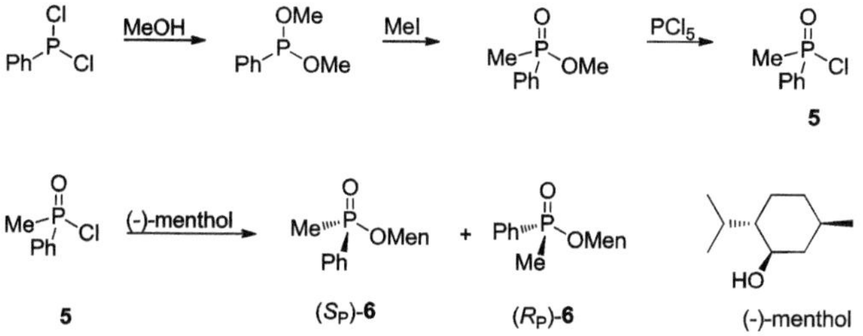

Scheme 4: Stereoselective synthesis of menthyl phosphinates, separable by fractional crysallization.[41-43]

This method found broad application in academic research as well as in industry where it was used by Knowles and co-workers for the syntheses of P-stereogenic phosphine ligands for rhodium-catalyzed asymmetric hydrogenations (vide supra).[3-6] CAMP was the first man-made ligand that gave enzyme like selectivity, DIPAMP, the next generation of Knowles' ligands, was used in the Monsanto synthesis of L-DOPA, providing 95% *ee* in the enamide hydrogenation step.

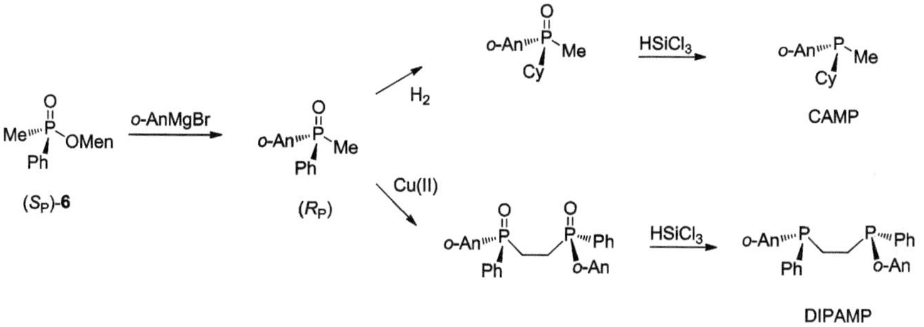

Scheme 5: Synthesis of the P-stereogenic ligands CAMP and DIPAMP as presented by Knowles.[3-6]

The use of menthyl phosphinates for the stereoselective synthesis of phosphine oxides was generally applied, although the reactions appeared to be very sensitive to variations of groups at phosphorus, oxygen and magnesium. Mislow, Horner, and others provided a broad variety of these menthyl phosphinates and also phosphine boranes[30, 45] and phosphinothioates[46-47] were successfully resolved.

Later, Imamoto and co-workers[48-49] discovered the possibility of cleaving the P-O bond between phosphorus and the menthyloxy group using single electron reducing agents such as Li-NH$_3$, lithium naphthalenide, lithium biphenylide or lithium 4,4'-di-*tert*-butylbiphenylide (LDBB). They quenched the reaction mixture with methanol or alkyl halides, providing the corresponding phosphine oxide, as shown in Scheme 6. Good selectivities (*ee* > 90%) were observed when performing the reaction at low temperature, whereas raising the temperature led to racemization. The same synthetic strategy could also be applied to phosphine boranes as demonstrated by Imamoto.

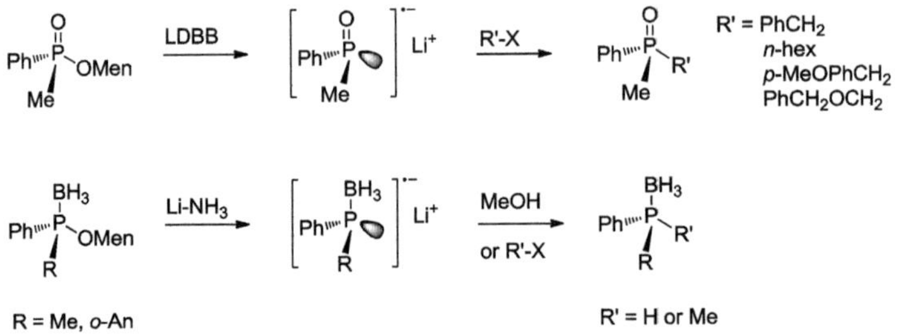

Scheme 6: Reductive cleavage of the menthyl group and subsequent reaction with electrophiles such as MeOH or alkyl halides.[48-49]

Contrary to the method described by Mislow, in which the substitution of the menthyl group occurred with inversion of the phosphorus center, retention of configuration was observed in the substitution method presented by Imamoto. Thus, two complementary methods were available allowing the synthesis of both enantiomers of the P-stereogenic phosphine from the same precursor.

1.3.3.2 Heterobifunctional Chiral Auxiliaries; Ephedrine and Campher Derivatives

The fact that alkoxy groups on phosphorus can be nucleophilically substituted with high stereoselectivity led to the idea of using a bifunctional chiral auxiliary with two different leaving groups. This should allow the stepwise introduction of two different nucleophiles with high stereoselectivity. In general, such a two-step procedure in which two nucleophiles are introduced sequentially provides a method allowing the synthesis of both enantiomers from the same precursor by simple variation of the nucleophiles that are introduced in step one or two, respectively.

Jugé and co-workers[50-51] used (-)-ephedrine as auxiliary in their powerful method for the preparation of P-stereogenic phosphines as outlined in Scheme 7.

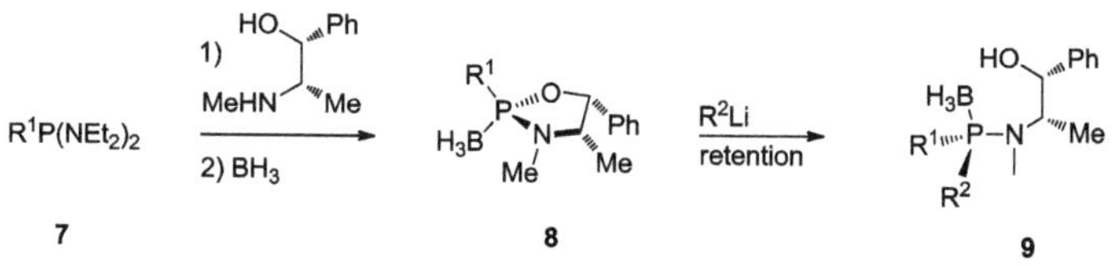

1 P-Stereogenic Phosphines

Scheme 7: Stereoselective synthesis of P-stereogenic phosphine boranes using (-)-ephedrine as chiral auxiliary, part 1.[50-51]

Starting from bis(diethylamino)phosphine **7** and (-)-ephedrine, oxaazaphospholidine borane **8** was obtained in diastereomerically pure form in 80% yield after borane protection. The X-ray structure of **8** showed that the R_P-isomer is formed in which the substituent on the phosphine and the methyl and phenyl groups on the auxiliary are in trans position to each other. The oxazaphospholidine ring could then be opened regioselectively with alkyl and aryl lithium reagents, such that the P-O bond was broken to form the corresponding phosphamide borane **9**. This opening reaction was observed to occur with high stereoselectivity ($de > 85\%$) and with retention of configuration at phosphorus. This stands in sharp contrast to substitution reactions on other acyclic phosphorus compounds, where a "normal" S_N2 mechanism with inversion at the stereocenter is observed (vide supra). Jugé and co-workers proposed the formation of a pentacoordinate intermediate **10** after attack of the incoming nucleophile at the sterically less hindered site next to the P-O bond, as illustrated in Scheme 8.

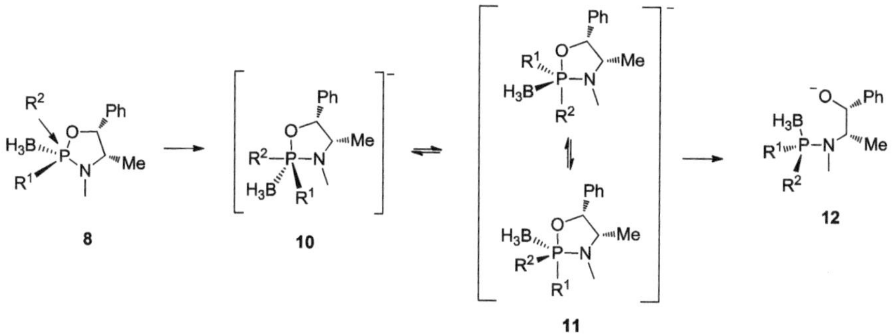

Scheme 8: Proposed mechanism for the opening of the oxazaphospholide ring, which occurs with retention.[51]

This intermediate can stereopermute into another one (**11**) having the oxygen group in apical position, which allows the cleavage of the P-O bond to form phosphamide **12**. This proposed mechanism explains the retention of configuration at phosphorus and is in agreement with a mechanism proposed for similar nucleophilic substitution reactions at trivalent phosphorus compounds.[52] A wide variety of alkyl and aryl lithium reagents were

1 P-Stereogenic Phosphines

used, affording good yields and stereoselectivity, even though some limitations were observed. Van Leeuwen and co-workers[53] found that 1,1'-dilithioferrocene (**13**) reacted with **8** in only low yields and an estimated diastereoselectivity of about 65:35, which was explained by the steric hindrance of the ferrocene moiety. Mezzetti's group[54] made the observation that, *o,o'*-disubstituted aryllithium compounds (**14-16**) did not react at all with **8**, concluding that phosphines bearing bulky aryl substituents are difficult to obtain in this way (Figure 3).

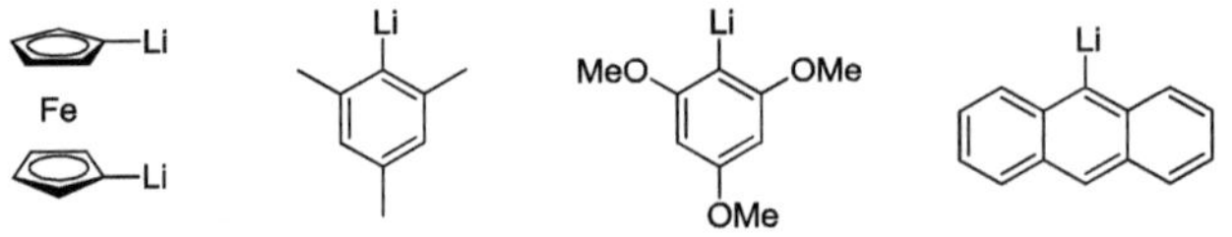

Figure 3: Aryl lithium compounds that gave low yield and diastereoselectiviy (**13**) or did not react at all (**14-16**) with **8**.[53-54]

Jugé and co-workers used an acidic methanolysis to convert the phosphamide **9** into the phosphinite boranes **17** as depicted in Scheme 9. High yields and stereoselectivity with inversion at the stereocenter are normally observed in this step, although one exception is known. Rippert et al. observed that phosphamides with a phenyl and a bulky *tert*-butyl substituent could not be converted into the corresponding phoshinite with this method, even at reflux temperatures.[55]

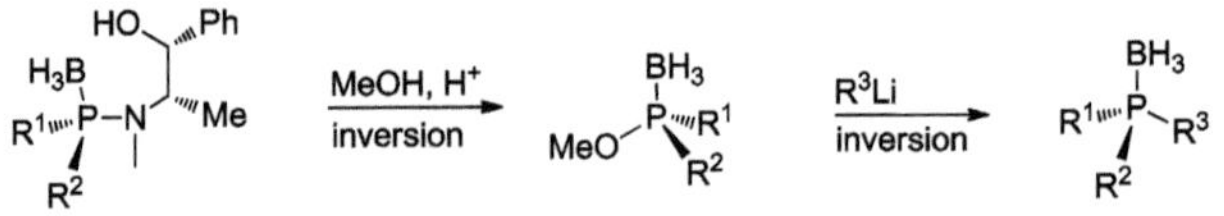

Scheme 9: Stereoselective synthesis of P-stereogenic phosphine boranes using (-)-ephedrine as chiral auxiliary, part 2.[50-51]

Addition of organolithium reagents to the phosphinite boranes thus obtained affords the borane protected P-stereogenic phosphines **18** with inversion at the stereocenter in good yields and with high stereoselectivity. Also this step is quite general for both aryl and alkyl lithium reagents, but with the same limitations as in the second step of the sequence. 2,6-Disubstituted aryllithium compounds did not react in this step either, as observed by Mezzetti's group.[54] Contrary to this observation, 1,1'-dilithioferrocene, which showed low reactivity and selectivity when used in the first substitution step of the sequence, reacted cleanly with **17** to the desired ferrocenyl diphosphines. Some loss of enantiomeric excess and monosubstituted byproduct were observed, but after chromatographic purification

1 P-Stereogenic Phosphines

enantiomerically pure product could be isolated.[53] Similarly to the methanolysis of the phosphamide borane, Jugé and co-workers[56-57] developed a method to convert **9** into chlorophosphine boranes **19** as shown in Scheme 10. These chlorophosphine boranes are very reactive compounds and are sensitive towards moisture, thus purification is not generally possible. Therefore, in most of the cases, in situ formed chloro phosphines are directly reacted to the corresponding P-stereogenic phosphine boranes. Additionally, careful control of the reaction conditions is crucial in order for high stereoselectivity to be obtained. However, excellent results as in the case of phosphinite boranes could not be achieved.[58-59]

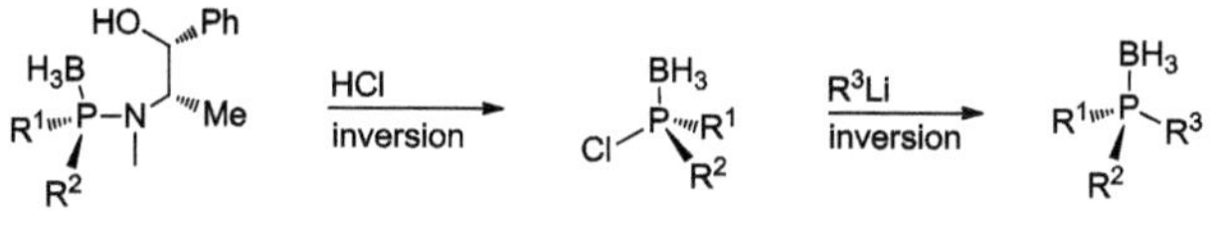

Scheme 10: Method for the stereoselective synthesis of chlorophoshine boranes as developped by Jugé and co-workers.[56-59]

(-)-Ephedrine was also used as chiral auxiliary in the synthesis of phosphine oxides of high entiomeric purity by Jugé and co-workers,[60] as outlined in Scheme 11. Phenyloxaazaphospholidine (**20**) reacts with electrophiles such as methyl or benzyl halides to the corresponding phosphinamides with retention of configuration, involving a Michaelis-Arbuzov[61-63] rearrangement. Some loss of enantiopurity at the phosphorus center was observed, but crystallization led to diastereomerically pure products, which then could be reacted further to the P-stereogenic phosphine oxides **21**.

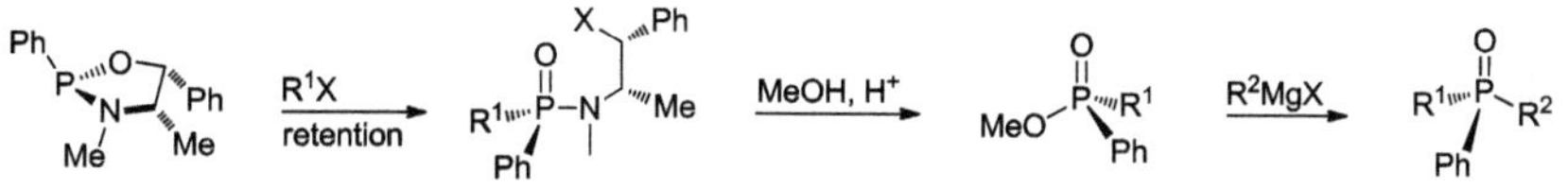

Scheme 11: Synthesis of P-stereogenic phosphine oxides, using a Michaelis-Arbuzov-Grignard reaction sequence.[60]

Corey and co-workers[64] used the camphor derivative **22** for the stereoselective synthesis of mono and bidentate P-stereogenic phosphines such as DIPAMP (Scheme 12). Auxiliary **22** and dichlorophenylphosphine were converted stereospecifically into the oxathiaphospholidine **23** in a thermodynamically controlled reaction and subsequently oxidized to the corresponding thiophosphonate **24** using elemental sulfur. NOE experiments clearly showed a *cis* arrangement of the phenyl group on phosphorus and the two hydrogen atoms at C(4) and C(5), respectively.

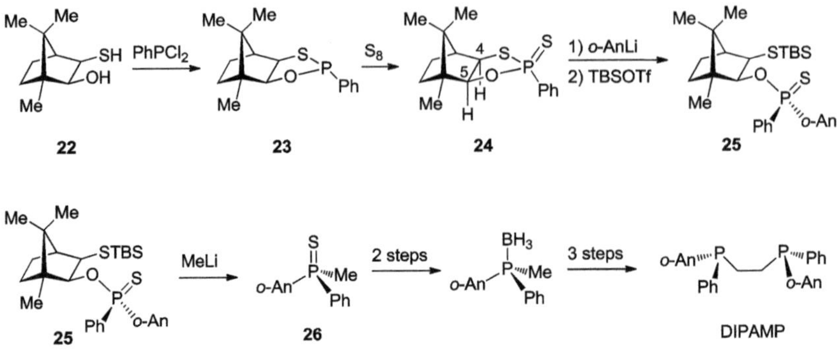

1 P-Stereogenic Phosphines

Scheme 12: Corey's synthetic approach towards mono- and bidentate P-stereogenic phosphines using a camphor derivative as chiral auxiliary.[64]

Nucleophilic substitution on phosphorus with an aryl lithium reagent and in situ protection of the thiolate group with TBSOTf afforded the thiophosphinate ester **25**. This displacement reaction occurs with retention at the stereocenter, which can be explained by a mechanism similar to the one of the first substitution step in the sequence using (-)-ephedrine as the auxiliary (vide supra). The chemoselectivity of the reaction, that is breaking the P-S bond preferentially than the P-O bond, can be explained by the argument that P-S bonds are slightly weaker than P-O bonds. The chiral auxiliary could then be substituted with methyllithium under inversion at phosphorus to give the P-stereogenic phosphine sulfide **26** in 80% yield and enantiomerically pure. The auxiliary was recovered from the reaction mixture after in situ deprotection of the TBS group and chromatographic purification in greater than 80% yield.

Few other examples, where different chiral auxiliaries such as cinchonine and cinchonidine[65] or chiral diols[66] were used are known in the literature. Despite these methods provide high stereoselectivity in certain cases, they met with less interest than those described above.

1.3.3.3 Enantioselective Deprotonation Using Chiral Bases

It was known that methylphosphine oxides[5, 44] as well as methyl phosphine boranes[30, 45] can be deprotonated at the methyl group using strong bases such as LDA or *sec*-BuLi. The astonishingly high acidity of these methyl groups is caused by the electron-withdrawing character of the oxo- or the borane group at phosphorus, respectively. These highly nucleophilic carbanions can then be further reacted with a variety of electrophiles to yield α-functionalized products as demonstrated in Scheme 13. Mislow[44] was the first to show that the deprotonation-carboxylation-decarboxylation-sequence proceeds without any epimerizaton at the phosphorus center, whereas Knowles[5] and Imamoto[30, 45] used the lithiated intermediates **27** and **28** for the synthesis of DIPAMP derivatives **29**.

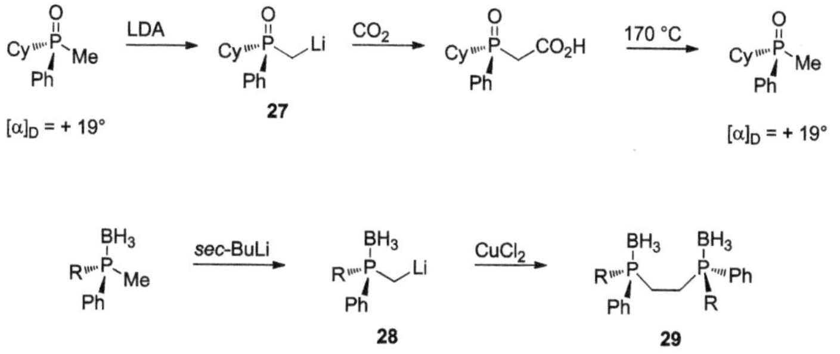

Scheme 13: Deprotonation of P-stereogenic phosphine oxides and boranes.[5, 30, 44-45]

Evans and co-workers[67] used a chiral auxiliary to selectively deprotonate one of the two enantiotopic methyl groups in **30** as illustrated in Table 2. They chose to use *sec*-BuLi as the base and (-)-sparteine as the auxiliary, as this is known to effectively coordinate to lithium, generating a chiral environment around the base.[68-69]

Table 2: Enantioselective deprotonation of prochiral phosphine boranes using (-)-sparteine as chiral auxiliary.[67]

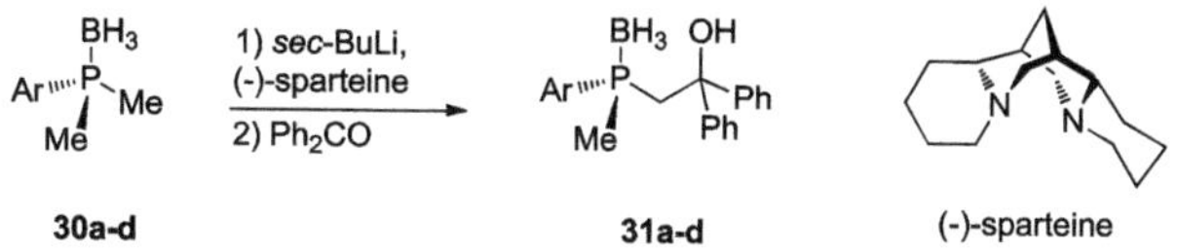

substrate	Ar	yield [%]	ee [%]
30a	phenyl	88	79
30b	*o*-anisyl	81	83
30c	*o*-tolyl	84	87
30d	1-naphthyl	86	82

When using benzophenone as electrophile, high yields and enantioselectivities of around 80% were obtained. Evans[67] and Imamoto[70-71] applied this method to the synthesis of P-stereogenic bidentate phosphines as depicted in Scheme 14. Whereas Evans used the classical oxidative coupling method to generate the diphosphine borane, Imamoto trapped the generated alkyl lithium species with a dichlorophosphine to obtain **32**. This was then further reacted with methylmagnesiumbromide and after, borane protection, diphosphine borane **33** and its meso form were isolated as a 1:1 mixture. After separation via crystallization and boron deprotection, the desired ligand, called MiniPHOS, is obtained in low yield (13-28%).

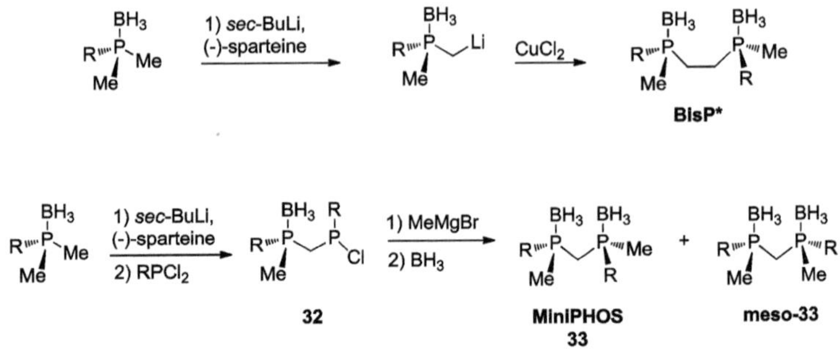

Scheme 14: Stereoselective synthesis of bidentate P-stereogenic phosphines.[67, 70-71]

In 2001, Imamoto[72-73] and co-workers published a sequence towards a non C_2-symmetrical bidentate ligand bearing two P-stereogenic phosphines. To do so, two intermediates had to be prepared, both using an enantioselective deprotonation as the key step as outlined in Scheme 15. The first intermediate, the secondary phosphine borane **34**, was synthesized in a two-step procedure from **35**. After introduction of a hydroxy group using molecular oxygen as the electrophile, the P-C bond was oxidatively cleaved to yield the secondary phosphine borane in good yields and high enantiomeric excess. As second intermediate, the phosphine borane **36** with an alkyl chain bearing a good leaving group, such as mesylate or tosylate, was prepared in a four step synthesis. Only moderate yields but enantiomeric excess higher than 90% were achieved in this sequence. **34** was then deprotonated with *n*-BuLi and reacted with the electrophile **36**, affording the desired C_1-symmetric diphosphine borane **37** in good yield.

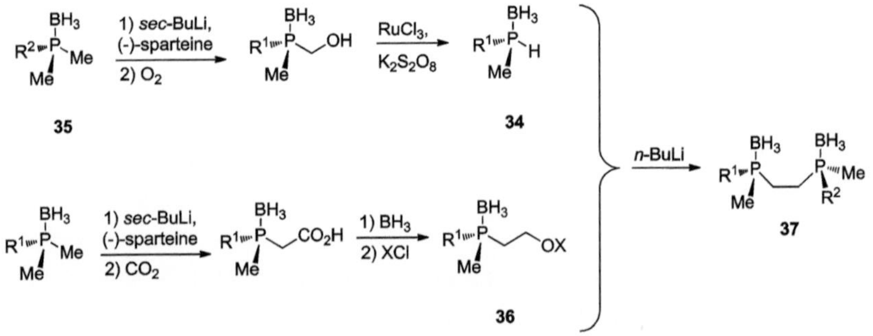

Scheme 15: Synthesis of C_1-symmetrical bidentate ligands containing two P-stereogenic phosphines.[72-73]

The secondary phosphine borane **38** was used to synthesize a large variety of mono- and bidentate phosphine ligands[74] among them the airstable diphosphine QuinoxP*,[75] an excellent ligand for asymmetric catalysis (Scheme 16).

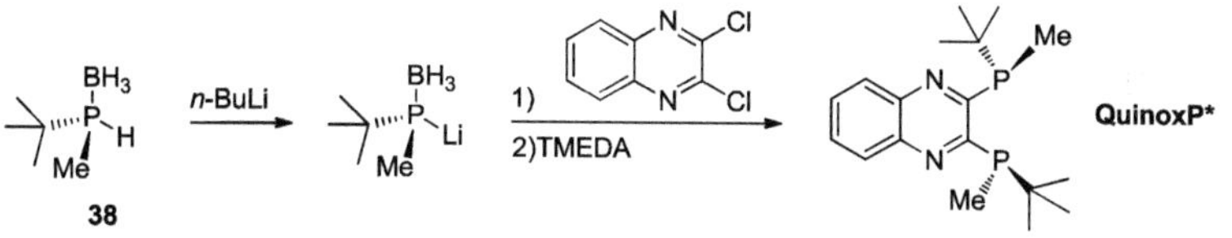

Scheme 16: Synthetic strategy towards the P-stereogenic bidentate ligand QuinoxP*.[20]

1.3.3.4 Dynamic Resolution of P-Stereogenic Phosphides

Enantioselective deprotonation using (-)-sparteine as chiral auxiliary was also applied to secondary phosphines to synthesize P-stereogenic mono- and bidentate ligands as can be seen in Scheme 17.

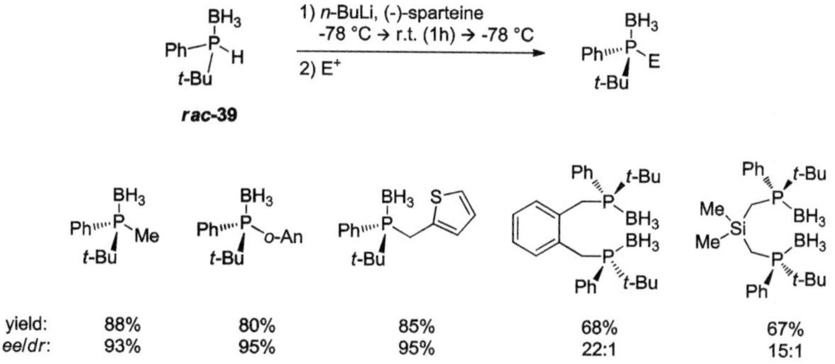

yield:	88%	80%	85%	68%	67%
ee/dr:	93%	95%	95%	22:1	15:1

Scheme 17: Synthesis of mono- and bidentate phosphines via dynamic resolution of P-stereogenic phosphides.[76]

Livinghouse and co-workers[76] showed that the phosphide prepared from *tert*-butylphenylphoshine borane (**39**) and *n*-BuLi could be dynamically resolved when (-)-sparteine was used as the chiral auxiliary, and that the observed enantioselectivity is time and temperature dependent. It turned out that stirring a mixture of **39**, *n*-BuLi and (-)-sparteine for one hour at room temperature before adding the electrophile at low temperature gave the best results in terms of selectivity. Enantiomeric excess up to 95% in the case of monodentate phosphines and diastereomeric ratios up to 22:1 in the case of diphosphines were observed.

1.3.4 Enantioselective Synthesis of P-Stereogenic Phosphines by Asymmetric Catalysis

Although most of the chiral auxiliaries discussed in the previous sections are easily available, cheap and recoverable after the reaction, the application of methods where only catalytic amounts of chiral compounds are used for the synthesis of P-stereogenic phosphines is highly desirable. To date, few examples are found in the literature compared to the large number of existing catalytic reactions. Most of the applications use dynamic resolution of chiral racemic secondary phosphines with a suitable metal ligand system and subsequent arylation or alkylation of the metal phosphido complex.

In 2000, Glueck's group[77] demonstrated the catalytic activity of a platinum complex in the hydrophosphination of acrylates. However, the activity and enantioselectivity achieved in these reactions were still low. Better results were obtained in the platinum-catalyzed alkylation of secondary phosphines as shown in Scheme 18.[78]

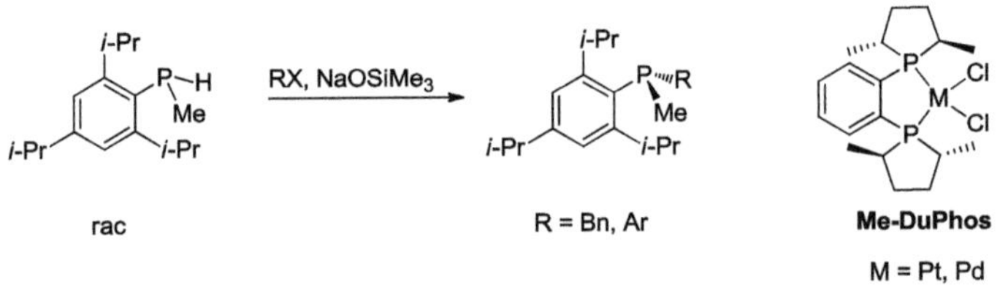

Scheme 18: Asymmetric phosphination of benzyl and aryl halides using a chiral Pt or Pd catalyst.[78-79]

Starting from a racemic secondary phosphine, up to quantitative yields and enantioselectivities higher than 80% *ee* could be achieved. A similar system with palladium instead of platinum was used for the coupling of racemic secondary phosphines to aryl halides or triflates, providing the enantioenriched tertiary phosphines in high yields with up to 88% *ee*.[79] Fast pyramidal inversion at the phosphorus center of the phosphido metal complex was proposed to account for this kinetic resolution of secondary phosphines.[80] In 2004, Helmchen and co-workers[81] presented the synthesis of PHOX-type ligands containing a P-stereogenic phosphine via palladium-catalyzed coupling of an aryl halide to a secondary phosphine, as outlined in Scheme 19. As a model system, ortho-substituted iodobenzene and a variety of diarylphosphines were used, affording the desired products in moderate to good yields with up to 90% *ee*. Long reaction times (1-3 days) and a remarkable influence of the aromatic groups on yield and selectivity were observed. In general, electron poor arenes gave lower yields but better enantioselectivities.

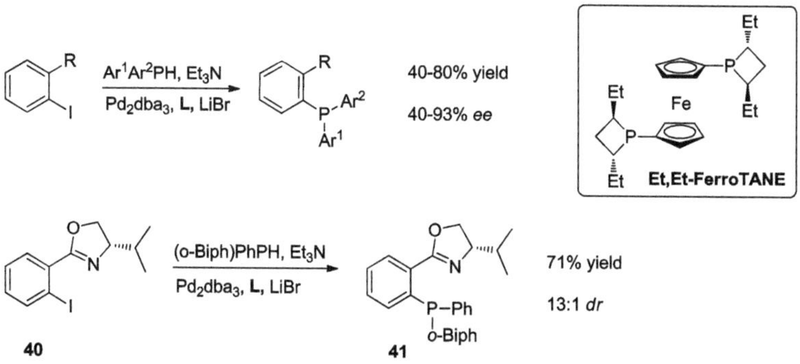

Scheme 19: Palladium-catalyzed symmetric synthesis of P-stereogenic PHOX type ligands.[81]

When the chiral iodoarene **40** was used, the P-stereogenic PHOX ligand **41** was obtained in good yield and with a 13:1 diastereomeric ratio. Toste and Bergman applied a triisopropylsilyl substituted tertiary phosphine instead of a secondary phosphine in a similar reaction and achieved moderate to good yields and enantioselectivities up to 98% *ee*.[82]

In 2006, the same investigators published the use of a ruthenium catalyst for the preparation of P-stereogenic tertiary phosphines as outlined in Scheme 20.[83-84] Racemic methylphenylphosphine was reacted with benzylic chlorides using 10 mol% of ruthenium catalyst **42** to afford the products in high yields and up to 87% *ee*.

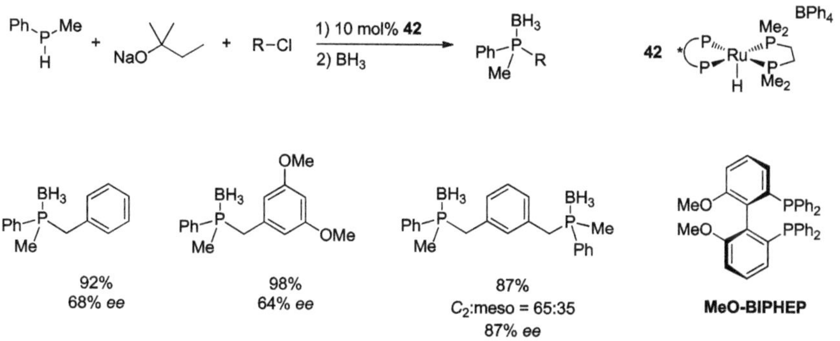

Scheme 20: Ruthenium catalyzed alkylation of secondary phosphines.[83-84]

1.3.5 P-Stereogenic Phosphines with a Ferrocene Backbone

Since the 1970ies, it has been shown that chiral diphosphine ligands with a ferrocene moiety in the backbone provide high activity and enantioselectivity in a wide variety of transition-metal-catalyzed reactions.[85] Several structurally different families of this type of ligands have been developed, such as Josiphos,[86] Taniaphos,[87] Mandyphos,[88] Walphos,[89] Trap[90] and Bophoz.[91] One of the most important features of these compounds is their high modularity, as electronic and steric properties of the substituents attached to the phosphines are normally easy to adjust. The ferrocene backbone exhibits special electronic and steric properties and provides a rigid structure, which results in highly stable transition metal complexes. Most of the aforementioned ligands incorporate central carbon and planar chirality, which allow for high selectivity when applied in catalysis. Compared to the vast number of chiral ferrocenyl phosphines known in the literature, ligands combining carbon, planar and phosphorus chirality are relatively rare. Nevertheless, the success of chiral ferrocene based ligands and the auspicious prospects of P-stereogenic phosphines motivated several groups to combine these areas. Some of the results are discussed below.

1.3.5.1 Central Phosphorus Chirality

Nettekoven et al.[53, 92] as well as Mezzetti and co-workers[93] independently developed the synthesis of P-stereogenic bidentate phosphines (**43**) using 1,1'-dilithioferrocene and the phosphite **44**, which was synthesized according to the method of Jugé (Scheme 21).

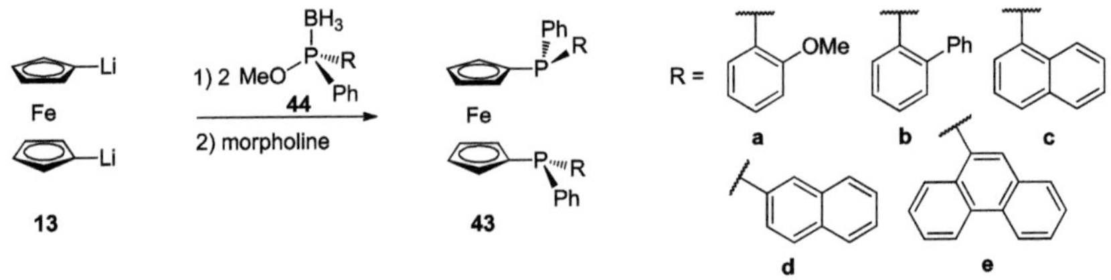

Scheme 21: Stereoselective synthesis of P-stereogenic bidentate phosphines with a ferrocene backbone.

A similar ligand, bearing a methyl and a *tert*-butyl group on the phosphine as depicted in Scheme 22 was introduced by Imamoto and co-wokers[39] in 1999. Instead of applying a stereoselective method, the enantiomers formed were separated using HPLC with a preparative chiral column. High yields and *ee* values of up to 92% were obtained in the rhodium catalyzed hydrosilylation of ketones using these ligands.

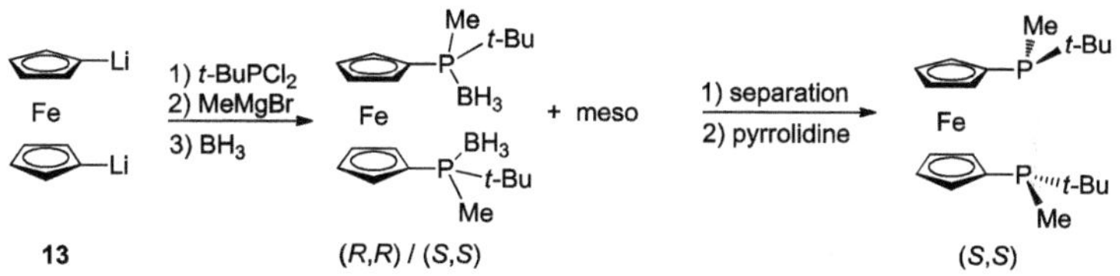

1 P-Stereogenic Phosphines

Scheme 22: Synthesis of a ferrocenyl diphosphine as reported by Imamoto in 1999.[39]

In 2000, Nettekoven et al.[94] reported the synthesis of a P-stereogenic phosphine containing two ferrocene units in the backbone as depicted in Scheme 23. Starting from lithioferrocene, the tertiary phosphine oxide **45** was synthesized in three steps via substitution at phosphite **44**, borane deprotection and oxidation. Diastereoselective ortho deprotonation using an amide base and subsequent addition of iodine afforded **46** in a diastereomeric ratio up to 97:3.

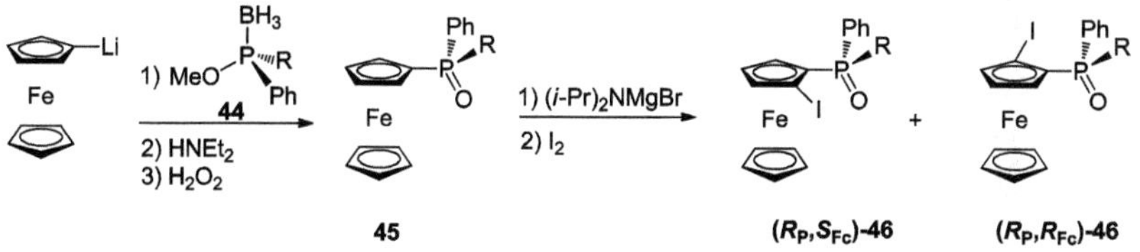

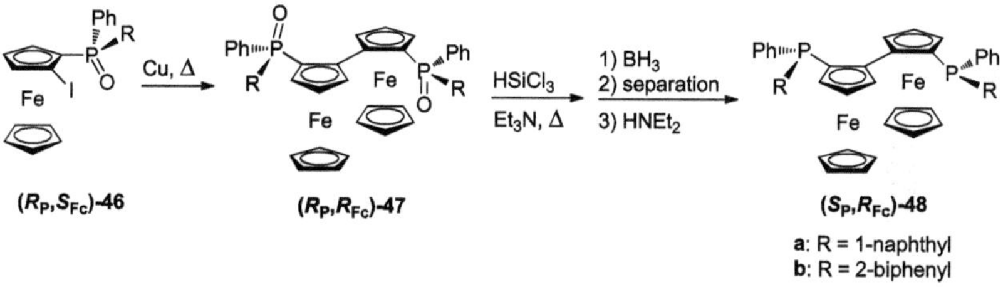

Scheme 23: Synthesis of bidentate P-stereogenic phosphine ligands containing two ferrocene units in the backbone.[94]

Copper mediated Ullmann coupling at elevated temperature transformed the iodoferrocene species into the desired diferrocene product **47**. Reduction of the phosphine oxide required rather harsh conditions under which partial racemization of the stereogenic phosphorus centers was observed. The diastereomeric mixture was therefore borane protected and the isomers separated using flash column chromatography. After deboranation, the desired P-stereogenic diphosphine **48** was obtained as a single diastereoisomer in good yield. The same group also successfully demonstrated the synthesis of bidentate P-stereogenic ligands

1 P-Stereogenic Phosphines

containing three ferrocene units as it can be seen in Scheme 24.[95] Starting from enantiopure **49**, which was synthesized according to the method of Jugé, the methoxy group on phosphorus is substituted by 1,1'-dilithioferrocene and the desired product **50** obtained after deboranation in moderate yields. Both types of ligands were successfully applied to several palladium catalyzed allylic substitution reactions giving high yields and *ee* values up to 93%.

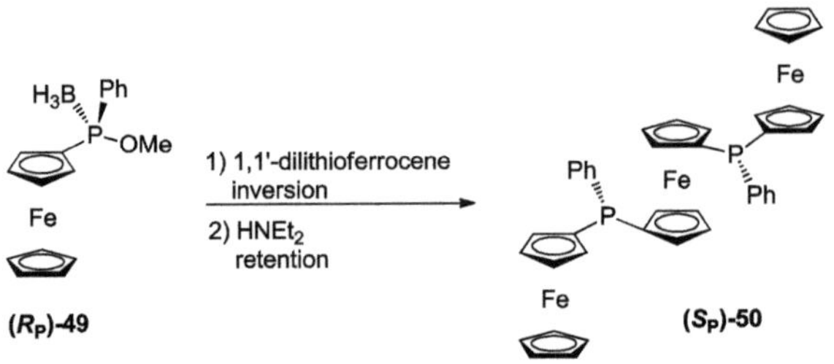

Scheme 24: Synthesis of a bidentate P-stereogenic phosphine containing three ferrocene units.[95]

Imamoto and co-workers[96] used the diastereoselective deprotonation method, described by Evans for the synthesis of a diphosphine with an ethylene backbone and ferrocene substituents as depicted in Scheme 25. Starting from lithioferrocene, the phosphine borane **51** was synthesized by a one-pot procedure in 38% overall yield. Diastereoselective deprotonation using *s*-BuLi and (-)-sparteine as chiral auxiliary followed by copper(II) mediated oxidative coupling yielded the diphosphine borane with only small amounts of the *meso*-byproduct. Recrystallization from hot toluene afforded the enantiopure product in 33% yield, which was then borane deprotected using pyrrolidine to yield the desired bidentate P-stereogenic phosphine ligand **52** in good yields and without any loss of enantiopurity.

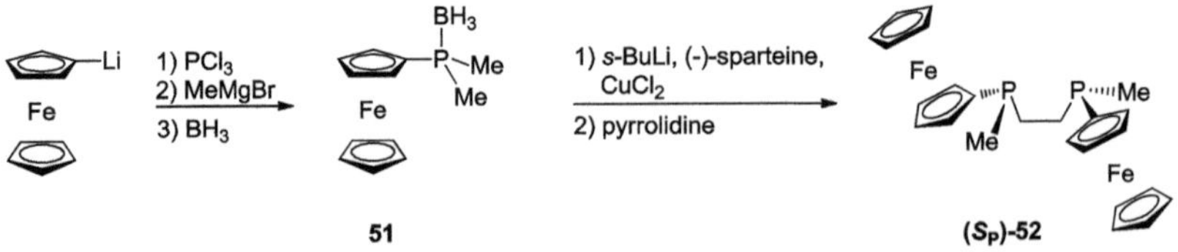

Scheme 25: Synthesis of a P-stereogenic diphosphine ligand with an ethylene backbone and ferrocene substituents.[96]

1.3.5.2 Central Carbon, Planar Carbon and Central Phosphorus Chirality

The first ferrocene-based diphosphines containing central carbon, planar carbon and central phosphorus chirality were published by Togni and co-workers. Diphosphines of type **53**[97] with the chiral phosphorus atom directly attached to the Cp-ring as well as bidentate ligands **54**[86] with the P-stereogenic phosphine on the side chain were successfully synthesized (Figure 4). Analogously to the synthesis of Josiphos **54**[12] was synthesized by reacting PPFA with racemic ArPhPH, which gave the product as a 1:1 mixture of isomers. Since separation of the diastereoismers by flash column chromatography failed, they were resolved using a chiral palladium complex as described in chapter 1.3.2.2. The ligands thus obtained showed disappointingly moderate selectivity in the hydrogenation of carbon carbon double bonds. Similarly, diphosphine **53** was never obtained in diastereomerically pure form and separation of the diastereoisomers proved to be very difficult, therefore **53** was not applied as ligand in asymmetric catalysis.

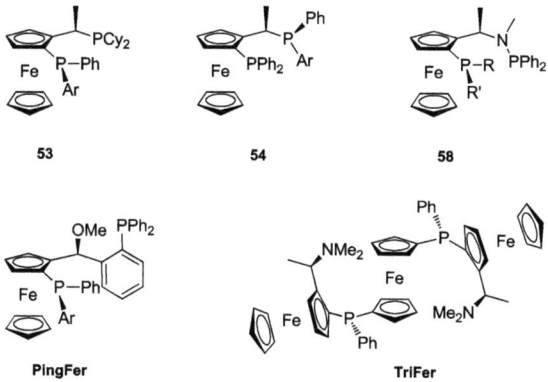

Figure 4: Selected diphosphine ligands combining central carbon, planar and central phosphorus chirality.[86, 97]

Chen and co-workers[98] recently published a method of introducing P-stereogenic phosphines into several ferrocene systems by the use of a chiral directing group as depicted in Scheme 26. Starting from lithiated Ugi's amine (**55**), slow addition of RPCl₂, followed by an aryl or alkyl Grignard reagent afforded the amine-phosphines **56** in high yields and, in most cases, as a single diastereoisomer. The high stereoselectivity of this reaction was explained by the formation of a tertiary ammonium salt as an intermediate. In this five-membered ring the substituent on the phosphorus points away from the unsubstituted Cp-ring and attack of the incoming organometallic reagent occurs from the front to open the ring and give the product.

1 P-Stereogenic Phosphines

These amino phosphines were then transformed to the corresponding diphosphine of the Josiphos (**57**) or the BoPhoz[91] type (**58**) without observing any epimerization.

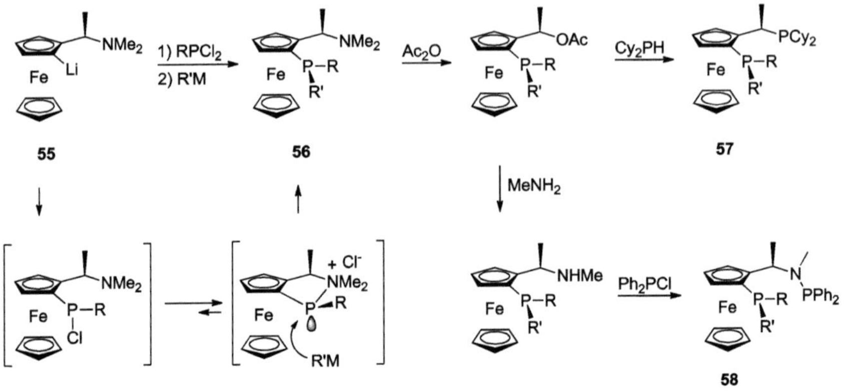

Scheme 26: Stereoselective synthesis of diphosphine ligands combining central carbon, planar carbon and central phosphorus chirality.[98]

PingFer,[99] a P-stereogenic version of second generation Taniaphos,[87] was synthesized starting from planar chiral aldehyde **59**, which was developed by Kagan and co-workers.[100] Addition of aryl magnesium bromide **60** afforded the desired alcohol in 98% yield as a single diasteromer, which was subsequently converted to the methylether **61**. Bromine-lithium exchange was performed using *t*-BuLi and the lithiated species quenched with PhPCl$_2$ followed by addition of an aryllithium reagent. Diphosphines **62** could be isolated in high yield and with moderate to excellent diastereoselectivity.

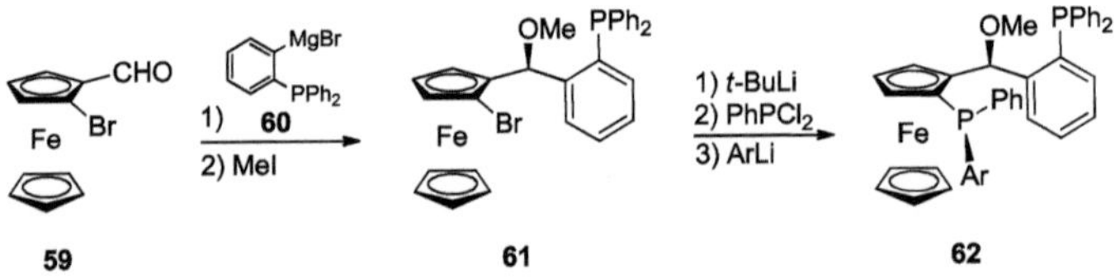

Scheme 27: Synthesis of a P-stereogenic version of Taniaphos as published by Chen and co-workers.[99]

The same group also prepared a C_2 symmetric diphosphine ligand combining central carbon, planar carbon and central phosphorus chirality.[101] Starting from Ugi's amine (**63**), the desired ligand TriFer was easily synthesized in 84% yield and with a diastereoselectivity of 10:1 (Scheme 28).

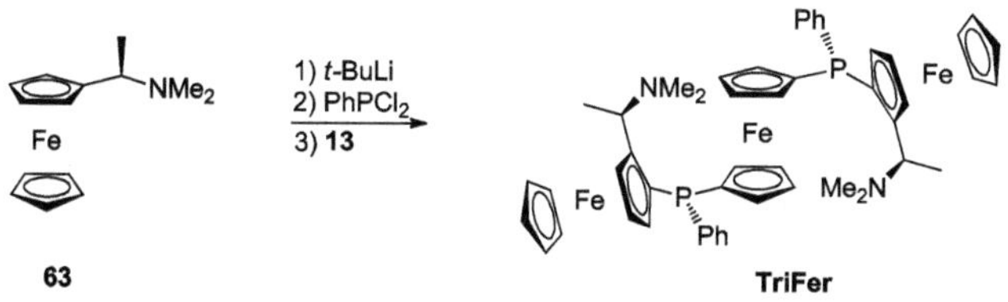

Scheme 28: Synthetic strategy towards a C_2 symmetric ferrocene based diphosphine ligand with carbon, planar and phosphorus chirality.[101]

1.4 Applications of P-Stereogenic Phosphines in Asymmetric Catalysis[13, 22]

This last section of this introduction will address the main goal of the synthesis of P-stereogenic phosphines: their application in asymmetric catalysis. Since the beginning of this field, chemists applied these phosphines in certain catalytic reactions to test their performance. Over the years some standard reactions and standard substrates were established, which allow an easy and accurate comparison between the different ligands and ligand classes. In the next subchapters some of these applications will be briefly discussed. The most important cases, such as hydrogenation of carbon carbon double bonds, the reduction of ketones, and allylic alkylations will be adressed as well as some newer examples of organocatalytic applications.

1.4.1 Asymmetric Hydrogenation Reactions

Since the first successful applications of chiral phosphine ligands in the early 1970s by Knowles[3-6] and Kagan,[7-8] the rhodium-catalyzed hydrogenation of functionalized olefins is one of the most useful and best studied catalytic reactions. In the late 80ies, Noyori[9, 102] developed his BINAP-Ru-system, which showed excellent selectivity in the hydrogenation of carbonyl groups. Since then a huge variety of chiral phosphine ligands have been successfully utilized in these two catalytic applications and the mechanisms of these reactions are well understood. Nowadays, the two above mentioned hydrogenation reactions belong to the most used methods for the generation of stereogenic carbon centers in academic research as well as in industrial processes. The methods are highly general and usually full conversion and high

1 P-Stereogenic Phosphines

enantioselectivities are achieved. This, together with the fact that the methods are cheap and atom economic, as H_2 gas is used as hydrogen source, explains the great success of these reactions.

1.4.1.1 Rh-Catalyzed Hydrogenation of Dehydroamino Esters and Enamides

Dehydroamino esters are one of the most important classes of substrates used in asymmetric hydrogenation, as the products are α-amino acid derivatives, important building blocks for the synthesis of pharmaceuticals or pesticides. Standard substrates to test the activity and selectivity of given ligand systems are, among others, derivatives of the α-acetamido cinnamic acid (ACA). Some selected examples are listed in Table 3.

Table 3: Selected examples of ligands used in the hydrogenation of dehydroamino esters.

entry	ligand	R^1	R^2	S/C	ee [%]
1	CAMP[6]	Ph	H	-	88
2	DIPAMP[4]	Ph	H	2000	95
3	DIPAMP[5]	Ph	Me	900	96
4	t-BuBisP*[70]	Ph	Me	500	99.9
5	t-BuBisP*[70]	Ph	H	500	98
6	t-BuMiniPHOS[71]	H	H	500	>99.9
7	t-BuMiniPHOS[71]	H	Me	500	>99.9
8	TangPhos[103]	Ph	H	100	>99
9	TangPhos[103]	Ph	Me	100	99
10	(S_P)-PingFer[99]	Ph	Me	200	99.6
11	(R_P)-PingFer[99]	Ph	Me	200	69
12	(S_P)-BoPhoz*[98]	H	Me	200	>99
13	(R_P)-BoPhoz*[98]	H	Me	200	90
14	43e[53]	Ph	H	100	99
15	43e[53]	Ph	Me	100	99

CAMP,[6] one of the very first ligands used in asymmetric catalysis and the only monodentate ligand in this table, gave 88% *ee*, which was an impressive result at that time. When changing from the monodentate ligand to the bidentate DIPAMP, enantioselectivity could be increased up to 96%.[4-5] *t*-BuBisP*[70] and MiniPHOS,[71] both developed by

Imamoto and co-workers showed high activity and enantioselectivity, even when β-unsubstituted dehydroamino esters were used as substrates (entry 6 and 7). Zhang and co-workers[103] developed the highly rigid bidentate ligand TangPhos and achieved excellent hydrogenation selectivity. PingFer,[99] a ferrocene based diphosphine ligand combining carbon-centered, planar and phosphorus-centered chirality afforded enantioselectivities above 99% ee (entry 10). Significant matched-mismatched effects were observed; changing the chirality on phosphorus from S to R drastically decreased the ee to 69% ee (entry11). PingFer gave slightly better selectivity (+2-3% ee) than the non P-stereogenic version (Taniaphos). Chen et al.[98] used a P-stereogenic version of BoPhoz in the rhodium-catalyzed hydrogenation of ACA derivatives. In this case significant matched-mismatched effects were also observed (compare entry 12 and 13), therefore choosing the right combination of the three different stereocenters is crucial in order to obtain high selectivities. The P-stereogenic diphosphine ligands with a rigid ferrocene backbone **43e** also gave enantioselectivities of up to 99% ee (entry 14-15).[53]

Some of the ligands discussed above were also tested in the rhodium-catalyzed asymmetric hydrogenation of enamides as shown in Table 4. With 1-acetamido-1-phenylethene as substrate, *t*-BuBisP*[104] and TangPhos[103] gave excellent enantioselectivity (entries 1 and 6), whereas *t*-BuMiniPHOS[104] achieved only 66% ee (entry 3). In the case of substrates with larger steric bulk, *t*-BuMiniPHOS was also very selective (entry 5). When ortho-substituted arenes such as *o*-anisoles were used, the enantioselectivity decreased drastically as can be seen in entries 2 and 4.

Table 4: Selected examples of ligands used in the hydrogenation of enamides.

entry	ligand	R^1	R^2	ee [%]
1	*t*-BuBisP*[104]	Ph	H	99
2	*t*-BuBisP*[104]	*o*-An	H	50
3	*t*-BuMiniPHOS [104]	Ph	H	66
4	*t*-BuMiniPHOS [104]	*o*-An	H	47
5	*t*-BuMiniPHOS [104]	Ph	Me	98
6	TangPhos [103]	Ph	H	>99
7	TangPhos [103]	2-Np	H	>99

1 P-Stereogenic Phosphines

In contrast to that observation, meta-substituted aromatic systems usually gave high enantioselectivity and therefore *t*-BuBisP* could be applied in the asymmetric hydrogenation of enamide **64**, a precursor of the acetylcholinesterase inhibitor SDZ-ENA-173 (Scheme 29).

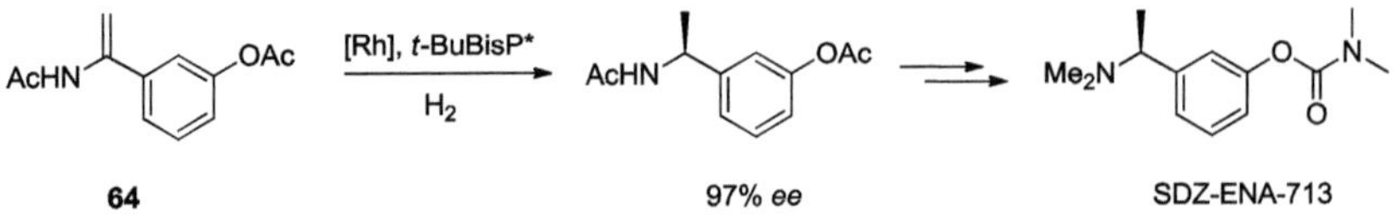

Scheme 29: Asymmetric hydrogenation of enamide **64** towards to the acetylcholinesterase inhibitor SDZ-ENA-713.[104]

1.4.1.2 Ru-Catalyzed Hydrogenation of β-Ketoesters and Ketones

The asymmetric hydrogenation of β-ketoesters provides a highly efficient and practical method for the synthesis of chiral β-hydroxy esters, which are important building blocks in natural product synthesis. Besides the well-known and highly effective Ru-BINAP system,[102] several P-stereogenic phosphines were applied to this specific reaction in recent years. Some selected examples are listed in Table 5.

Table 5: Asymmetric ruthenium-catalyzed hydrogenation of β-ketoesters and ketones.

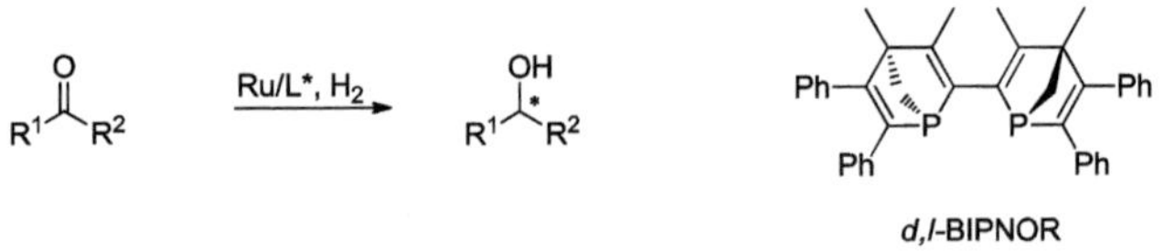

entry	ligand	R^1	R^2	S/C	conv. [%]	ee [%]
1	*t*-BuBisP*[105]	Me	CO_2Me	200	86	97
2	*t*-BuBisP*[105]	Ph	CO_2Et	200	100	89
3	TangPhos[106]	Me	CO_2Me	1000	100	99.8
4	TangPhos[106]	Ph	CO_2Et	1000	100	90
5	(*R*,*R*)-QuinoxP*[107]	Me	CO_2Me	500	>99	99.7
6	(*R*,*R*)-QuinoxP*[107]	Ph	CO_2Et	500	89	99.3
7	*d,l*-BIPNOR[108]	Ph	Me	200	30	60
8	*d,l*-BIPNOR[108]	2-Np	Me	500	65	81

Excellent enantioselectivity was observed with alkyl substituted β-ketoesters using *t*-BuBisP*[105] and TangPhos,[106] respectively (entries 1,3). In the case of aryl substituted substrates, which in general give lower *ee*'s than the alkyl substrates, up to 90% *ee* were achieved with each of these ligands (entries 2 and 4). This accounts for an increase in enantioselectivity of approximately 5% compared to the Ru-BINAP system. The relatively rigid bidentate ligand QuinoxP*[107] gave similar results to TangPhos in the hydrogenation of alkyl substrates. In the case of aryl substituted β-ketoester *ee*'s of more than 99% were achieved, which is, to our knowledge, the highest value for this substrate reported in the literature. *d,l*-BIPNOR[108] was applied to the asymmetric hydrogenation of acetophenone and acetonaphthone, but only moderate conversion and enantioselectivity was observed.

1.4.2 Palladium-Catalyzed Asymmetric Allylic Alkylation

Pd-catalyzed asymmetric allylic alkylation is a typical enantioselective carbon-carbon bond formation reaction.[109-111] Like asymmetric hydrogenation this catalytic reaction is used as a standard method to test the performance of phosphine ligands.

Imamoto and co-workers used monodentate P-stereogenic phosphine ligands (**65**) in the asymmetric allylic alkylation of 1,3-diphenylallyl acetate and achieved quantitative yields and moderate to high enantioselectivity (Table 6).[112]

Table 6: Palladium-catalyzed-asymmetric allylic alkylation with P-stereogenic phosphine ligands.

entry	ligand	isolated yield [%]	*ee* [%]
1	*t*-Bu-**65**[112]	>99	80
2	Cy-**65**[112]	>99	91
3	(S_P)-**52**[96]	99	95
4	**43f**[95]	84	81
5	(S_P,R_{Fc})-**48a**[95]	87	88
5	(*R,R*)-QuinoxP*[107]	85	92

# 1 P-Stereogenic Phosphines

The ligand bearing a cyclohexyl substituent gave slightly higher *ee*'s than the larger *t*-Bu-analog whereas the phosphine with Ph instead of Cy gave only racemic product (entries 1-2). The same group also used the bidentate ligand **43f** in this reaction which resulted in high yields and excellent enantiomeric excess (entry 3).[96] Similar, ferrocene-based ligands were applied by Nettekoven et al.[95] as shown in entry 4 and 5. Yields and enantioselectivities between 80% and 90% were achieved with both ligands. In this catalytic application (*R,R*)-QuinoxP*[107] also showed its significant potential as a ligand for transition-metal-catalyzed asymmetric reactions, since 85% yield and 92% *ee* were achieved (entry 6).

1.4.3 Miscellanous Examples

1.4.3.1 Palladium-Catalyzed α-Arylation and α-Vinylation

In 2009, Buchwald and co-wokers[113] described the palladium-catalyzed α-arylation and α-vinylation of oxindoles using phosphine ligands as depicted in Scheme 30. Among the ligands they screened only biaryl monodentate phosphines were found to give appreciable yields or enantioselectivities in this reaction. Moderate to good yields and *ee*'s of up to 78% were achieved applying KenPhos or (*i*-Pr)$_2$-MOP as ligands. A significant improvement was achieved when the P-stereogenic ligand **66** was applied, with the enantioselectivity raising to 97% *ee*. The method was expanded to a broad variety of aryl bromides and vinyl bromides and in general good yields and enantioselectivities were achieved. These results demonstrate the importance of P-stereogenity in combination with other stereogenic units.

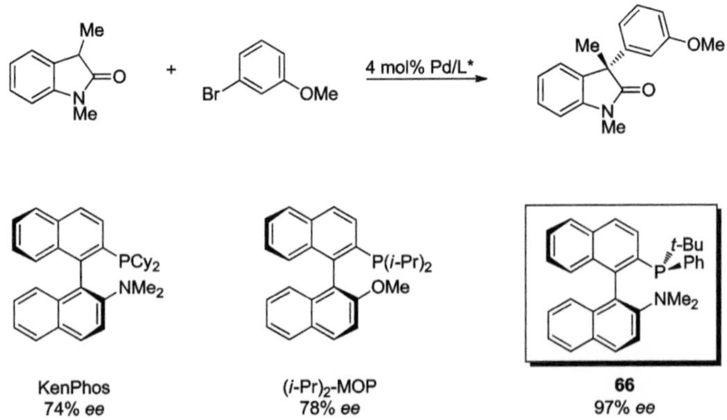

Scheme 30: Pd catalyzed asymmetric α-arylation of oxindoles.[113]

1.4.3.2 Palladium-catalyzed Asymmetric Epoxide Ring-Opening

In 2007, Imamoto and co-wokers[114] developed a ligand similar to BisP* (**67**) but with alkinyl substituents instead of the methyl group. They reasoned that this alkinyl group should, in close proximity to phosphorus, be smaller than a methyl group, since the A-value of the former is 0.41-0.52 kcal/mol compared to 1.74 kcal/mol for methyl.[115] This increase in steric discrimination compared to BisP* should than result in an increase in selectivity. Moreover, as alkinyl groups are electron-withdrawing, the electron density on the phosphine is decreased, which should change the catalytic behavior of these ligands. The ligands were tested in rhodium-catalyzed hydrogenation and conjugate boration as well as in the palladium-catalyzed epoxide ring-opening and achieved excellent enantioselectivities in each case. Results of the latter catalytic reactions are shown in Table 7. Dimethyl- as well as diethylzinc as the alkylating agent worked well and also small substituents on the aromatic systems were tolerated in this type of reaction. Ligands with small residues on the alkinyl substituents such as a hydrogen atom or a methyl group gave up to 99.9% *ee*'s, whereas more bulky residues resulted in somewhat lower selectivity. These values are the highest reported so far for this type of reaction.

Table 7: Palladium-catalyzed epoxide ring-opening using diphosphine ligands with alkinyl substituents.[114]

entry	R^1	R^2	R^3	yield [%]	ee [%]
1	H	Me	H	93	99.9
2	H	Me	Me	94	99.8
3	H	Et	H	90	99.9
4	H	Et	Me	89	99.2
5	MeOCH$_2$O	Me	H	89	98.4
6	MeOCH$_2$O	Me	Me	92	97.8

1.4.3.3 P-stereogenic Phosphines as Reagents in Asymmetric Synthesis/Organocatalysis

In addition to their use as ligands in organometallic catalysis, chiral phosphorus compounds are more and more applied to asymmetric organocatalytic reactions.[116-117] However, compared to the huge variety of chiral phosphines used as ligands in transition-metal catalysis, the number of reports dealing with chiral phosphines in organocatalysis is rather small. In 1996, Vedejs et al.[118] reported the first enantioselective acyl transfer catalyzed by chiral phosphines. Six years later, the same group developed the P-stereogenic phosphine **68**, which afforded high acitivity and selectivity in the same reaction.[119] The same phosphine was then also successfully applied to an organocatalyzed Steglich rearrangement as depicted in Scheme 31.[120] When applying 10 mol% of the P-stereogenic phosphine **68**, enol carbonate **69** was converted to the corresponding lactone **70** in 87% yield and 92% *ee*.

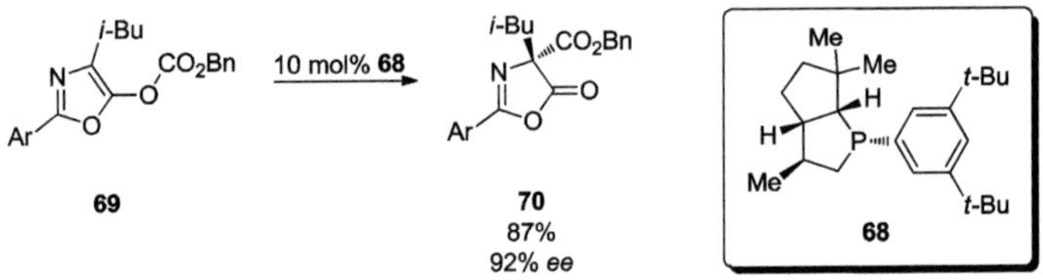

Scheme 31: Organocatalyzed Steglich rearrangement using P-stereogenic phosphine **68**.[120]

In 2007, Marsden and co-workers published the application of P-stereogenic phosphines as reagents in the asymmetric aza-Wittig reaction as it can be seen in Scheme 32. The best yields and enantioselectivities were observed with phosphine-borane **71**, after stepwise or in situ deboranation with DABCO. Although the *ee*'s are so far the highest reported for this reaction, in general only moderate yields and moderate to good enantioselectivities were observed. It also has to be mentioned, that this reaction is not catalytic in phosphine as phosphine oxide is formed as a byproduct and usually a slight excess of the phosphine has to be used.

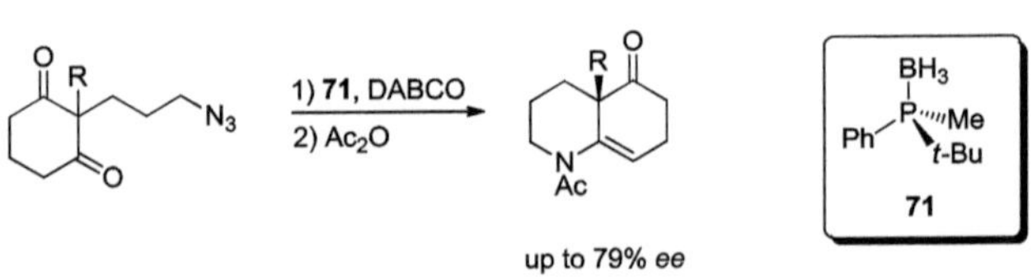

Scheme 32: Phosphine promoted asymmetric aza-Wittig reaction using P-stereogenic compound **71**.

2 Secondary Phosphines and Secondary Phosphine Oxides

2.1 General Aspects of Planar Chiral Ferrocene Systems

A ferrocene derivative bearing two different substituents in 1,2- or 1,3-position of the cyclopentadienyl ring is planar chiral. A broad variety of methods to introduce planar chirality are known and most of them make use of a ferrocene precursor containing a stereogenic directing group. As many excellent reports about this topic exist, we will not go more into detail here.[121-123] The chiral starting material for all of the chiral ferrocene molecules presented in the next subchapters was Ugi's amine (Figure 5),[124] which was generously provided by Solvias AG.[125]

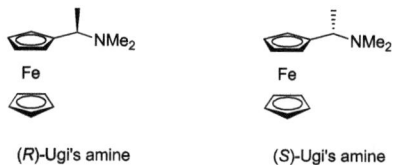

Figure 5: (R)- and (S)-Ugi's amine.

The success of Ugi's amine as precursor for the synthesis of planar chiral systems relies mainly on two properties: the directed ortho-lithiation, which is highly diastereoselective and the nucleophilic substitution of the amino group, which proceeds with complete retention of configuration.

2.1.1 Diastereoselective *ortho*-Lithiation

The amino group in Ugi's amine facilitates deprotonation of one of the two diastereotopic ortho positions preferentially upon addition of a strong base such as *n*-BuLi through coordination to lithium. Deprotonation of H^5 leads to unfavorable steric repulsion between the methyl group and the ferrocene core in the metallated intermediate as shown in Scheme 33. Therefore, lithiation in position 2 is favored and quenching of the lithiated intermediate with an electrophile affords the two diastereoisomers in a ratio of about 96:4. Hayashi used chlorodiphenylphosphine as electrophile in the synthesis of the first ferrocenyl phosphine ligand, which was later named PPFA (1-(diphenylphosphino)ferrocenyl-2-(1-ethylamine).[126]

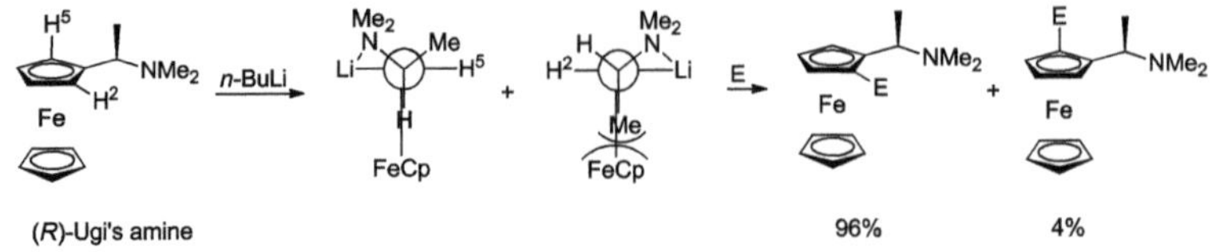

Scheme 33: Diastereoselective *ortho*-lithiation of (*R*)-Ugi's amine.

2.1.2 Substitution of the Dimethylamino Group with Retention of Configuration

The amino group in Ugi's amine can be substituted by nucleophiles with complete retention of configuration of the stereogenic carbon. A non-classical S_N1 mechanism via a configurationally stable ferrocenium intermediate, as shown in Scheme 34, is responsible for this highly selective transformation. Both, the departure of the dimethyl amino group as well as the nucleophilic attack are occurring at the exo site (above) of the ferrocene unit, giving rise to retention of configuration. This highly selective substitution was used for the synthesis of a broad variety of chiral ligands such as Josiphos,[12] and the tridentate ligand Pigiphos.[127]

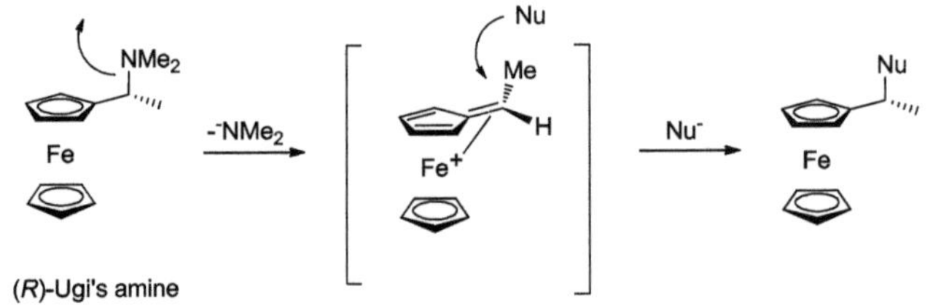

Scheme 34: Nucleophilic substitution of the amino group in Ugi's amine with complete retention of configuration.

2.1.3 Stereochemistry of Chiral Ferrocenes

As the classical CIP rules cannot be simply applied to planar chiral systems, Schlögl[128] introduced a simple rule, which is the currently accepted standard for ferrocene systems displaying planar chirality. The rule states:

> "The observer looks along the principal axis of the molecule so that the more highly substituted ring is directed towards him, whereby the priority of the groups is decisive. The substituents are then, as usual, arranged in decreasing order of priority according to the sequence rule. The choice of symbol (R) or (S) depends on the resulting direction (clockwise or counterclockwise)."

2 Secondary Phosphines and Secondary Phosphine Oxides

This rule is well illustrated using (R_C,S_{Fc})-PPFA as shown in Figure 6. In order to differentiate between central chirality and planar chirality the symbol for the first is tagged with the index C and that for the latter with the index Fc, respectively. Additional phosphorus stereocenters are denoted with indices P1 and P2, respectively. P1 is the phosphorus atom directly attached to the ferrocene, P2 the one on the side chain. The symbols for the different stereogenic centers are written in brackets in front of the name of the molecule starting with phosphorus-centered chirality, followed by carbon-centered and planar chirality. An example containing twofold phosphorus, carbon and planar chirality is shown in Figure 6.

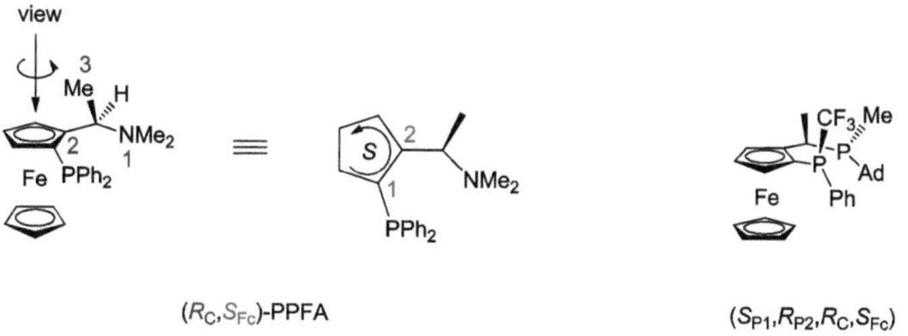

Figure 6: Application of the sequence rule to planar-chiral ferrocene systems as proposed by Schlögl.

It has to be mentioned that in our group we work with (R)- as well as with (S)-Ugi's amine, which allows the synthesis of both enantiomers of certain ligands. To simplify matters all molecules presented in this thesis are drawn with (R)-configuration at the stereogenic carbon, although in some cases both enantiomers or only the (S)-enantiomer were synthesized. Crystal structures of (S)-derivatives were inverted to reflect the configuration drawn in the schemes, but this is always noted in the caption below the figure if needed.

2.2 General Aspects of Secondary Phosphine Oxides (SPO)

2.2.1 Introduction[129-130]

In transition-metal catalysis where phosphorus compounds are used as ligands, tertiary phosphine and phosphoramidites are the dominant structures. Secondary phosphines as well as secondary phosphine oxides (SPO) and their heteroatom substituted analogs (HASPO) are rarely used, though they have gained in importance in the last few years. SPOs are of interest due to their stability toward air and moisture, easy synthesis and the possibility to form secondary interactions upon coordination to a transition metal. In solution, an equilibrium between the pentavalent form (SPO **72**) and the trivalent form (phosphinous acid **73**) exists, lying predominantly to the side of the secondary phosphine oxide (Scheme 35). Complexation to a transition metal occurs in most cases by coordination to phosphorus, providing complex **74**, although oxygen coordination to the metal is also known. If two SPOs are coordinated to one metal center, loss of a proton can occur, forming complex **75**, in which the two SPOs form an anionic bidentate ligand through hydrogen-bond stabilization.

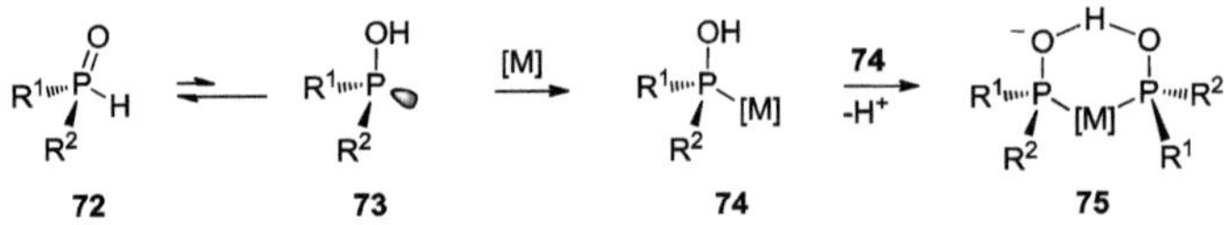

Scheme 35: Secondary phosphine oxide **72**, its tautomeric form **73** and their transition metal complexes **74** and **75**.[129-130]

2.2.2 Synthesis of P-Stereogenic Secondary Phosphine Oxides

The synthesis of secondary phosphine oxides is usually accomplished by hydrolysis of the corresponding chlorophosphine or by oxidation of a secondary phosphine. Different routes to P-stereogenic secondary phosphine oxides are known, but most of them are based upon resolution of racemic mixtures. Haynes and co-wokers[131-132] made use of chiral amines to resolve phosphanylthioic acid **76** by fractional crystallization of the ammonium salt as shown in Scheme 36.

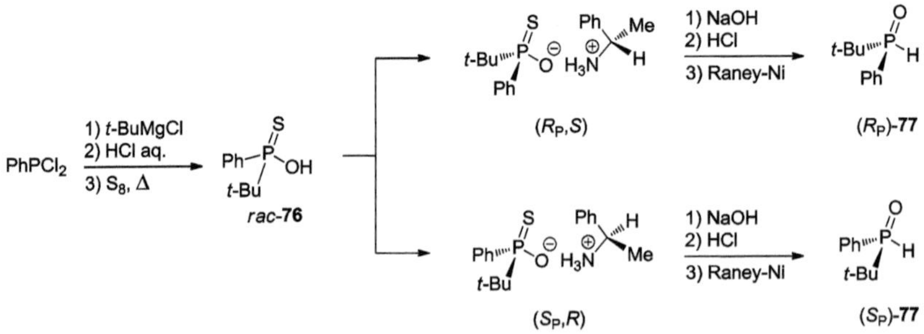

Scheme 36: Synthesis and resolution of P-stereogenic secondary phosphine oxide **77**.[131-132]

rac-**76** could be synthesized in three steps and 55% overall yield on large scale. Addition of (*S*)-(-)-α-methylbenzylamine precipitated the (R_P,*S*)-ammonium salt, which was separated from the mother liquor. The (S_P)-isomer remaining in the mother liquor was separated from the chiral amine and then precipitated using (*R*)-(+)-α-methylbenzylamine. The resolved

2 Secondary Phosphines and Secondary Phosphine Oxides

phosphanylthioic acids were liberated from the chiral amine and reduced to the desired SPOs **77** using Raney-Nickel.

Buono and co-workers[133] made use of a chiral amino alcohol as a chiral auxiliary for the stereoselective synthesis of P-stereogenic SPO **77** as depicted in Scheme 37. Compound **78** was readily synthesized from (*S*)-(+)-prolinol and PhP(NMe$_2$)$_2$ in good yields. The authors could show that the oxazaphospholidine ring of **78** could be opened using Grignard reagents with retention of configuration and with high stereoselectivity.

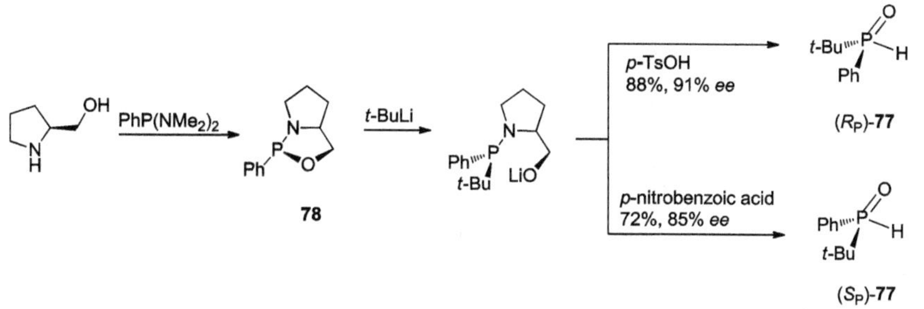

Scheme 37: Stereoselective synthesis of P-stereogenic SPO **77** using prolinol as chiral auxiliary.[133]

More interestingly, both enantiomers of **77** could be obtained by simple variation of the Brønstedt acid used in the hydrolytic cleavage of the auxiliary. Strong acids such as *p*-toluene sulfonic acid (pK_a ~1) gave preferentially (*R*$_P$)-**77**, whereas acids with low acidity (pK_a ~3-5) provided the *S*-enantiomer. No exact mechanistic explanation for this behavior is given by the authors. Thus, this quick and easy method allows access to both enantiomers of P-stereogenic secondary phosphine oxide **77** from the same starting materials.

2.2.3 Application of Secondary Phosphine Oxides in Transition Metal-Catalysis

Secondary phosphine oxides have mainly found application in palladium catalyzed cross coupling reactions of various kinds (Stille, Suzuki, Heck, Sonogashira, etc.). In general, high activity is observed also with aryl chlorides as substrates and in some cases even better results are observed using SPO ligands than with standard phosphine ligands.[134-135] One example, a Suzuki reaction of chloropyridine **79** towards a COX-2 selective inhibitor is depicted in Scheme 38. With POPd, one of the most common SPO palladium catalysts, the cross coupling afforded the product in 98% yield after four hours in refluxing dimethoxyethane.[135]

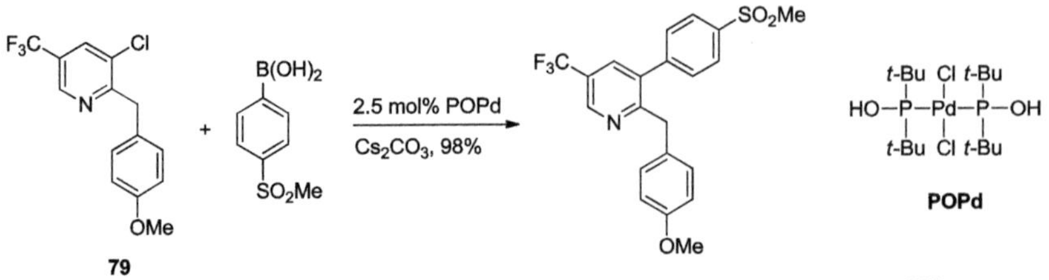

2 Secondary Phosphines and Secondary Phosphine Oxides

Scheme 38: Suzuki cross coupling of pyridinyl chloride 79 using SPO catalyst POPd.[135]

Dai, Haynes and co-workers[136] used the P-stereogenic SPO (R_P)-77 in palladium-catalyzed allylic alkylation with malonate as shown in Scheme 39. The yield as well as enantioselectivity proved to be highly dependent on the solvent and the ligand to palladium ratio. Best results were obtained in THF with an in-situ formed catalyst using 2 mol% of the palladium precursor and 5 mol% of (R_P)-77.

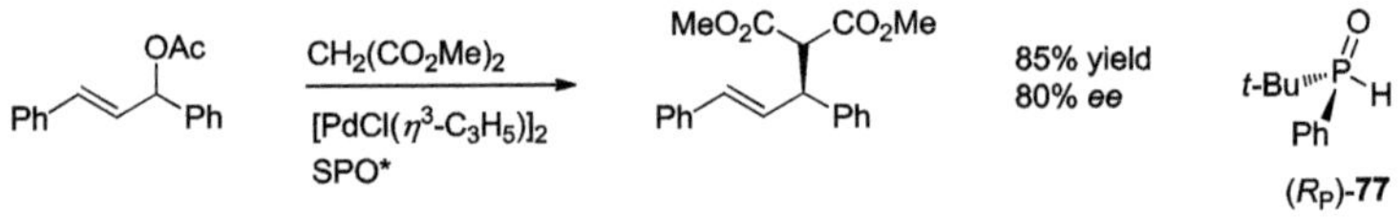

Scheme 39: Palladium catalyzed allylic alkylation with the P-stereogenic SPO (R_P)-77.[136]

SPO 77 was also applied to iridium-catalyzed asymmetric hydrogenation of imines by de Vries and co-workers[137] as shown in Scheme 40. Full conversion and enantioselectivities of up to 80% were obtained when pyridine was used as additive. The best performance was obtained with 25 bar of hydrogen and the ratio [IrCl(COD)]$_2$/SPO/pyridine/imine = 2.5:10:10:100.

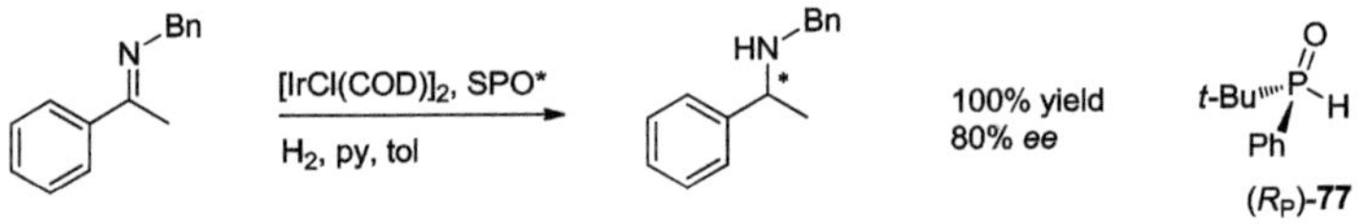

Scheme 40: Iridium catalyzed hydrogenation of benzylic imines using SPO (R_P)-77.[137]

2.2.4 Secondary Phosphine Oxides with a Ferrocene Backbone

In 2010, Pfaltz and co-workers in collaboration with Solvias published the synthesis of a bidentate ligand consisting of an SPO moiety and a tertiary phosphine connected over a ferrocene backbone.[138] They postulated that moderate *ee*'s found for hydrogenations with Ir and Rh originate from an insufficient affinity of the SPO for these metals. Therefore, a combined phosphine/SPO ligand with a somewhat rigid tether should not only lead to a better coordination of the SPO to the metal, but to better defined complexes in general. Two straightforward routes for the synthesis of the ligands were developed starting from Ugi's amine as it can be seen in Scheme 41. In route 1, the SPO was installed first, followed by the introduction of the tertiary phosphine moiety. Surprisingly, the SPO withstood the harsh acidic conditions of the second step. The absolute configuration of the SPO was dependent on the residues (R) and the conditions used in the hydrolysis step. In route 2, the tertiary phosphine was installed after bromination of Ugi's amine, followed by halogen lithium exchange, phosphination and hydrolysis of the chlorophosphine.

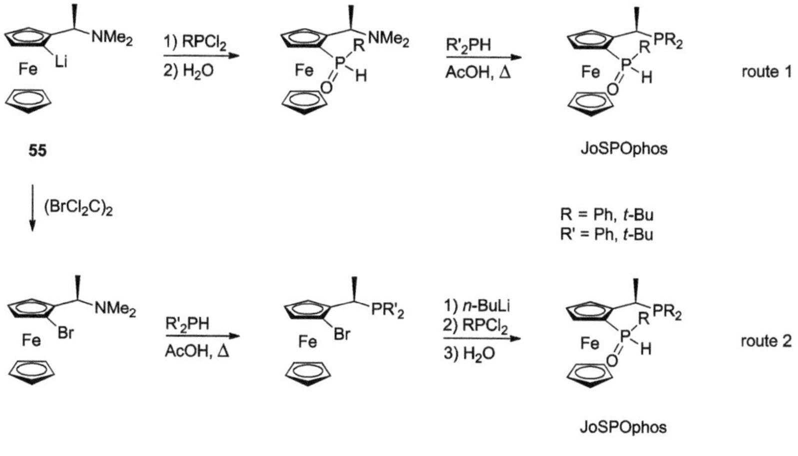

Scheme 41: Synthesis of a bidentate SPO/phosphine ligand with a ferrocene backbone.[138]

Application of the JoSPOphos ligands in rhodium-catalyzed hydrogenation of carbon-carbon double bonds gave excellent results; enantioselectivities of up to 99% and turnover frequencies (TOF) from 2'000-20'000 were achieved. The best ligand, providing 90% - 99% *ee* with all substrates tested was the one in which R = Ph and R' = *t*-Bu. It appears that the dominant factor for induction is the absolute configuration of the secondary phosphine moiety. Thus in nearly all cases tested, the product configuration was inverted when going from the R_P-isomer to the S_P-isomer of the ligand.

2 Secondary Phosphines and Secondary Phosphine Oxides

A former member of our group, Pietro Butti, isolated from the Pigiphos-synthesis small amounts of a side product, which was attributed to the secondary phosphine oxide **80**, according to an X-ray crystal structure he obtained (Figure 7).[139]

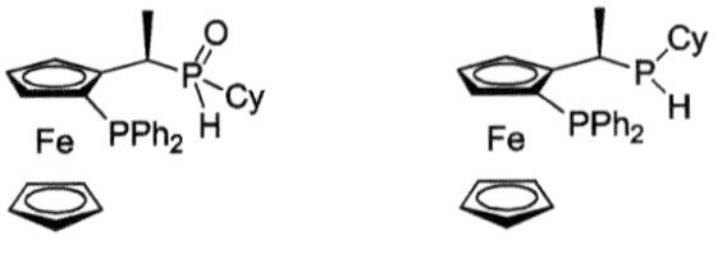

Figure 7: Secondary phosphine oxide **80** and secondary phosphine **81** as observed in our group.[139-140]

Later, it was found that the side product was in fact the unoxidized phosphine **81**, which had been described previously by Barbaro et al.[140] Therefore, it was assumed that oxidation of the secondary phosphino group took place during the crystallization process. A similar process was observed when the secondary phosphine **82** was mixed with [PtCl(COD)]$_2$, yielding single crystals of the SPO-complex **83**, which was also analyzed by X-ray diffraction (Scheme 42).

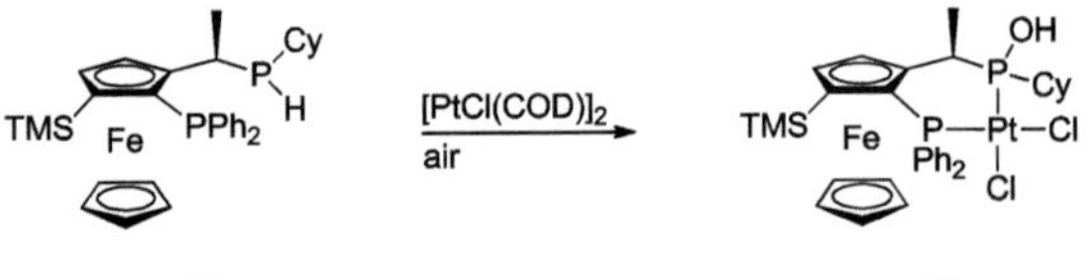

Scheme 42: Complexation and in situ oxidation of secondary phosphine **82** yielded SPO-complex **83**.[140]

Neither secondary phosphine oxide ligand **80**, nor the platinum complex **83** were applied to asymmetric transition-metal catalysis.

2.3 Aim of this Work

The above-mentioned findings that ferrocene-based ligands combining a tertiary phosphine moiety and an SPO group are excellent ligands for the hydrogenation of alkenes led us to further explore this type of ligands. Based on the observation that SPO ligands **84** are able to form stable bidentate complexes with late transition metals such as platinum, one of the goals of this thesis was to develop ligands with an SPO moiety on the side chain and a tertiary phosphine unit on the ring. The retrosynthetic strategy is depicted in Scheme 43.

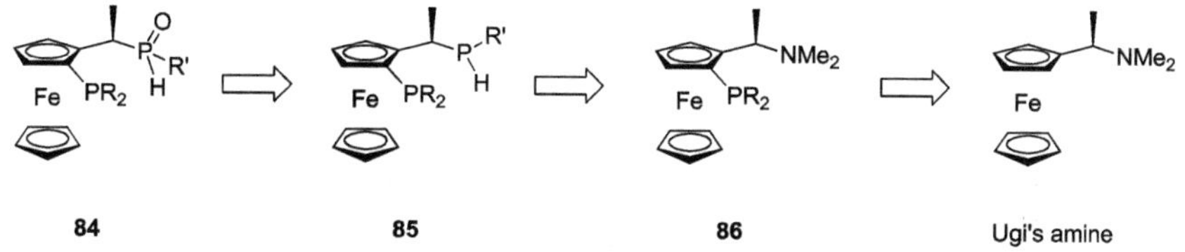

Scheme 43: Retrosynthetic strategy for the synthesis of SPO ligand **84**.

The first step, from a retrosynthetic perspective, the oxidation of the secondary phosphine **85**, should occur during work up or chromatography as secondary phosphines are known to be very sensitive to oxidation. It was planned to introduce the secondary phosphine unit via the well-known[12] acid promoted substitution of the dimethylamino group. The aminophosphine **86** would be obtained via stereoselective ortho-lithiation of enantiomerically pure Ugi's amine,[124] followed by addition of a chlorophosphine, a method very well described in literature.[141] This strategy should provide access to a broad library of SPO-P ligands with different substituents on the phosphines allowing steric and electronic tuning of the ligands. Synthesis and characterization of transition metal complexes of these ligands as well as their application to asymmetric catalysis is also a goal of this thesis.

2.4 Synthesis of Ferrocene-Based Secondary Phosphines

The planned synthesis of the secondary phosphine oxides started as discussed with a stereoselective ortho lithiation of Ugi's amine at -78 °C. Addition of chlorodiphenylphosphine to the lithiated intermediate yielded aminophosphine PPFA as depicted in Scheme 44.

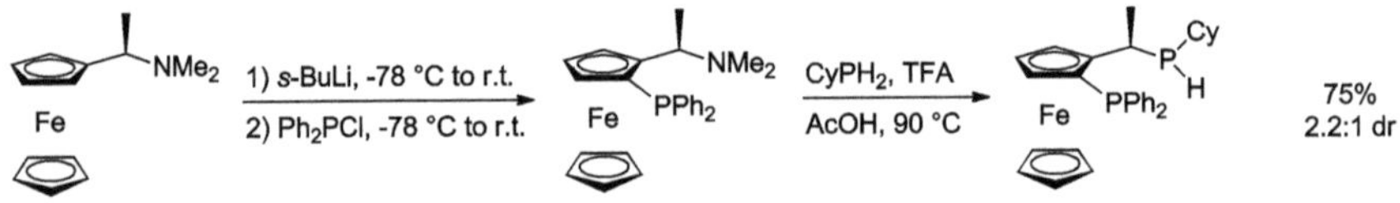

Scheme 44: Two-step synthesis of secondary phosphine **81** starting from Ugi's amine.

For the introduction of the phosphine unit on the side chain, an improved method for the Pigiphos-synthesis developed in our group[139, 142] was applied but using an excess of the primary phosphine (1.5-2.5 eq.) rather than 0.5 eq. The diphosphine product **81** was thus obtained in good yield and as a diastereomeric mixture of 2.2:1. All attempts to separate the diastereoisomers by flash column chromatography, crystallization or HPLC failed.

To our surprise, no formation of a secondary phosphine oxide was observed during synthesis and work up or on the column used for purification. This implies that secondary phosphines of type **85** are very stable toward oxidation, including those with electron-rich substituents. The stability toward oxidation of certain phosphines is not well understood and several factors have to be considered as discussed for primary phosphines by Gilheany and co-workers.[143] Besides the electronic nature of the substituents, steric factors, possible conjugation with a π-system as well as interactions with other functional groups have to be taken into account. Which factors explain the stability of **81** remains unclear, but other similar primary or secondary ferrocenyl phosphines are known to be stable against oxidation.[140, 144-145]

We assumed that the 2.1:1 diastereomeric ratio found in the synthesis of **81** is the result of a thermodynamically controlled equilibrium since the introduction of the phosphine occurs at high temperature. Therefore, the size of the substituents on the primary phosphine used in this step should be crucial to the diastereomeric outcome of the reaction. To verify this assumption, several secondary phosphine derivatives were synthesized as shown in Table 8.

2 Secondary Phosphines and Secondary Phosphine Oxides

Table 8: Synthesis of secondary phosphine compounds with different substituents.

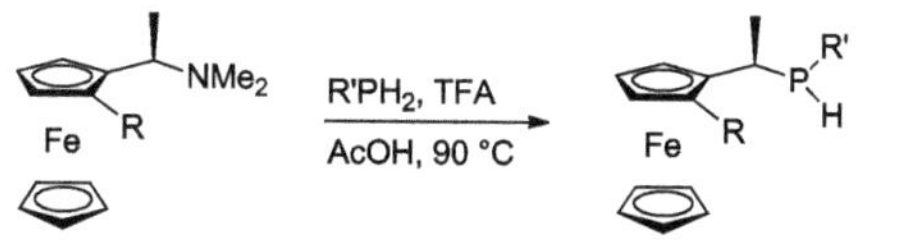

entry	R	R'	product	yield [%]	d.r.
1	PPh$_2$	Cy	81	75	2.2:1
2	PPh$_2$	Ph	87	89	2.1:1
3	PPh$_2$	2-Np	88	68	2.1:1
4	PPh$_2$	1-Ad	89	90	10:1
5	P[3,5-(CF$_3$)$_2$Ph]$_2$	1-Ad	90	86	5:1
6	PCy$_2$	1-Ad	91	70	3:1
7	Br	1-Ad	92	97	5:1

Small primary phosphines such as cyclohexyl- and phenylphosphine yielded diastereomeric ratios of about 2:1 as indicated in entries 1 and 2. To our surprise, when larger aromatic phosphines such as 2-naphthylphosphine were used the ratio did not change (entry 3). The second aromatic ring in 2-naphthylphosphine seems to be too far away from the reaction center to influence the diastereoselectivity. The best diastereomeric ratios were obtained with 1-adamantylphosphine, clearly indicating that bulky groups on the phosphine favor the formation of one diastereomer over the other (entry 4). The ortho-substituents on the cyclopentadienyl ring also showed a certain influence on the diastereomeric ratio although no clear correlation between size and d.r. could be inferred as can be seen from entries 5-7. The best diastereomeric ratio was obtained with R = PPh$_2$ and R' = Ad, giving a 10:1 ratio. **89** is highly crystalline and insoluble in polar solvents, two factors that gave rise to high yields and a fast and easy purification. When the reaction was carried out under high concentration conditions, secondary phosphine **89** crystallized from the reaction mixture. Filtration and washing with water and methanol yielded the pure product in over 90% yield.

Single crystals of **89** and **90** suitable for X-ray diffraction measurements could be obtained by slow cooling of hot concentrated hexane solutions or slow diffusion of methanol into concentrated ether solutions, respectively. An ORTEP-representation of **89** and **90** are shown in Figure 8. **90** was synthesized from the (S_C, R_{Fc})-precursor, therefore the structure shown in Figure 8 was inverted for better comparison with **89**. The exact position of an H-atom cannot be accurately determined by X-ray due to the small electron density associated with H-atoms. The electron cloud of the lone pair on phosphorus and the one of a P-H bond cannot be

2 Secondary Phosphines and Secondary Phosphine Oxides

differentiated by X-ray diffraction methods. Therefore, although X-ray crystal structures of **89** and **90** were obtained, the exact position of the P-H hydrogen and thus the absolute stereochemistry of the secondary phosphine could not be determined.

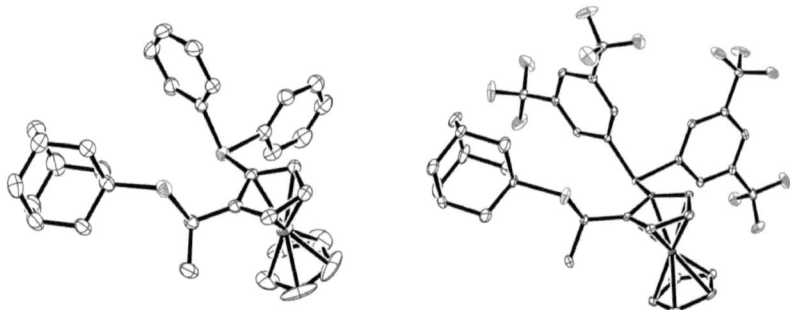

Figure 8: ORTEP representation of **89** and **90** (inverted); hydrogen atoms are omitted for clarity and thermal ellipsoids are set to 30% probability.

Interestingly, the ^{31}P-NMR spectrum of **89** obtained by dissolving single-crystal material still showed both diastereoisomers in the same ratio as observed after the reaction, which confirms the assumption that the two diastereoisomers are in equilibrium. Moreover, it seems that this specific diastereomeric ratio does not result from the long heating at 90 °C as speculated above but is reached even at room temperature. If pyramidal inversion were the source of epimerization then it should not occur at room temperature and can therefore be excluded.[1, 146] Thus, we believe that a fast proton exchange on the secondary phosphine accounts for this epimerization as was observed in the case of (-)-menthylmesitylphosphine by Wild and co-workers.[147] Secondary phosphines are amphoteric, therefore protonation of **93** (e.g. by another secondary phosphine) can occur to form a phosphonium ion intermediate (**94**) which is now achiral (Scheme 45). After deprotonation, either starting material **93** or its enantiomer **95** is formed depending on which proton is removed.

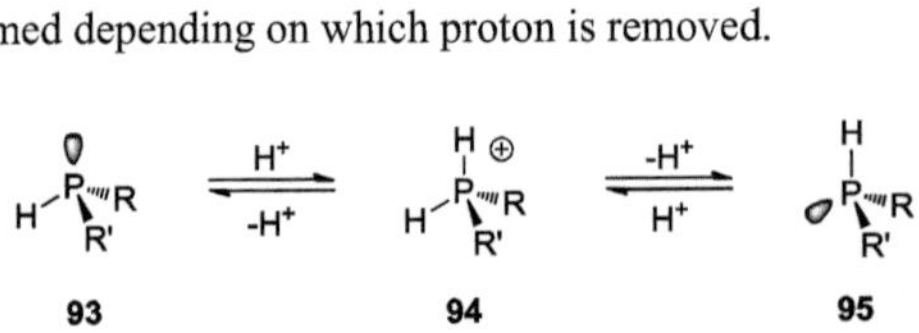

Scheme 45: Racemization of secondary **93** phosphine through a protonation/deprotonation mechanism.

In order to substantiate this proposed mechanism, we designed an experiment in which a proton transfer could be observed using NMR spectroscopy. First we synthesized **92-D**, deuterated on the secondary phosphine by stirring **92** in an acidic D_2O/C_6D_6 mixture. By this method 90% deuterium incorporation was obtained, as assigned by a large 1:1:1-triplet in the ^{31}P NMR spectrum. This material was then mixed with **89-H** in order to verify if proton/deuterium scrambling could be observed (Scheme 46).

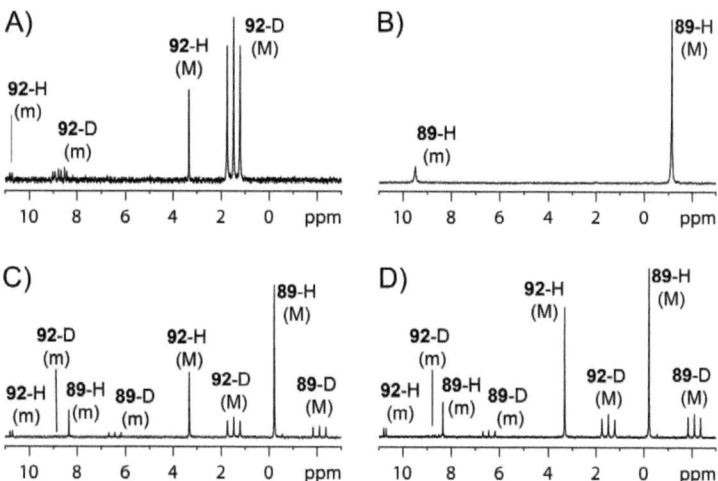

Scheme 46: Deuterium scrambling experiment of secondary phosphines **89** and **92**.

Indeed, after five minutes **92-H** and **89-D** were observed by ^{31}P-NMR spectroscopy and after one hour full scrambling was observed. Figure 9 depicts the ^{31}P NMR spectra of the secondary phosphine region before and after mixing.

Figure 9: ^{31}P NMR spectra of a H/D scrambling experiment for **89** and **91**. A) shows 90% deuterated **92-D** and B) undeuterated **89-H**. C) 5 min after mixing **92-D** and **89-H**, all 8 species can be seen. D) 70 min after mixing, full deuterium scrambling is observed.

(M) and (m) denote major and minor isomers of the secondary phosphines, respectively. In A) the spectrum of 90% deuterated **92-D** and in B) undeuterated **89-H** is shown. C) shows the spectrum 5 minutes after mixing of **92-D** and **89-H**. The signals of all 8 species can be observed. In D) 70 minutes after mixing: integration of **89-H** and **92-H** confirms roughly full scrambling of deuterium. During the scrambling experiments the ratio of the two diastereoisomers of **89-H** remained constant. Thus, with this experiment it was clearly shown

2 Secondary Phosphines and Secondary Phosphine Oxides

that fast proton exchange between the secondary phosphines takes place and that this could account for the formation of two diastereoisomers. Moreover, since the diastereomeric ratio stays constant, we can conclude that this specific d.r. is the result of an equilibrium situation.

2.5 Synthesis of Secondary Phosphine Oxides

As the secondary phosphines turned out to be very stable against air oxidation, we planned to transform them into the SPOs using oxidation reagents. **89** was chosen as substrate as it is easy to synthesize und purify and has the highest diastereomeric ratio, which, as we thought, may be reflected in the SPO product. An overview is given in Table 9.

Hydrogen peroxide was our first choice, because it is cheap, commercially available and has been used successfully for the oxidation of phosphines. When using one equivalent of peroxide no reaction was observed even after heating at 40 °C. When applying an excess, namely three equivalents of H_2O_2, only **96** in which both phosphine units were oxidized could be isolated. The use of t-butyl peroxide and peracetic acid resulted in a mixture of starting material **89**, doubly oxidized product **96**, desired product **97**, and **98** with intact an secondary phosphine moiety and a tertiary phosphine oxide group. As no selective oxidation of the secondary phosphine unit with classical oxidation reagents was achieved, we decided to investigate other synthetic strategies.

Table 9: Attempts of selective oxidation of the secondary phosphine moiety in **89**.

entry	[Ox]	eq. [Ox]	observed products
1	H_2O_2	1	89
2	H_2O_2	3	96
3	t-BuOOH	1	89 + 96 + 97 + 98
4	Peracetic acid	1	89 + 96 + 97 + 98

2 Secondary Phosphines and Secondary Phosphine Oxides

Another possibility to transform a secondary phosphine into its oxide is to first synthesize the corresponding chlorophosphine and hydrolyze this reactive intermediate as depicted in Scheme 47. The use of phosgene as chlorinating reagent afforded, after treatment with aqueous NaOH solution either decomposition or a mixture of product and doubly oxidized material.

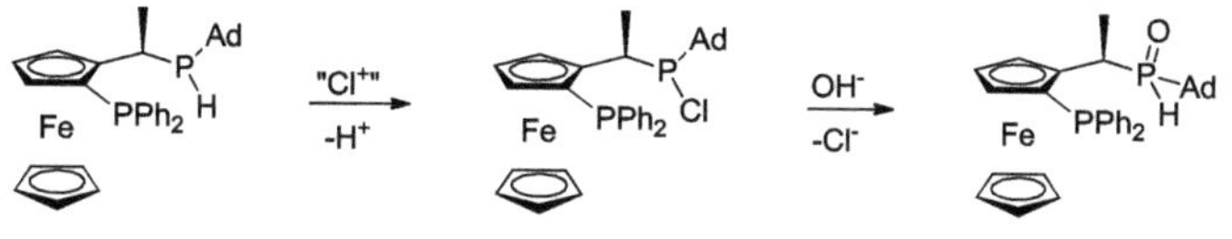

Scheme 47: Attempted synthesis of SPO **97** via chlorophosphine intermediate.

With N-chlorosuccinimide (NCS) and t-butyl hypochlorite no product formation was observed, independently of the conditions applied or bases used. The use of NCS together with catalytic amounts of a titanium catalyst ($CpTiCl_3$ in our case), as was used for the synthesis of α-chloro-β-ketoesters[148] and trifluoromethyl substituted chlorophosphines[149] resulted in a wild mixture of products without any formation of the desired SPO. Softer chlorinating agents such as PCl_3 or CCl_4 which are often used to transform R_2PH into R_2PCl were not able to chlorinate substrate **89**. These results and those obtained from the oxidation attempts clearly indicate that the nucleophilicity of both phosphine moieties are too similar to allow a selective reaction solely on the secondary phosphine unit. Thus, it seems to be crucial to carry out the oxidation of the secondary phosphine in the absence of the tertiary phosphine moiety or use a protecting group strategy to avoid parallel oxidation of both phosphino groups.

One possible approach would be to oxidize the phosphine prior to introduction into the ferrocene system. Transformation of adamantylphosphine into its oxide **99** can be performed using one equivalent of hydrogen peroxide or by air oxidation. **99** was then reacted with PPFA or **100** in acetic acid, but no product formation was observed (Scheme 48). It seems that primary phosphine **99** is not nucleophilic enough for this kind of substitution, at least under the conditions used.

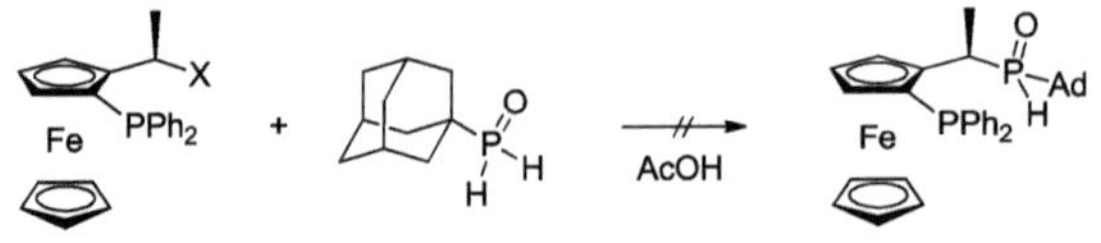

PPFA: X = NMe₂
100: X = OAc

99

97

Scheme 48: Planned synthesis of SPO **97** via substitution with primary phosphine oxide **99**.

The complementary strategy would be to install the tertiary phosphine moiety in the very last step to avoid its oxidation. Therefore, we decided to use a brominated ferrocene derivative, which was planned to be transformed, after bromine-lithium exchange, into the desired product. Secondary phosphine **92** was synthesized in good yields as listed in Table 8 and was smoothly oxidized to the corresponding SPO **101** as shown in Scheme 49. Bromine lithium exchange and subsequent addition of chlorodiphenylphosphine to the lithiated intermediate should then give the desired product. Unfortunately, no conditions were found to accomplish this last step successfully; starting material and unidentified decomposition products were observed in all attempted experiments.

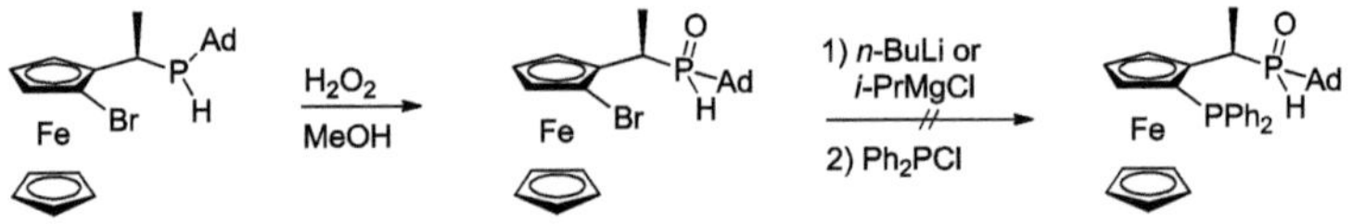

92

101

97

Scheme 49: Unsuccessful attempt to synthesize SPO **97** via brominated intermediate **101**.

A completely different approach to the introduction of the phosphine on the side chain was reported by Knochel and co-workers.[150] Instead of using a nucleophilic substitution at the α-carbon, they generated an α-ferrocenyl lithium derivative and quenched it with electrophiles such as chlorodiphenylphosphine. An example of this formal Umpolung of ferrocene reactivity is shown in Scheme 50.

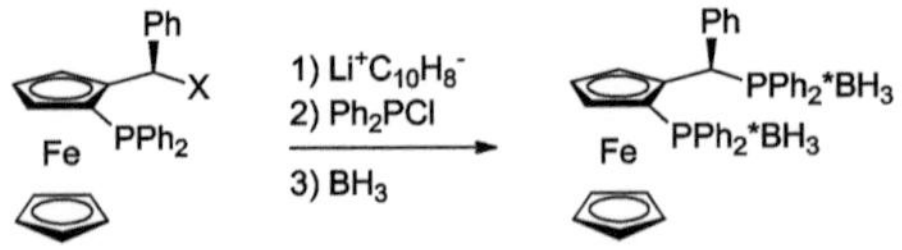

102
X = OMe, SPh

103
60%

Scheme 50: Introduction of a phosphino group using an Umpolung reaction as reported by Knochel and co-workers.[150]

The reductive lithiation of (thio)ether **102** was accomplished using lithium naphthalide and the borane protected diphosphine **103** was obtained in 60% yield.

We planned to apply this method for the synthesis of P-SPO compound **104** as outlined in Scheme 51. Starting from PPFA, thioether **105** was readily prepared in good yield by applying the usual substitution conditions. We intended to perform the reductive lithiation with lithium naphthalide and quench the intermediate with dichlorophenylphosphine, which should give the desired product after basic hydrolysis.

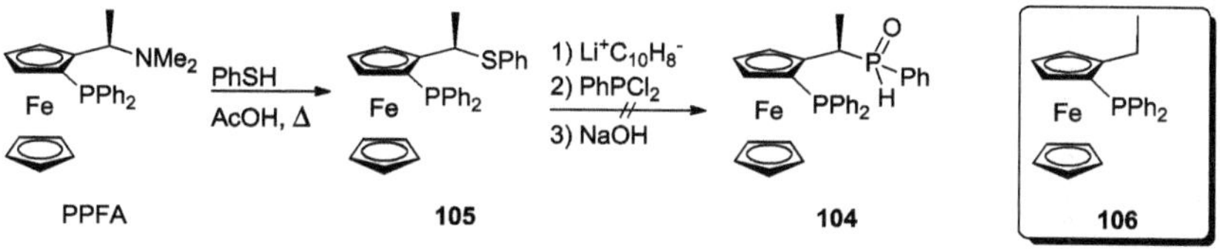

Scheme 51: Synthetic strategy for the preparation of SPO **104** using an Umpolung methodology.

Unfortunately, no product formation was observed and in all cases thioether **105** or reduction product **106** was isolated after the reaction. This Umpolung methodology seemed not to be suited for the synthesis of SPO **104**. Therefore, we had to look for other synthetic strategies.

The use of protecting groups for certain reactive functionalities in order to avoid side reactions is a widely applied strategy in organic synthesis. Therefore, it seems self-evident to apply such a strategy in our case. A protected diphenylphosphino group should allow the selective oxidation of the secondary phosphine moiety and afford after deprotection the desired product **97**. Borane is the most often used protecting group for phosphines, therefore, we started our sequence with the boranation of PPFA as depicted in Scheme 52.

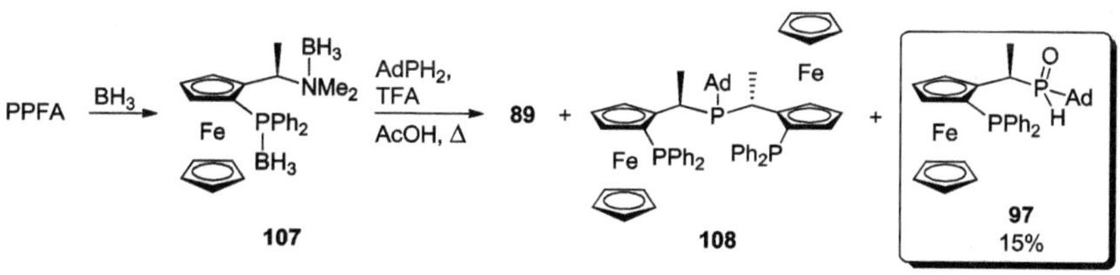

Scheme 52: Synthetic approach towards **97** using double borane protected amino phosphine **107** as intermediate.

The reason why we did not apply this method to **89** is due to the previously observed similarity in nucleophilicity of both phosphine groups which should result in unselective or

2 Secondary Phosphines and Secondary Phosphine Oxides

double boranation with this diphosphine. It turned out that under the conditions used, phosphine as well as amine boranation occurred. Selective phosphine protection with PPFA is possible when working in diluted solutions as shown by Gischig.[151] However, in our case the nitrogen boranation was not seen as a problem since it should favor the substitution of the amino group in the following step. Heating borane protected PPFA **107** together with 1-adamantylphosphine in degassed acetic acid overnight yielded a mixture of several products. Either the acidic conditions or the excess of primary phosphine lead to deboranation since no signals of a borane-protected diphenylphosphino group could be observed in the ^{31}P NMR spectrum. After purification by means of column chromatography three major products were isolated: secondary phosphine **89**, triphosphine Pigiphos **108** and in 15% yield the desired SPO **97**. The d.r. of **97** was around 3:1, which is much less than the one of secondary phosphine **89**. Separation of the isomers by column chromatography was not successful but crystallization from dichloromethane/hexane provided the major isomer in pure form. In the ^{31}P NMR spectrum, the signal for P2 of SPO **97** arises at 55.9 ppm, which represents a large downfield shift compared to the corresponding signal for **89** (δ = 3.4 ppm). The coupling constant between phosphorus and the adjacent hydrogen is with $^{1}J_{\text{P-H}}$ (**97**) = 454 Hz more than two times larger than that of **89** ($^{1}J_{\text{P-H}}$ = 213 Hz). These two observations suggest a pentavalent structure of the SPO.

Single crystals suitable for X-ray diffraction could be obtained and an ORTEP representation of **97** is shown in Figure 10. As the (S_C,R_{Fc})-isomer was used for crystallization the structure shown in Figure 10 is inverted for convenience.

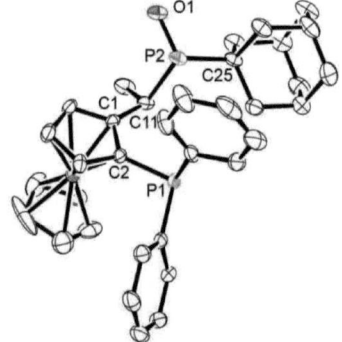

atoms	length [Å]	angle [°]
P2-O1	1.484 (1)	
C11-P2	1.8364 (17)	
C25-P2	1.837 (2)	
C11-P2-C25		110.48 (9)
C11-P2-O1		113.32 (11)
C25-P2-O1		113.65 (11)

Figure 10: ORTEP representation of SPO **97** (inverted). Hydrogen atoms are omitted for clarity and thermal ellipsoids are set to 30% probability.

From the X-ray structure one can infer that in the major isomer the SPO unit adopts the *S*-configuration. The phosphorus oxygen distance is 1.484(2) Å, typical for a P-O double bond, which confirms the pentavalent structure of the SPO. Selected bond length and angles are listed in a table in Figure 10. The angles around phosphorus are similar to the ones reported for *t*-butylphenylphosphine oxide.[152]

The reason for SPO formation, even in low yield, during this reaction under reducing (BH_3), rather than oxidative conditions, remains unclear and the results could never be repeated. As we thought some traces of oxygen together with the acidic conditions could be the oxidant, prolonged heating of **89** in acetic acid under air was performed. No formation of oxides could be observed confirming the high stability of **89** against air oxidation. The low yield and irreproducibility observed in this reaction led us to once again look for other synthetic strategies.

Another approach to overcome the problem of double oxidation is to lower the nucleophilicity of the tertiary phosphine moiety by introduction of electron deficient substituents. Aside from fluorine the trifluoromethyl group has one of the strongest electron-withdrawing characters available, therefore we decided to use bis(trifluoromethyl)phosphine **109** as precursor for the synthesis of P-SPO ligands. **109** can be synthesized from Ugi's amine in three consecutive steps as depicted in Scheme 53.[153]

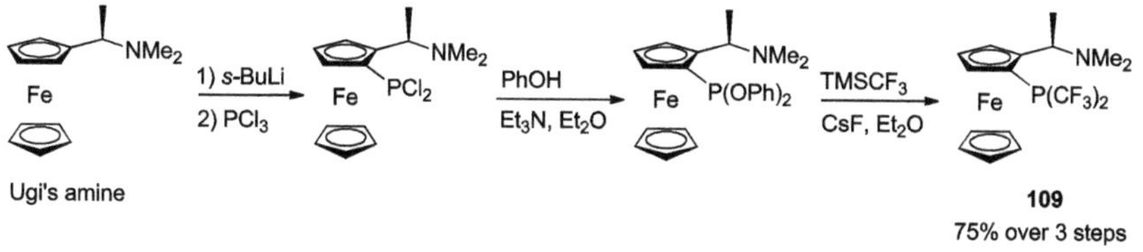

Scheme 53: Synthesis of a bis(trifluoromethyl)phosphine as reported in our group.[153]

Trifluoromethyl substituted phosphines are, to date, rather uncommon ligands in organometallic chemistry and homogeneous catalysis.[154-157] Substitution of the amine function with 1-adamantylphosphine was performed using the standard protocol but with a longer reaction time, as the bis(trifluoromethyl)phosphino group slows down the reaction compared to the more electron-rich diphenylphosphino group. Diphosphine **110** was obtained in excellent yield as a mixture of two diastereoisomers with a d.r. of 6:1. Selective oxidation of the secondary phosphine unit was performed using one equivalent of hydrogen peroxide as depicted in Scheme 54. Product **111** was obtained in high yield as a 20:1 mixture of diastereoisomers. Interestingly, in this case the d.r. of SPO **111** was much higher than the one

2 Secondary Phosphines and Secondary Phosphine Oxides

of the corresponding secondary phosphine, suggesting a dynamic kinetic resolution of **110** during the oxidation reaction.

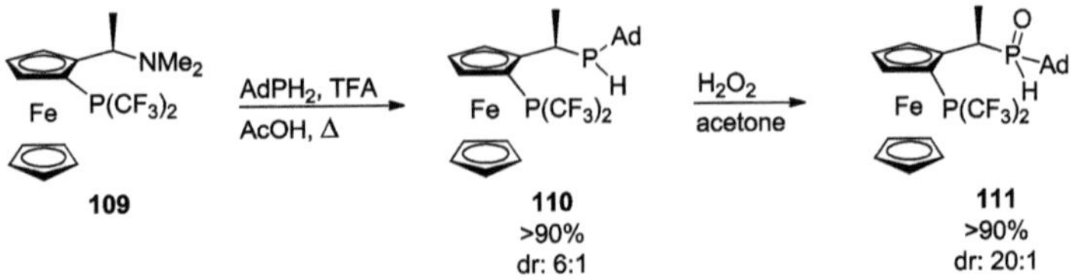

$$\underset{\textbf{109}}{\text{Fc-}{P(CF_3)_2}\text{-NMe}_2} \xrightarrow[\text{AcOH, } \Delta]{\text{AdPH}_2,\text{ TFA}} \underset{\substack{\textbf{110} \\ >90\% \\ \text{dr: 6:1}}}{\text{Fc-}{P(CF_3)_2}\text{-P(Ad)H}} \xrightarrow{\underset{\text{acetone}}{\text{H}_2\text{O}_2}} \underset{\substack{\textbf{111} \\ >90\% \\ \text{dr: 20:1}}}{\text{Fc-}{P(CF_3)_2}\text{-P(=O)(Ad)H}}$$

Scheme 54: Synthesis of secondary phosphine oxide **111** containing a bis(trifluoromethyl)phosphine unit.

However, isolation of the major isomer in pure form was not achieved, neither using chromatographic methods nor by crystallization.

2.6 Synthesis of Transition Metal-Complexes

Bidentate phosphine ligands are one of the most often used classes of ligands in asymmetric catalysis. The combination of diphosphines with transition metals such as rhodium, iridium, palladium or platinum provides highly efficient catalytic systems for a broad variety of transformations including hydrogenation, hydrosilylation, allylic alkylation, 1,4-additions, etc. Most of these systems have been extensively studied and very well characterized. A correlation between the structure of the complex and its catalytic performance is normally not straightforward and predictions for the outcome in catalysis remain difficult. Nevertheless, synthesis of transition-metal complexes with new ligands and their structural and spectroscopic analysis is of great importance in order to gain a more detailed insight into the coordination behavior of such ligands.

2.6.1 Palladium Complexes with Secondary Phosphine Ligands

In contrast to the vast amount of literature concerning metal complexes of tertiary phosphines, only very few reports of secondary phosphine complexes can be found.[158-162] One reason is the weakening of the P-H bond upon metal coordination, which leads to phosphido complexes and makes the study of the ligand properties of secondary phosphines for comparison with tertiary phosphines difficult. Palladium is the most often used metal in combination with secondary phosphines, therefore we decided to study the coordination behavior of ligand **89** to this metal.

2 Secondary Phosphines and Secondary Phosphine Oxides

Dichloropalladium phosphine complexes are the most common for several reasons. The metal precursors, in most cases [PdCl$_2$(COD)] or [PdCl$_2$(CH$_3$CN)$_2$] are air-stable and easily synthesized, in most cases clean reactions are observed and the byproducts can easily be separated. The phosphine dichloropalladium complexes display high stability toward air and are highly crystalline, thus enabling X-ray structure analyses in most cases. As dichloropalladium complexes of most of the important phosphine ligands exist, an easy comparison of the structural and spectroscopic properties of new complexes with the existing ones is possible.

Addition of secondary phosphine **89** to a solution of [PdCl$_2$(COD)] in dichloromethane yielded a deep red solution. The palladium complex **112** was isolated in high yield and purified by crystallization yielding the product as a 10:1 mixture of diastereoisomers as shown in Scheme 55. No formation of phosphido species was observed implying a rather high stability of **112** against deprotonation.

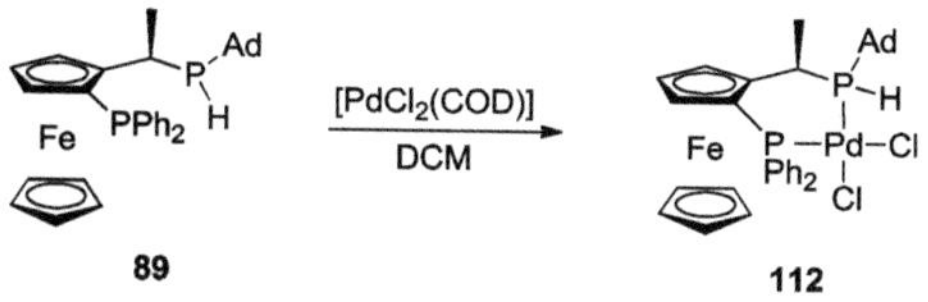

Scheme 55: Synthesis of a dichloropalladium complex **112** with secondary phosphine ligand **89**.

A rather small $^2J_{P-P'}$ cis coupling of 8 Hz (assuming $^4J_{P-P} = 0$ Hz) of the two phosphines and a large $^1J_{P-H}$ coupling of 403 Hz of the secondary phosphino group were observed. Slow diffusion of hexane into a concentrated chloroform or dichloromethane solution afforded single crystals suitable for X-ray diffraction measurement. An ORTEP representation of **112** and a table with selected bond lengths and angles are shown in Figure 11. The bond distance between P2 and Pd1 is slightly shorter than the one between P1 and Pd1, which can be explained with the higher s character of the lone pair in P2 compared to P1. The bond angles of secondary phosphines (~97° for HPPh$_2$) are about 5° smaller than the one of tertiary phosphines (~102° for PPh$_3$).[163] The smaller the angles between the substituents on phosphorus the higher the s character of the lone pair and therefore the shorter the bond to a coordinated metal. We assume that in **112** steric effects of the two phosphino groups can be neglected due to the small steric crowd around the metal center. In **112**, the bite angle (P1-Pd1-P2) is slightly larger than the typical 90°. However, the complex shows a nearly perfect square-planar geometry as the sum of all angles around the palladium center is very close to 360°.

2 Secondary Phosphines and Secondary Phosphine Oxides

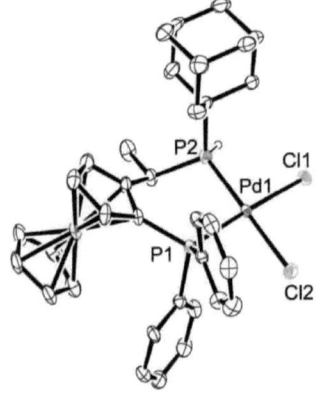

atoms	length [Å]	angle [°]
Pd1-P1	2.2629 (10)	
Pd1-P2	2.2257 (12)	
Pd1-Cl1	2.3605 (10)	
Pd1-Cl2	2.3505 (10)	
P1-Pd1-P2		94.23 (4)
Cl1-Pd1-Cl2		92.11 (4)
sum around Pd1		359.94 (8)

Figure 11: ORTEP representation of **112** and selected bond lengths and angles. Hydrogen atoms except the one at P2 and solvent molecules are omitted for clarity. Thermal ellipsoids are set to 30% probability.

Since the P-H bond of secondary phosphines coordinating to a metal is known to be reactive, we studied the behavior of complex **112** upon treatment with a base. **112** was mixed with one equivalent of potassium *tert*-butoxide in tetrahydrofuran affording a deep red solution. Isolation and crystallization yielded the pure product, a dimeric phosphido-bridged complex **113**, as shown in Scheme 56. In situ preparation, starting from diphosphine **89** gave the same product in somewhat better yields.

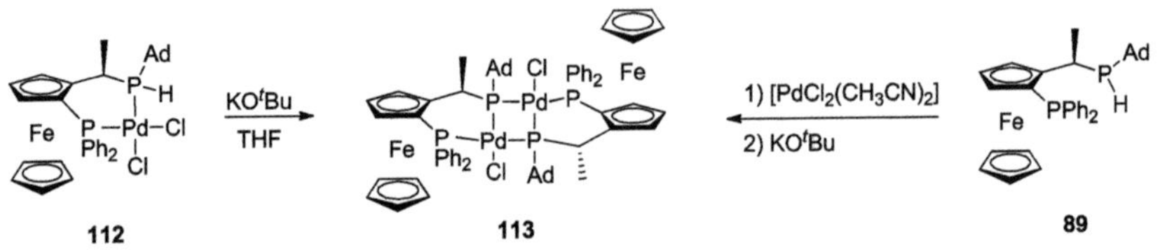

Scheme 56: Synthesis of a dimeric phosphido-bridged palladium complex **113** by proton abstraction from the secondary phosphine complex **112**.

In the ^{1}H NMR spectrum of complex **113**, one sharp set of signals for both ferrocene units was observed. This indicates that the complex is highly rigid or that a fast exchange between two different units is operating in solution. The ^{31}P NMR spectrum is somewhat more complicated and of higher order. An AA'XX' spin system with two sets of signals is observed, as shown in Figure 12.

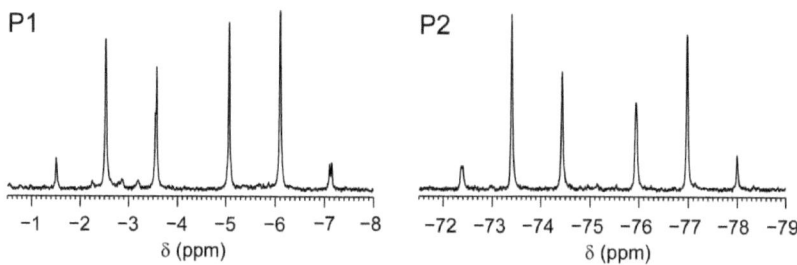

Figure 12: ^{31}P NMR spectrum of **113**, showing an AA'XX' spin system with two sets of signals for each phosphorus atom.

The signals around -4.5 ppm belong to P1 (PPh$_2$) and the other set of signals around -75 ppm is that of the phosphido group (P2). Spectra of similar appearance were reported by Glaser et al. for palladium and platinum complexes containing phosphine and bridging phosphido ligands.[164] We used iterative simulations for the determination of the coupling constants between the different phosphorus nuclei giving rise to such spectra. The simulated spectrum as well as the chemical shifts and the coupling constants giving the best fit of the simulation with the measurement are shown in Figure 13. It has to be mentioned that the relative errors of the coupling constants obtained from the simulation were rather high, especially for the small ones. However, the measured and simulated spectra are very similar in appearance, indicating that the calculated coupling constants match the real values remarkably well.

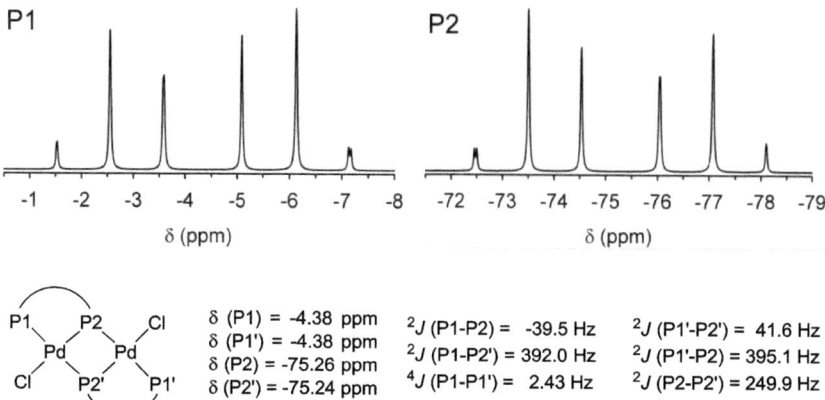

Figure 13: Simulation of the ^{31}P NMR spectrum of **113** and the chemical shifts and coupling constants iterative determined.

2 Secondary Phosphines and Secondary Phosphine Oxides

Whereas dichloropalladium complexes are widely used in coordination chemistry, their application range in catalysis is rather narrow. However, palladium allyl complexes are one example of organometallic compounds that have found broad application in coordination chemistry as well as in catalysis; among others they are used in asymmetric allylic alkylations.

The synthetic procedure we followed for the synthesis of a palladium allyl compound **114** is outlined in Scheme 57. Analysis of the crude product showed a significant amount of decomposed material and the formation of a diphosphine coordinated to a dichloropalladium complex. Recrystallization afforded a pure product, which was found to be complex **115**, as no secondary phosphine moiety was observed in ^{1}H and ^{31}P NMR spectra. **115** can be seen as the product of an intramolecular allylation reaction of a coordinated secondary phosphine. ^{31}P NMR studies of the reaction suggest the formation of an intermediate where ligand **89** coordinates to palladium in a bidentate fashion. Intramolecular allylation then releases the product, probably involving some redox processes as the allyl formally gets oxidized in the course of the reaction. We suggest that half of the palladium is reduced accounting for the low yield and the large amount of byproduct, although no palladium black was observed. However, the exact mechanism of this allylation remains unclear.

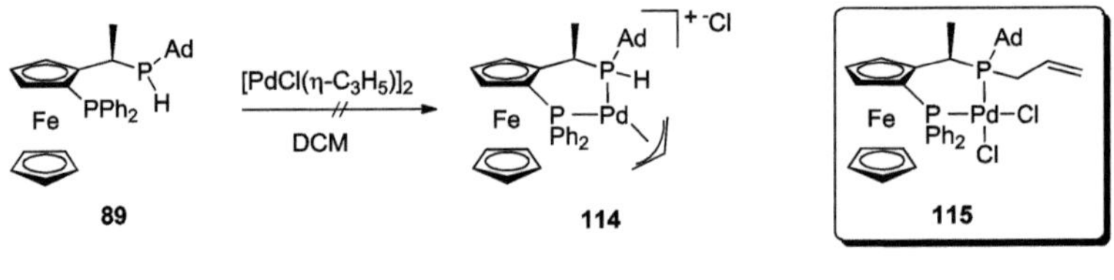

Scheme 57: Attempted synthesis of a palladium allyl complex **114** and structure of the isolated product **115** of this reaction.

Single crystals suitable for X-ray diffraction measurement were obtained by slow diffusion of hexane into a concentrated dichloromethane (DCM) solution of the metal complex. The postulated nature of **115**, was confirmed as shown in Figure 14, also collecting selected bond length and angles. Compared to palladium complex **112** P2-Pd1 is much longer in **115**, which is expected as P1 is part of a tertiary phosphine and not a secondary as in **112**. The bond lengths of Pd1 to P1, Cl1 and Cl2 as well as the bite angle P1-Pd1-P2 in **115** are very similar to those in **112**. A nearly perfect square planar geometry is also observed.

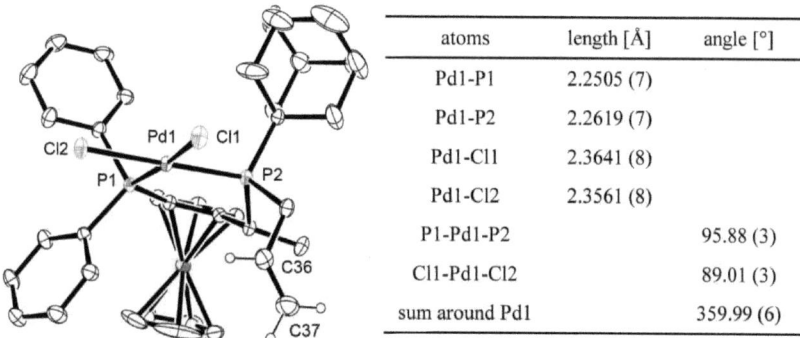

Figure 14: ORTEP representation of **115** and table with selected bond lengths and angles. Hydrogen atoms except the ones on C36 and C37 and solvent molecules were omitted for clarity. Thermal ellipsoids are set to 30% probability.

As olefins are known to act as ligands for palladium we wanted to verify if **115** would undergo olefin coordination after chloride abstraction. Disappointingly, no reaction or decomposition of complex **115** was observed upon addition of potassium hexafluorophosphate or silver hexafluoroantimonate, respectively.

2.6.2 Rhodium Complexes with Secondary Phosphine Ligands

Rhodium-catalyzed hydrogenation of olefins is a highly useful catalytic asymmetric reaction in industrial as well as in academic laboratories. Although in most of these applications the catalyst is prepared in situ, rhodium phosphine complexes are very well characterized spectroscopically as well as structurally. In most cases, cationic complexes with diolefin ligands such as norbornadiene or 1,5-cyclooctadiene serve as precatalysts in hydrogenation reactions. The advantage of these olefin complexes in catalysis is that the olefin is reduced under H_2 atmosphere and the active rhodium catalyst is generated. Under normal conditions these organometallic compounds are somewhat stable.

The secondary phosphine ligand **89** was reacted with [Rh(COD)$_2$]PF$_6$ to yield rhodium complex **116** in good yield as depicted in Scheme 58. In this case, the P-H bond of the secondary phosphine also remained untouched as confirmed by a large $^1J_{H-P}$ coupling constant of 337 Hz. Compared to the palladium complexes **112** and **115** the cis coupling of the two phosphorus atoms is considerably larger ($^2J_{P-P'}$ = 44 Hz) in the case of rhodium. The $^1J_{P-Rh}$ coupling constants of the phosphines to the metal were measured to be 139 Hz for P1 and 113 Hz for P2, respectively.

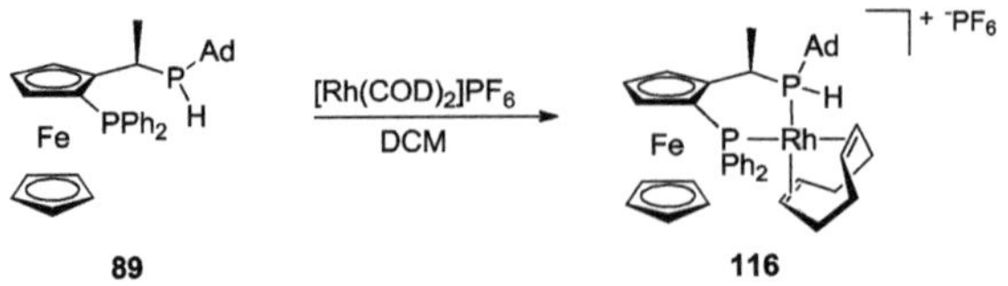

2 Secondary Phosphines and Secondary Phosphine Oxides

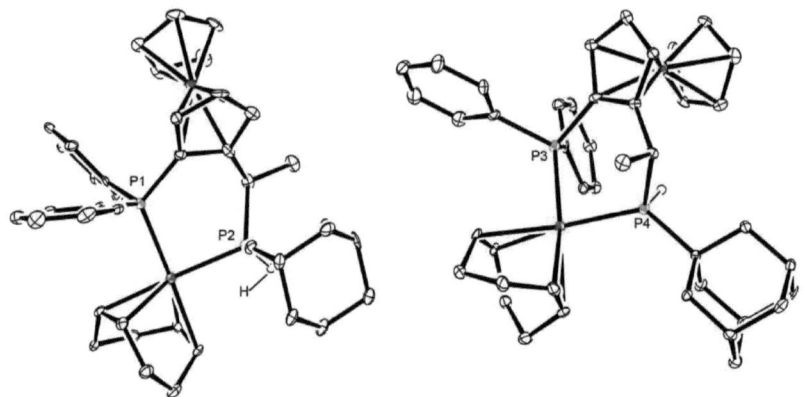

Scheme 58: Synthesis of a rhodium diolefin complex **116** with bidentate secondary phosphine **89**.

The X-ray diffraction measurement of a single crystal of **116** was performed and ORTEP representations are shown in Figure 15 and in Figure 16. Two crystallographically independent and conformationally different molecules were found in the asymmetric unit (Figure 15). The structures were inverted for convenience and the PF_6 counter ions and co-crystallizing solvent molecules were omitted for clarity in both cases.

Figure 15: ORTEP representation of **116** (inverted) with two crystallographically independent molecules in the unit cell. PF_6, solvent molecules and hydrogen atoms except the ones on P2 and P4, respectively are omitted. Thermal ellipsoids are set to 30% probability.

In Figure 16 the two different molecules are shown from the same perspective to allow a better comparison of their structural differences. The absolute configuration of the secondary phosphine is R in both molecules.

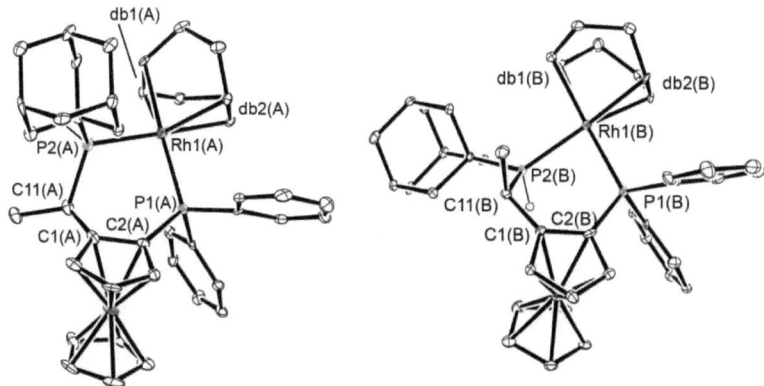

Figure 16: Inverted ORTEP representation of the two different molecules of **116** from the same perspective.

The difference between **116**-A and **116**-B lies in the conformation of the six-membered chelate ring. This ring can be regarded as a cyclohexene derivative adopting a distorted half chair conformation as represented schematically in Figure 17. In **116**-A the adamantyl substituent on P1 is in a pseudo axial position whereas the methyl group on C11 adopts a pseudo equatorial position. In **116**-B the ring is flipped and the methyl group in a pseudo axial position, whereas the adamantyl group now adopts a pseudo equatorial position.

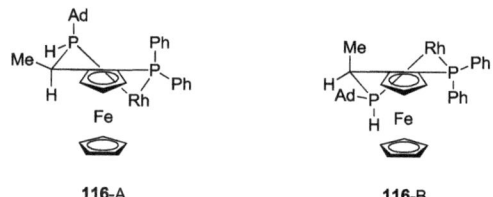

Figure 17: Schematic drawings of the two different half-chair conformations adopted by **116**-A and **116**-B.

The conformational differences between the two molecules in the asymmetric unit are also reflected in certain bond distances and angles as shown in Table 10. In **116**-A, the distance from the metal center to P1 is larger than to P2, whereas in **116**-B the opposite is the case. The bite angle (P1-Rh1-P2) is about 5° larger in A than in B, whereas the distance from rhodium to the centers of the double bonds is slightly shorter in A compared to B. The largest difference is observed for the torsion angle C2-C1-C11-P2. In conformation A the angle is

2 Secondary Phosphines and Secondary Phosphine Oxides

approximately -60° and in conformation B an angle of around 30° is found. This change in algebraic sign reflects the different arrangement of P2 in the halfchair conformation: exo in structure A and endo in structure B. The NMR spectroscopic detection of only one isomer leads to the assumption that in solution only isomer is present or a fast interconversion between A and B occurs.

Table 10: Selected bond lengths and angles of **116-A** and **116-B**. db1 and db2 indicate the center of the double bonds of the cyclooctadiene trans to P1 and P2, respectively.

atoms	116-A length [Å]	116-A angle [°]	116-B length [Å]	116-B angle [°]
Rh1-P1	2.313 (2)		2.295 (2)	
Rh1-P2	2.286 (2)		2.312(2)	
Rh1-db1	2.085		2.115	
Rh1-db2	2.094		2.120	
P1-Rh-P2		91.52 (7)		86.08 (6)
C2-C1-C11-P2		-59.2 (8)		31.0 (9)

2.6.3 Transition Metal Complexes with Secondary Phosphine Oxide Ligands

Palladium and platinum complexes with SPO ligands are well known and are often used in catalytic applications. Despite this fact, complexes with bidentate ligands combining a phosphine donor and a SPO are virtually unknown. To the best of our knowledge platinum complex **83** discovered in our group is the first example of such an organometallic compound. In order to study the coordination behavior of P-SPO ligands with such metals, we synthesized the palladium and rhodium complexes of ligand **111** as depicted in Scheme 59.

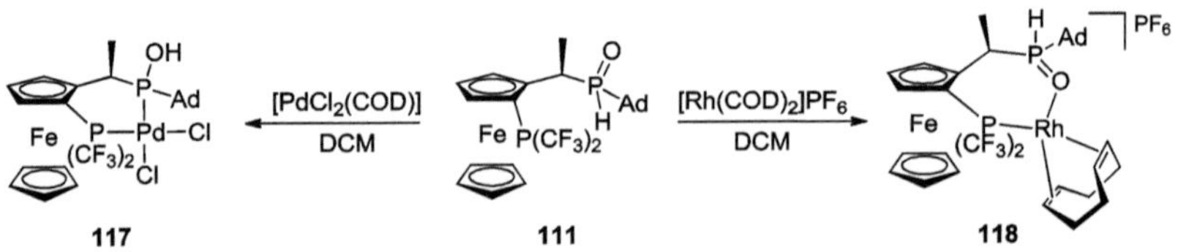

Scheme 59: Synthesis of palladium and rhodium complexes with P-SPO ligand **111**.

The synthesis of palladium complex **117** was straightforward. Mixing ligand **111** and [PdCl$_2$(COD)] afforded the corresponding complex in good yield and as a single diastereoisomer. ^{31}P NMR spectroscopy suggested the coordination of phosphorus to palladium since a high frequency shift was observed and the large P-H coupling,

2 Secondary Phosphines and Secondary Phosphine Oxides

characteristic for the ligand, completely vanished. The two phosphorus atoms showed a small $^2J_{P-P'}$ cis coupling of 10 Hz. Single crystals suitable for X-ray diffraction were obtained by diffusion of diisopropyl ether into a concentrated solution of **117** in a dichloroethane/dichloromethane mixture. An ORTEP representation and a table with selected bond lengths are shown in Figure 18.

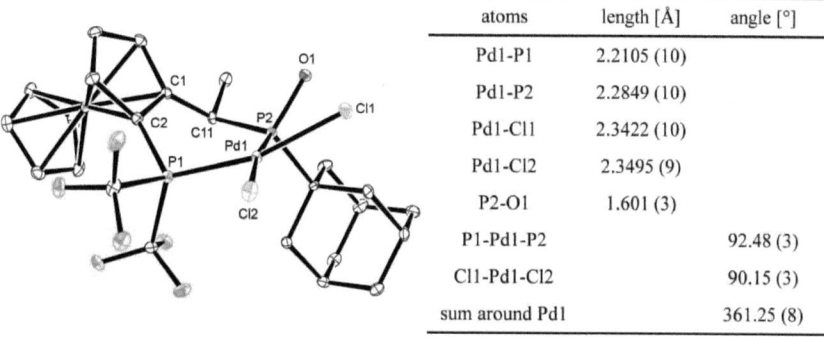

atoms	length [Å]	angle [°]
Pd1-P1	2.2105 (10)	
Pd1-P2	2.2849 (10)	
Pd1-Cl1	2.3422 (10)	
Pd1-Cl2	2.3495 (9)	
P2-O1	1.601 (3)	
P1-Pd1-P2		92.48 (3)
Cl1-Pd1-Cl2		90.15 (3)
sum around Pd1		361.25 (8)

Figure 18: ORTEP representation of **117** and selected bond lengths and angles. Hydrogen atoms are omitted for clarity and thermal ellipsoids set to 30% probability.

The distance between Pd1 and P1 is considerably shorter than the bond to P2, which can again be explained by the high s character of the lone pair on P1, typical for electronically poor phosphines. In addition to this, electron-withdrawing substituents on the phosphines increase their π-acidity, which leads to an increased π-back-donation from filled metal orbitals into empty ligand orbitals. In general, these two effects give rise to short phosphorus-metal bonds.[165] The distances Pd1-P1 and Pd1-P2 in **117** represent the shortest and longest phosphorus-palladium bonds, respectively observed for the palladium complexes presented in this chapter. The phosphorus-oxygen bond is 1.60 Å in length, therefore considerably longer than in the SPO **97** (1.48 Å) and in the range typical of a P-O single bond. A bite angle (P1-Pd1-P2) of slightly over 90° and a slightly distorted square planar geometry (361° around Pd1) is observed.

The SPO ligand **111** was reacted with [Rh(COD)$_2$]PF$_6$ to yield the corresponding rhodium complex **118** in good yield as depicted in Scheme 59. Careful study of ^{1}H as well as ^{31}P NMR spectra led to the assumption that the SPO is coordinating to rhodium via oxygen, rather than through phosphorus. There are several reasons for this assumption: 1) The chemical shift of P2 in rhodium complex **118** is very similar to the one of the free ligand. 2) A large P-H coupling ($^1J_{P-H}$ = 476 Hz) of similar value to that of **111** is observed, indicating an SPO species. 3) Neither coupling to rhodium nor to P1 is observed for P2 implying no direct interaction of the phosphorus atom to the metal. Unfortunately, no crystals suitable for X-ray analysis were obtained to prove this assumption. Oxygen coordination is known for oxophilic metals such as zinc but is rather uncommon for rhodium. If electronic reasons (special donor properties of bis(trifluoromethyl)phosphine) or steric repulsion, as assumed by Pfalz and co-workers[138] in a similar case, are responsible for this unusual behavior remains unclear.

2.7 Ferrocenyl Secondary Phosphines and SPOs in Asymmetric Catalysis

In order to investigate the potential of our diphosphine ligands in asymmetric catalysis, we decided to test them in the rhodium-catalyzed hydrogenation of dimethyl itaconate, a standard substrate for this type of reaction. The results with various ligands are listed in Table 11.

Table 11: Rhodium-catalyzed hydrogenation of dimethyl itaconate using SP and SPO ligands.

$$\text{CH}_2=\text{C}(CO_2Me)\text{CH}_2CO_2Me \xrightarrow[\text{TFE, r.t.}]{\text{1 mol\% [Rh(COD)]PF}_6,\ \text{1.1 mol\% ligand, 1 bar H}_2} \text{MeCH}(CO_2Me)\text{CH}_2CO_2Me$$

entry	ligand	conversion[%]	ee [%]	config. of the product
1	89	100	77	(R)
2	90	100	77	(R)
3	110	100	83	(R)
4	111	100	0	-
5	97	100	29	(S)
6	97 (dp)	100	90	(S)

In all catalytic reactions 1 mol% of catalyst and 1.1 mol% of chiral ligand were used. The reactions were performed in 2,2,2-trifluoroethanol (TFE) at room temperature and under 1 atmosphere of hydrogen pressure. Moderate *ee*'s of 77% were obtained with ligand **89** and **90** bearing aromatic substituents on P1. A slight improvement in selectivity was observed with the bis(trifluoromethyl)phosphine **110**, which provides 83% *ee*. Although in all three cases diastereomeric mixtures of the ligands were used the selectivities are quite high, suggesting the formation of one isomer of the active rhodium complex. It is important to note that with secondary phosphine ligands (entries 1-3) the (R)-isomer of the product is preferentially formed, whereas with typical Josiphos-ligands the (S)-isomer is preferred. This implies a different substrate coordination resulting from the different chiral environment of the ligand.

The bidentate ligand **111**, combining SPO and bis(trifluoromethyl)phosphine moieties disappointingly showed no asymmetric induction in this kind of catalytic reaction. In contrast, diphosphine **97** led to 29% *ee* when used as 3:1-diastereomeric mixture (entry 5). A drastic improvement in selectivity was obtained when using diastereomerically pure SPO **97** affording 90% *ee* (entry 6). These results impressively show the influence of the absolute configuration of the SPO unit on the selectivity. Additionally, the formation of the opposite enantiomer of the product was observed when changing from secondary phosphine **89** to SPO **97**, implying a completely different electronic and steric environment around the metal.

In the frame of a collaboration with Solvias AG,[125] the secondary phosphine ligand **89** was tested in order to clearly determine the reactivity of this ligand. In addition to dimethyl itaconate (DMI), methyl α-acetamido acrylate (MAA) was also used as substrate. The observed turnover frequencies (TOF) and enantiomeric excesses are shown in Table 12.

Table 12: Hydrogenation results using DMI and MAA as substrates and diphosphine **89** as ligand.

entry	R	TOF [1/h]	*ee* [%]	config. of the product
1	CH$_2$CO$_2$Me (DMI)	3000	86	(*R*)
2	NHAc (MAA)	8000	92	(*S*)

For DMI and MAA, high turnover frequencies of 3000 h^{-1} and 8000 h^{-1}, respectively, were observed. The enantiomeric excess for DMI (86%) is slightly higher in MeOH than that observed using TFE as solvent. For MAA a value of 92% *ee* was obtained, which is surprisingly high for a secondary phosphine ligand.

2.8 Conclusion and Outlook

In this subchapter we presented the high yield syntheses of a variety of ferrocenyl secondary phosphines and explored their reactivity. The synthesis of several palladium and rhodium complexes of ligand **89** showed its ability to form stable complexes as well as the reactivity of the secondary phosphine moiety toward certain metal complexes. Additionally, it was demonstrated that these special types of ligands provide in part surprisingly high enantioselectivities in rhodium-catalyzed hydrogenation of functionalized olefins. Unfortunately, to date, no selective oxidation conditions were found for the synthesis of the desired bidentate ligand **97**. In contrary, it was shown that the secondary phosphine moiety can be oxidized in the presence of electron-poor tertiary phosphines, thus accessing bidentate P-SPO ligand **111**. The coordination behavior of this ligand to transition metals was examined and it was found that in palladium complexes P-coordination is favored and in rhodium complexes O-coordination is favored. The high selectivity in the hydrogenation of itaconate with P-SPO ligand **97** shows the significant potential of such ligands. Therefore, a selective synthesis of these ligands still remains an important target for the future.

3 P-Stereogenic Trifluoromethyl Phosphine Ligands

3.1 Introduction

One of the major goals in ligand synthesis is to systematically tune the properties of ligands of a certain class. Tuning of the steric properties via introduction of small or bulky substituents is relatively easy and therefore a large variety of ligands with different steric demand around the donor atoms exist. To modify the electronic properties is rather difficult, but the use of alkyl or silyl substituents enables at least the synthesis of electron-rich ligands. Moderately electron deficient ligands can be synthesized using electron poor aromatic substituents, thus bis-[3,5-bis(trifluoromethyl)phenyl]phosphines or bis(pentafluorophenyl)-phosphines are widely used. For ligands with strong π-acceptor properties the situation is worse. The major drawback of strongly electron-withdrawing ligands such as CO, NO or PF_3 is their lack of a suitable site for modification. However, perfluoroalkylphosphines possess both highly electron-withdrawing character and the possibility for further tuning of the sterical demand and introduction of chirality. Ligands containing of the smallest perfluoroalkyl group (CF_3) have gained significant interest in recent years. However, in the expanding field of alkyl and aryl substituted phosphine ligands relatively few reports dealing with trifluoromethyl phosphines can be found.

In the 1980s, Grobe and co-workers published several reports on trifluoromethylphosphine ligands in chromium, molybdenum and tungsten carbonyl complexes.[166-167] Some examples are shown in Figure 19. The authors were interested primarily in the coordination behavior of these ligands and the electronic influence on carbonyl vibration frequencies. 20 years later, Hoge and co-workers reported platinum and molybdenum complexes of bidentate ligands containing two bis(trifluoromethyl)phosphino groups (Figure 19).[168] They focused on the synthesis of the ligands as well as on spectroscopic and structural properties of the metal complexes.

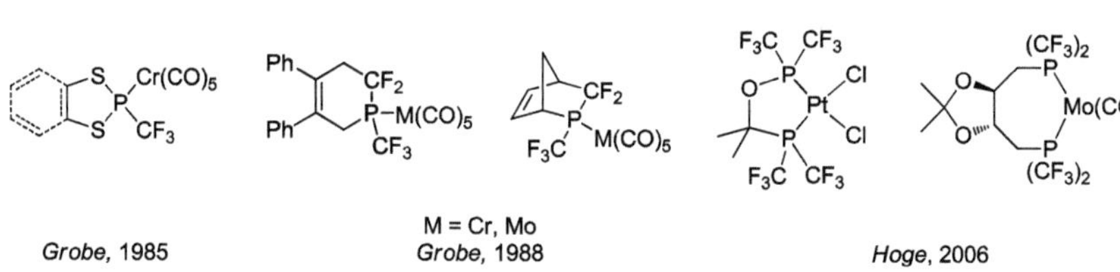

Figure 19: Selected examples of metal complexes with trifluoromethylphosphine ligands.[166-168]

3 P-Stereogenic Trifluoromethyl Phosphine Ligands

Grobe used $F_2C=PCF_3$ and $(CF_3)PI_2$ as precursors, whereas Hoge used $(CF_3)_2PP(CF_3)_2$ to introduce the trifluoromethyl group into his ligands. These precursors are very reactive and difficult to synthesize and purify. Thus, this is one of the major reasons why trifluoromethylphosphine ligands are still fairly rarely used in coordination chemistry and homogeneous catalysis.

3.1.1 Nucleophilic Trifluoromethylation of Phosphines

In 2001, Hoge and co-workers introduced the use of $TMSCF_3$, also known as Ruppert's reagent, for the synthesis of trifluoromethylphosphines as shown in Scheme 60.[169] They used cyanophosphines as substrates and catalytic amounts of ammonium cyanide for the activation of $TMSCF_3$.

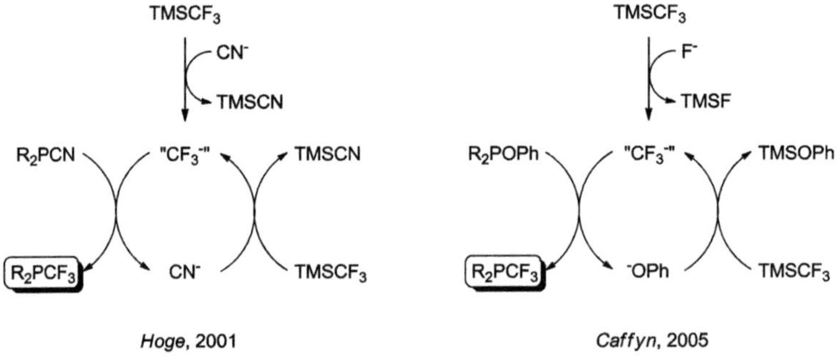

Scheme 60: Synthesis of trifluoromethylphosphines using $TMSCF_3$ as reported by Hoge[169] and Caffyn.[170]

In 2005, a similar method was reported by Caffyn and co-workers as shown in Scheme 60[170]. They used phenylphosphinites and phenylphosphonites instead of cyanophosphines and catalytic amounts of cesium fluoride in order to start the reaction. This methodology allowed the relatively facile synthesis of mono- and bis(trifluoromethyl)phosphines, although not all substrates are suited for this type of reaction. Roddick and co-workers applied Caffyn's method to the synthesis of platinum pincer complexes bearing two bis(trifluoromethyl)phosphino groups as shown in Figure 19.[154]

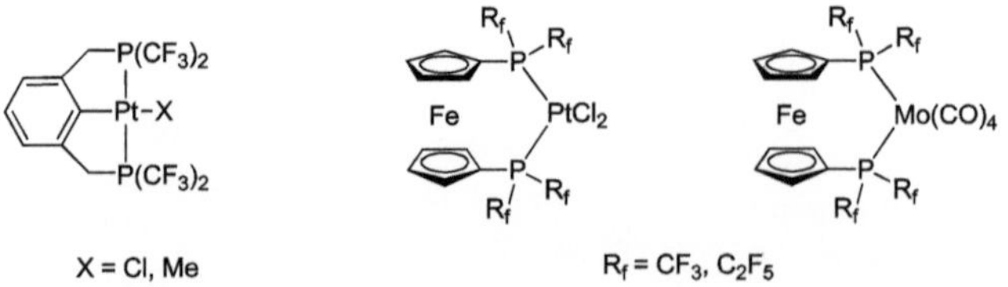

Figure 20: Platinum and molybdenum complexes of perfluoroalkylphosphine ligands as reported by Roddick[154] and Caffyn.[155]

3 P-Stereogenic Trifluoromethyl Phosphine Ligands

In 2008, Caffyn and co-workers reported the first bidentate perfluoroalkylphosphine ligand with a ferrocene backbone.[155] The corresponding dichloroplatinum and tetracarbonyl molybdenum complexes were synthesized, their X-ray crystal structures determined and their carbonyl stretching frequencies analyzed. It was concluded that these ferrocenyl ligands show a highly electron-withdrawing character along with large cone and bite angles, making these ligands some of the bulkiest bidentate (perfluoroalkyl)diphosphines synthesized to date.

In order to gain more detailed insight into the coordination behavior of bis(trifluoromethyl)phosphines, ligands analogous to Josiphos, Pigiphos and P,N-type ligands containing one or two bis(trifluoromethyl)phosphino group were synthesized in our group as shown in Scheme 61.[171] The amine **109** was prepared applying the methodology developed by Caffyn and the synthetic sequence is shown in Scheme 21 in Chapter 2.

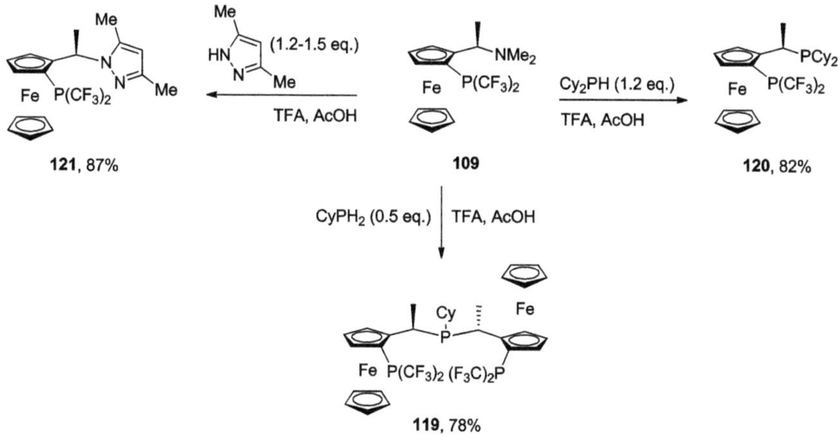

Scheme 61: Synthesis of bis(trifluoromethyl) analogues of Josiphos, Pigiphos and P,N-type ligands.[171]

A wide variety of different transition metal complexes (Pt, Pd, Ir, W) were synthesized and analyzed crystallographically as well as spectroscopically. In all cases, relatively short metal-phosphorus bonds were observed for phosphines bearing trifluoromethyl groups. Surprisingly, the synthesis of a Ni-Pigiphos complex failed with a ligand containing two bis(trifluoromethyl)phosphino groups (**119**). This finding suggests that these phosphines have such a strong electron-withdrawing character that the formation of stable complexes with certain transition metals such as nickel is hampered. Ligands **120** and **121** were applied in several examples of homogeneous catalysis such as rhodium-catalyzed hydrogenation and hydroboration, ruthenium-catalyzed transfer hydrogenation, as well as palladium-catalyzed

3 P-Stereogenic Trifluoromethyl Phosphine Ligands

allylic alkylation. Disappointingly, the activity and/or selecitivity obtained with these ligands in the reactions tested were far worse than the results observed with the original ligand systems. The combination of highly electron-withdrawing character together with small steric bulk seems not to be advantageous in the reactions tested so far. Nevertheless, the concept that bis(trifluoromethyl)phosphines in combination with other donors or phosphines bearing only one CF_3 group should provide good results in homogeneous catalysis remains plausible.

3.1.2 Electrophilic Trifluoromethylation of Phosphines

In 2008, our group published a mild and efficient method for the synthesis of trifluoromethylphosphines using electrophilic hypervalent iodine reagents.[172] Togni's reagents **122** and **123** proved to be suitable for the transformation of primary and secondary phosphines into their mono(trifluoromethyl) analogues as shown in Scheme 62. Later, it was shown that DBU promotes double trifluoromethylation of 2-naphthylphosphine, affording the corresponding bis(trifluoromethyl)-2-naphthylphosphine in moderate yields.[157]

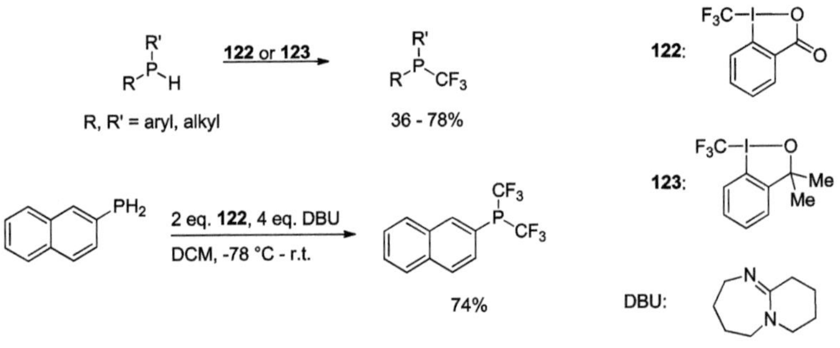

Scheme 62: Trifluoromethylation of primary and secondary phosphines using hypervalent iodine reagents.[157, 172]

This methodology was applied for the synthesis of a *P*-bis(trifluoromethyl) derivative of BINAP (**124**) in a five-step synthesis starting from (*R*)-BINOL as depicted in Scheme 63.[157]

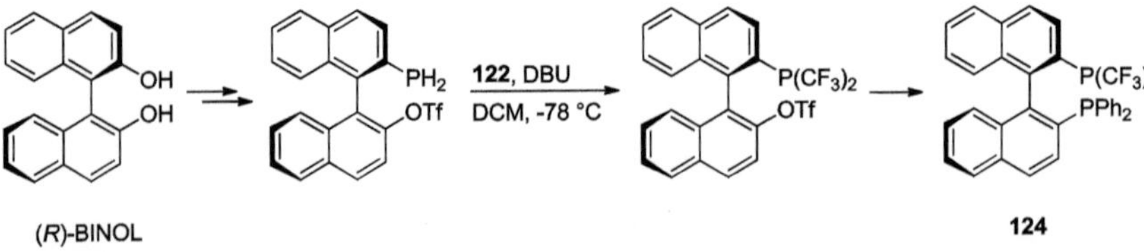

Scheme 63: Synthesis of a doubly *P*-trifluoromethylated analogue of BINAP.[157]

X-ray crystal structure analysis of the corresponding dichloropalladium complex showed a pronounced difference in bond length between the palladium center and the two different

3 P-Stereogenic Trifluoromethyl Phosphine Ligands

phosphorus atoms. The bond to the phosphine bearing the trifluoromethyl groups is significantly shorter (by 4.5 pm) than the one to PPh$_2$. This can be explained with an increased s character of the lone pair orbital of P(CF$_3$)$_2$ resulting from the high electron-withdrawing character of the trifluoromethyl groups or by the difference in steric demands of the two phosphines. Similar results were reported in the literature as well as in Section 2.2.5.3 of this thesis. To date, this type of BINAP ligand has not been tested in asymmetric transition-metal-catalyzed reactions.

The electrophilic trifluoromethylation of primary phosphines using Togni reagent **122** was also used for the synthesis of P-stereogenic ferrocene-based trifluoromethylphosphines as shown in Scheme 64.[149, 173] Phenylphosphine was converted to the corresponding trifluoromethylphenylphosphine and used as nucleophile in the substitution of the protonated hydroxy group in **125**. Separation of the diastereoisomers (**126**) by recrystallization and subsequent replacement of the dimethylamino group with diphenylphosphine afforded the P-stereogenic diphosphines (R_P)-**127** and (S_P)-**127**. Complexes of **127** with Ir, Rh and Pd metal centers were synthesized and crystallographically analyzed. In all three cases, the bond from the metal to the phosphorus bearing the electron-withdrawing trifluoromethyl group was significantly shorter than that to the diphenylphosphino group. This indicates that the substitution of only one phenyl group by a trifluoromethyl group has a significant influence on the donor and acceptor properties of these phosphines towards transition metal centers.

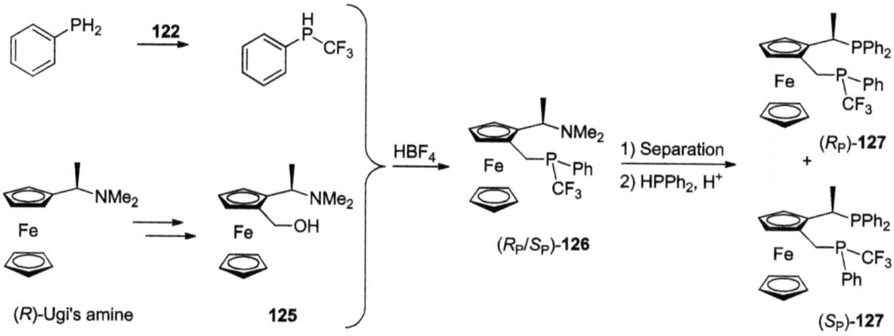

Scheme 64: Synthesis of a ferrocene-based P-stereogenic trifluoromethylphosphine.[149, 173]

(R_P)-**127** and (S_P)-**127** were tested in the rhodium-catalyzed hydrogenation of dimethyl itaconate providing full conversion in chlorinated solvents such as dichloromethane and chloroform. Interestingly, no reaction was observed when the reaction was carried out in typical hydrogenation solvents such as methanol or TFE. The best selectivity (76% *ee*) was observed with (S_P)-**127** in DCM at 0 °C, whereas with (R_P)-**127** in the same solvent only

13% *ee* were achieved. In chloroform, the influence of the absolute conformation of the P-stereogenic phosphine is even more pronounced; (S_P)-**127** gave the *R*-enantiomer of the product in 69% *ee* and (R_P)-**127** provided only 37% *ee* of the *S*-enantiomer. It was suggested that the methylene bridge between the ferrocene ring and the phenyl(trifluoromethyl)-phosphino group provides extensive conformational freedom, thus explaining the modest selectivities observed in the reactions. The use of an analogous ligand bearing two diphenylphosphino groups gave slightly lower selectivities than (S_P)-**127**, supporting the suggestion that not the trifluoromethyl group but a certain flexibility of the complex accounts for the moderate selectivity. Several attempts to synthesize a Josiphos analogue containing a P-stereogenic phenyltrifluormethylphosphino group were unsuccessful. Therefore, the above mentioned suggestion could not be further substantiated.

The relatively high activities observed with the system described above and the finding that the flexibility and not the properties of the trifluoromethylphosphine are responsible for the modest enantioselectivities led to the conclusion that CF_3 remains an interesting group for the electronic and steric tuning of phosphines. Thus, finding efficient and selective methods for the synthesis of effective trifluoromethylphosphine ligands as well as their applications in homogenous catalysis remains an important task.

3.2 Intramolecular Substitution of Trifluoromethyl Groups

3.2.1 Stereoselective Synthesis of 1,2-Diphospholes

In Chapter 2 it was shown that ferrocene-based secondary phosphines not only form stable ligands, but also can react further with the metal precursor under certain conditions. As secondary phosphines (SPs) are used for the synthesis of tertiary ones, the idea of selectively functionalizing the SP moiety in order to obtain Josiphos-type ligands is somewhat self-evident. Thus, we planned to transform secondary phosphine **110** into the phenyl-substituted phosphine **128** as depicted in Scheme 65. Deprotonation of the secondary phosphine moiety using potassium *tert*-butoxide and 18-crown-6 in tetrahydrofuran, followed by addition of fluorobenzene as electrophile should afford diphosphine **128**. The formation of a dark red, reaction mixture immediately after the addition of the base was observed. The addition of phenylfluoride had no influence on the color and after aqueous work up a dark reddish brown oil was formed. Purification by column chromatography yielded a highly crystalline solid, which showed no aromatic signals in the ^{1}H NMR spectrum, indicating that the nucleophilic aromatic substitution of fluorobenzene was unsuccessful. ^{1}H NMR as well as ^{31}P NMR measurements clearly showed the absence of a secondary phosphine moiety and in the ^{19}F NMR spectrum only one CF_3 group was observed. Together with the large phosphorus-phosphorus coupling of $J_{P-P'}$ = 208 Hz, which is typical for two directly connected phosphorus atoms, this strongly indicates the formation of a 1,2-diphosphole **129**.

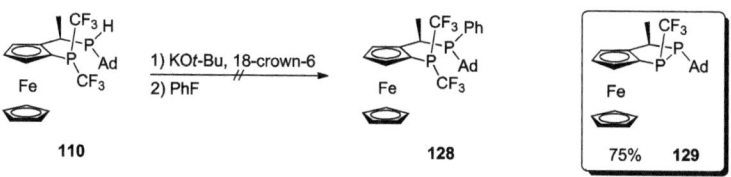

Scheme 65: Synthesis of 1,2-diphophole **129** by intramolecular substitution of a CF_3 group.

This suggestion was then confirmed by elemental analysis, high resolution MALDI mass spectrometry as well as crystal structure analysis (vide infra). The formation of diphosphole **129** is explained by a formal intramolecular substitution of a CF_3 group by the phosphide that was generated after proton abstraction from the secondary phosphine unit. This reaction also indicates that a trifluoromethyl anion may act on P-centers as a leaving group, this being rather surprising as such substitutions are very uncommon. A literature search reveals only one report dealing with the breaking of a P-CF_3 bond by nucleophilic substitution. Roddick

3 P-Stereogenic Trifluoromethyl Phosphine Ligands

and co-workers described the slow hydrolysis of a P-CF$_3$ substituent of a platinum pincer complex **130** as shown in Scheme 66.

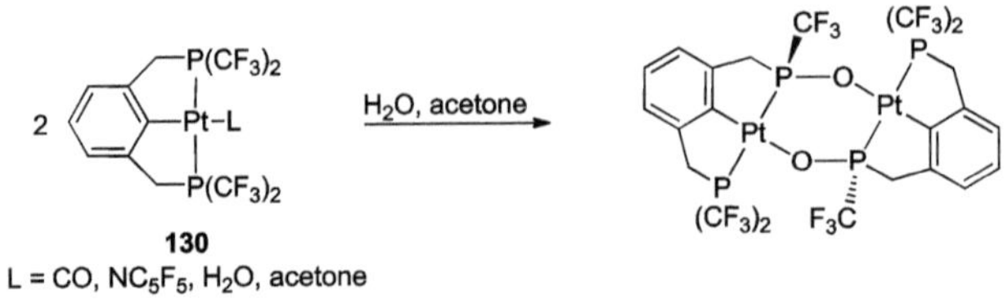

Scheme 66: Hydrolytic cleavage of a P-CF$_3$ bond in platinum phosphine complexes as observed by Roddick.[156]

In their case the P-CF$_3$ bond is activated by coordination of the phosphine to platinum and the formation of HCF$_3$ gas as the by-product facilitates this hydrolysis. In our case, under basic conditions no formation of HCF$_3$ as driving force for the P-CF$_3$ bond cleavage should occur and a free CF$_3$ anion should be formed in the solution. In order to prove this, the reaction was carried out in the presence of benzophenone as an efficient trapping agent, and the corresponding trifluoromethyl alcohol was isolated in 75% yield (Scheme 67). This reaction was repeated with 2-naphthaldehyde, in order to see if the chiral environment of **110** leads to an asymmetric addition of the CF$_3$ group to a prochiral substrate. Unfortunately, the corresponding alcohol was isolated in low yield and in racemic form, indicating no asymmetric induction by the enantiomerically pure starting material.

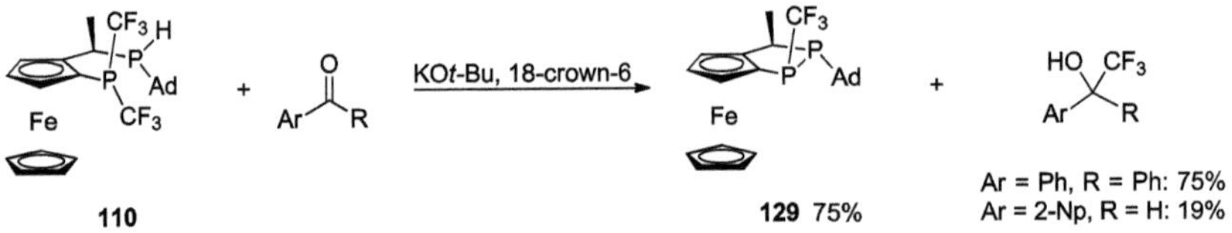

Scheme 67: Nucleophilic trifluoromethylation of aryl ketones and aldehydes upon formation of diphosphole **129**.

The scope of this cyclization reaction was extended to phosphines bearing a cyclohexyl group instead of the adamantyl group (Scheme 68). Starting from the amine **109** the secondary phosphine **131** was isolated in good yields as a 10:7 mixture of diastereoisomers. Treatment of this mixture with potassium *tert*-butoxide and crown ether in tetrahydrofuran yielded diphosphole **132** in 40% yield as a single diastereoisomer.

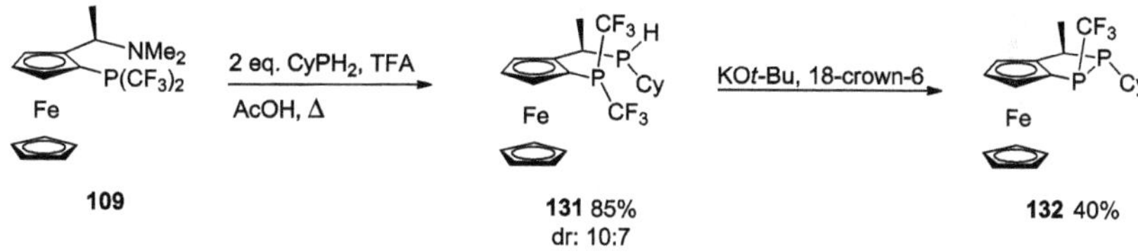

3 P-Stereogenic Trifluoromethyl Phosphine Ligands

Scheme 68: Synthesis of a cyclohexyl substituted 1,2-diphosphole **132**.

The cyclization reaction was accompanied by decomposition and the formation of undefined side products, which explain the rather low yield. Variation of solvents, temperature as well as the nature and amounts of base had no positive influence on the outcome of the reaction. Since we suggested the trifluoromethyl anion to be responsible for the formation of decomposition and side products, the reaction was repeated in the presence of benzophenone, however without any improvement in the yield. Therefore, it remains unclear if the low diastereomeric ratio of starting material **131**, some minimal amounts of impurities or other reasons are responsible for the low yield.

In both cases only one of the four theoretically possible stereoisomers of the 1,2-diphosphole is formed, indicating a very selective substitution of one of the diastereotopic CF_3 groups. This shows that the two stereocenters in **129** and **132** (carbon and planar chirality) direct the formation of the two new stereocenters at P1 and P2 in a highly selective manner. The isomer drawn in the schemes is the most reasonable one, as the steric interactions between all substituents on the diphosphole ring (Fc, CF_3, Ad, Me) are minimized. This arrangement is also confirmed by the crystal structure analysis of **129** and **132** as shown in Figure 21.

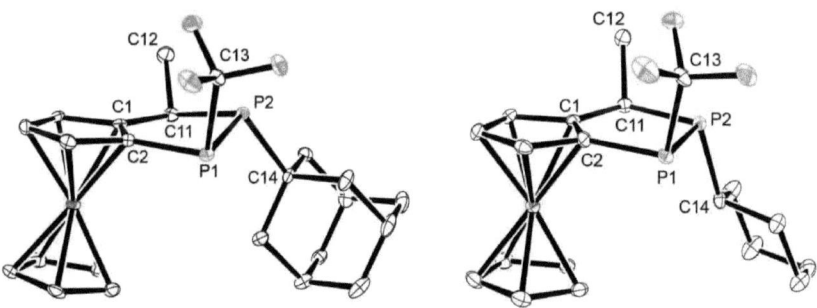

Figure 21: ORTEP representations of **129** and **132**. Of the two crystallographically independent molecules of **132** present in the asymmetric unit, only one is shown. Hydrogen atoms are omitted for clarity and thermal ellipsoids are set to 30% probability

3 P-Stereogenic Trifluoromethyl Phosphine Ligands

As expected, the bulky adamantyl or cyclohexyl groups on P2 are in trans position to both the trifluoromethyl group on P1 as well as the methyl group on C11. This minimizes steric interactions between these substituents and explains the highly stereoselective outcome of the cyclization. Over the course of the reaction two new stereogenic P centers are formed with absolute configuration *S* for P1 and *R* for P2, respectively. Selected bond lengths and angles are listed in Table 13. The bond lengths as well as the angles between the atoms of the 1,2-diphosphole ring are very similar in both structures. The only significant difference between **129** and **132** is the position of P2 in relation to the plane defined by the other four atoms of the diphosphole ring. This can be seen in the last two entries of Table 13 that show the torsion angles between the ferrocene plane and the planes through P1 and P2 (C1-C2-P1-P2, 7°) and C11 and P2 (C2-C1-C11-P2, 16°), respectively. This difference is also apparent by inspection of the ORTEP drawings; the diphosphole ring in **129** is slightly puckered, whereas the one in **132** is perfectly planar. In **129**, P2 deviates by 0.93 Å from the plane formed by the other four atoms of the diphosphole ring.

Table 13: Selected bond lengths and angles of **129** and **132**.

atoms	129 length [Å]	129 angle [°]	132 length [Å]	132 angle [°]
P1-P2	2.2178 (12)		2.2171 (13)	
C2-P1	1.808 (3)		1.810 (4)	
C11-P2	1.885 (3)		1.893 (4)	
P1-C13	1.878 (3)		1.872 (4)	
P2-C14	1.856 (3)		1.855 (4)	
C1-C2-P1		117.8 (2)		117.9 (3)
C2-C1-C11		121.6 (3)		121.2 (3)
C2-P1-P2		92.04 (10)		93.65 (12)
P1-P2-C11		97.25 (11)		97.00 (12)
C1-C2-P1-P2		13.1 (2)		6.0 (3)
C2-C1-C11-P2		-8.8 (4)		7.2 (4)

3.2.2 Ring Opening of 1,2-Diphospholes

The chemistry of 1,2-diphosphacyclopentanes is relatively underdeveloped and these phosphines are rarely used in catalysis.[174] The lone pairs of both phosphines in **129** as well as **132** point in opposite direction, therefore a bidentate coordination to a transition metal seems impossible. In fact, no metal complexes of **129** and **132** were synthesized and both diphospholes were not used in asymmetric catalysis. We reasoned that breaking of the P-P bond and the introduction of two new substituents on the phosphines would be more valuable as a precursor for Josiphos type ligands containing two P-stereogenic phosphino groups.

3.2.2.1 1,2-Diphosphole Chemistry in Literature: A Short Introduction

In 1985, Kauffmann and co-workers presented the ring opening of 1,2-diphosphacyclopentanes and 1,2-diphosphacyclohexanes by nucleophilic attack of a Grignard reagent followed by electrophilic trapping by alkyl bromides as shown in Scheme 69.[175]

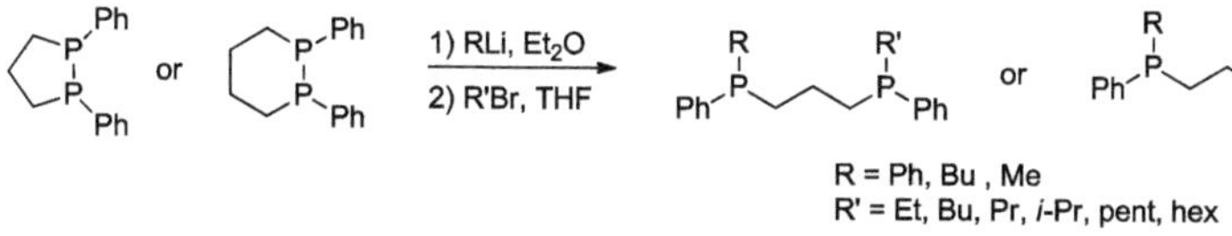

Scheme 69: Cleavage of P-P bonds in 1,2-diphosphacycloalkanes as reported by Kauffmann and co-workers.[175]

Good yields were obtained in all cases reported when the solvent was changed from diethylether to tetrahydrofuran after the Grignard addition. The products were spectroscopically analyzed and complexes with nickel and ruthenium were prepared. In 1992, Alder and co-workers showed that the bond between two phosphorus atoms can be broken by alkylation of both P centers with a strong electrophile and subsequent hydrolytic cleavage of the P-P bond (Scheme 70).[176]

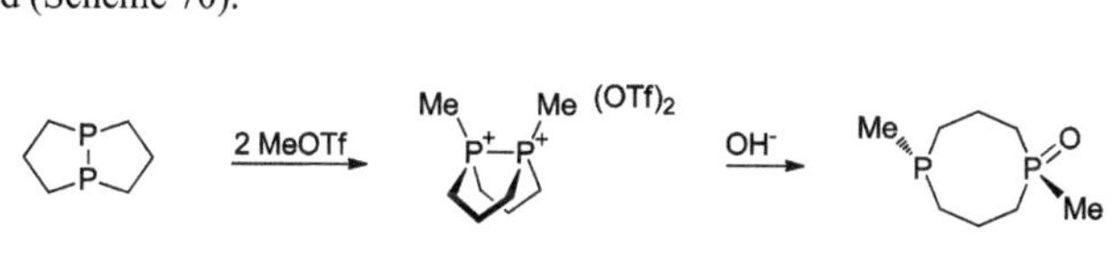

Scheme 70: P-P bond cleavage by double alkylation and subsequent hydrolysis as reported by Alder.[176]

Some years later, the same authors reported a similar reaction but with single phosphorus alkylation and P-P bond cleavage using suitable Grignard reagents as shown in Scheme 71.[177-178] This strategy can be seen as complementary to the reaction sequence reported by Kauffmann. First, an electrophilic alkylation using methyl iodide, methyl triflate or benzyl bromide was conducted giving the corresponding phosphonium salts in good yields. In a

3 P-Stereogenic Trifluoromethyl Phosphine Ligands

second step, the P-P bond was broken by nucleophilic attack of a Grignard reagent such as methyl or phenyl lithium or benzyl magnesium chloride.

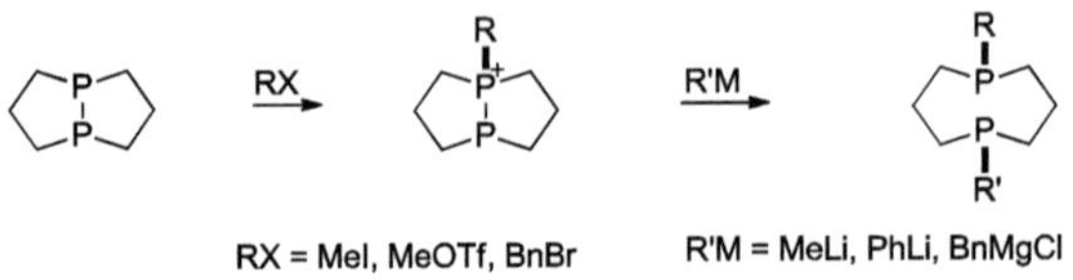

RX = MeI, MeOTf, BnBr   R'M = MeLi, PhLi, BnMgCl

Scheme 71: Diphosphole-ring-opening by electrophilic alkylation followed by Grignard addition.[177-178]

The products were isolated in good yields as a single isomer and their structures were confirmed by crystal structure analysis to be the cis isomers.

3.2.2.2 Transformation of Ferroceno-1,2-Diphospholes into Josiphos-Type Ligands

Encouraged by the promising results reported above, we applied the methodology developed by Alder and co-workers to our system. In a first trial, diphosphole **129** was treated with methyl triflate in dichloromethane followed by a solvent change to tetrahydrofuran and addition of methyl lithium at low temperatures as shown in Scheme 72. The Josiphos-like product was isolated as a single stereoisomer in moderate yield, indicating highly stereoselective alkylation reactions.

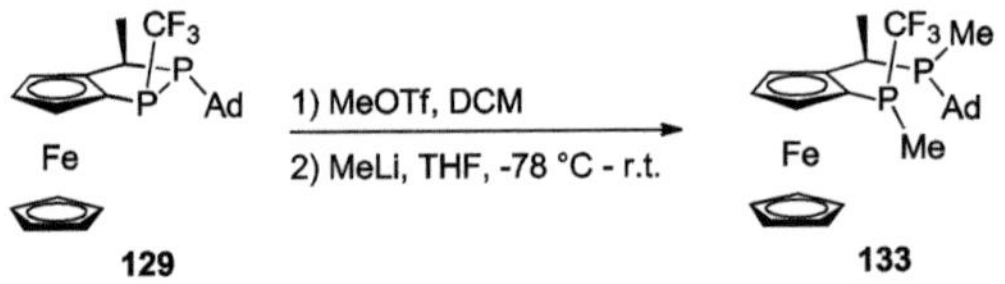

1) MeOTf, DCM
2) MeLi, THF, -78 °C - r.t.

129 → **133** 52%

Scheme 72: Ring-opening-sequence of 1,2-diphosphole **129** using MeOTf and MeLi.

However, the regioselectivity of the electrophilic alkylation was not clear, as the same product should be formed, regardless of which phosphine is alkylated in the first step. Nevertheless, we were strongly convinced that the electrophilic alkylation should occur preferentially at electron-rich P2.

Thus we wanted to extend the scope of this reaction sequence using both 1,2-diphospholes in hand and applying different electrophiles and Grignard reagents in order to prepare a small library of twofold P-stereogenic diphosphine ligands. All newly synthesized Josiphos-type ligands are collected in Table 14.

Table 14: Synthesis of twofold P-stereogenic Josiphos-type ligands from 1,2-diphospholes **129** and **132**.

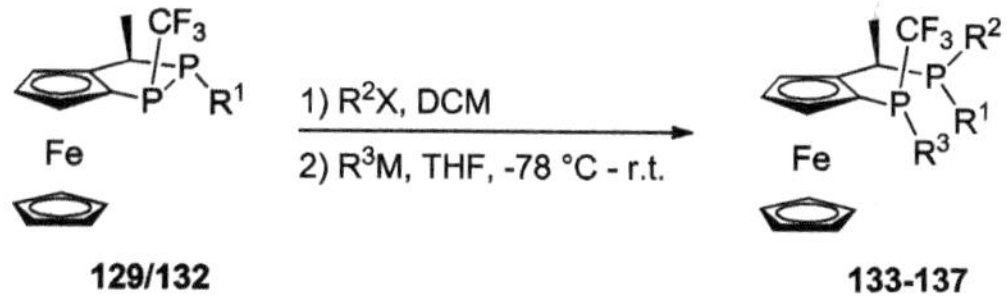

entry	R^1	R^2X	R^3M	product	yield [%]
1	Ad	MeOTf	MeLi	**133**	52
2	Ad	MeOTf	PhMgCl	**134**	60
3	Ad	MeOTf	i-PrMgCl	**135**	60
4	Cy	MeOTf	PhMgCl	**136**	79
5	Cy	MeOTf	i-PrMgCl	**137**	62

Unfortunately, methyl triflate was the only electrophile that could be successfully applied in the reaction sequence shown above. Methyl or ethyl iodide were ineffective and ethyl triflate and Meerwein salts (Me$_3$OBF$_4$, Et$_3$OBF$_4$) also failed to afford the desired products under the conditions tested. Whether the source of this limitation is of steric or electronic nature remains unclear. The scope of Grignard reagents was much broader. Whereas the use of phenyl lithium resulted in a mixture of products, application of phenyl magnesium chloride and *iso*-propyl magnesium chloride provided the products in moderate to good yields. We are convinced that a wide range of alkyl- and aryl-Grignard reagents are suitable for this type of reaction, allowing the tuning of the steric as well as electronic properties of P1. As suggested above, the reaction sequence turned out to be highly regioselective as only one isomer was isolated in all cases. Thus, the electronic and steric differences between the two phosphino groups lead to an electrophilic alkylation occurring exclusively at the more electron-rich P1.

The absolute configuration of both newly formed P-stereogenic phosphines is assigned to be *R* for P1 and *S* for P2 based on the X-ray crystal structure of complex **138** (vide infra). This is in good agreement with the configuration of diphospholes **129** and the reaction mechanism of the ring opening sequence. The electrophilic alkylation at P2 as well as nucleophilic attack at P1 should occur with inversion, thus affording *R* configuration at P1 and *S* configuration at P2 in **133**, respectively.

3.2.3 Transition-Metal Complexes

To assess the coordination behaviour of the new diphosphines with late transition metals, **133** and **134** were reacted with [PtCl$_2$(COD)] or [PdCl$_2$(COD)] in dichloromethane to obtain the corresponding platinum and palladium complexes in good yield (Scheme 73).

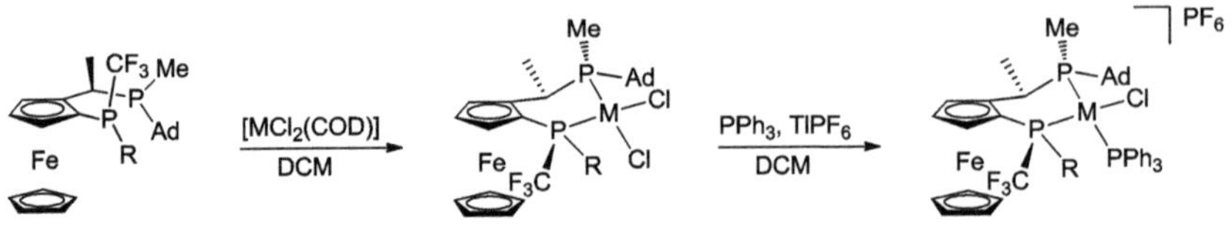

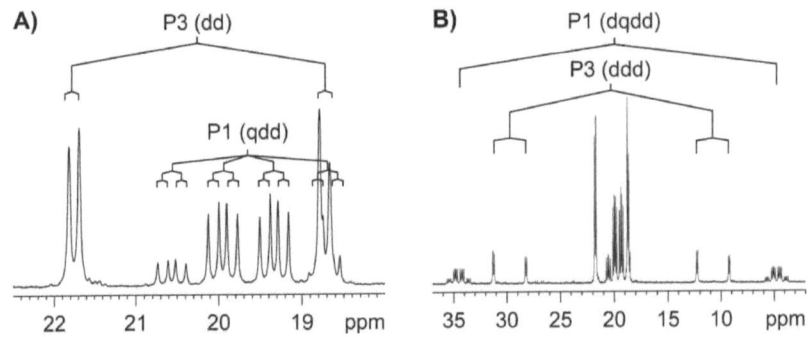

133: R = Me
134: R = Ph

139: R = Me, M = Pt
138: R = Ph, M = Pd

140: R = Me, M = Pt

Scheme 73: Synthesis of palladium and platinum complexes of ligands **133** and **134**.

Platinum complex **139** turned out to be nearly insoluble in all solvents tested and no spectroscopic measurements were obtained. It was therefore converted to the cationic complex **140** upon chloride abstraction and addition of triphenylphosphine. The product was obtained in moderate yield and analyzed NMR spectroscopically. P1, the PCF$_3$ phosphino group appears as quartet of doublets of doublets (qdd) in the ^{31}P-NMR as shown by spectrum A) in Figure 22.

Figure 22: Selected parts of the ^{31}P-NMR spectrum of platinum complex **140**.

The quartet with a coupling constant of $^2J_{P-F}$ = 74.7 Hz arises from the coupling with the fluorine atoms of the CF$_3$ group. The doublet of doublets comes from coupling to P2 and P3 (PPh$_3$) and shows $^2J_{P-P}$ values of 26.6 and 15.4 Hz, respectively, typical for a cis arrangement. In the same region of the phosphorus spectrum the signal of the triphenylphosphino group (P3) can be seen. It arises as a doublet of doublets with a large coupling constant of 368 Hz, typical for a trans arrangement and a small coupling of 15.3 Hz, equal to the one to P1. This clearly indicates the arrangement drawn in Scheme 73 in which PPh$_3$ is cis to P1 and trans to P2. The total regioselectivity achieved in the substitution reaction is the result of the significantly weaker trans effect due to the PCF$_3$ donor on chloride dissociation compared to that of the alkyl substituted P2. The corresponding complex with Josiphos as ligand gave an 8:1 mixture of regioisomers with preference for the isomer with PPh$_3$ trans to P1, indicating the special properties of the trifluoromethylphosphino group in this type of ligand. In B) a larger part of the ^{31}P NMR spectrum shows the satellite signals of P1 and P3 arising from coupling with the ^{195}Pt-nucleus. The satellite of P1 has a multiplicity of dqdd with a coupling constant of $^1J_{P-Pt}$ of 3614 Hz, whereas the coupling of P3 to ^{195}Pt shows a significantly smaller $^1J_{P-Pt}$ value of 2304 Hz. This is indicative of the high s character of the lone pair at P1, resulting from the high electron-withdrawing character of the trifluoromethyl group.

Single crystals of **138**, suitable for X-ray analysis were obtained by diffusion of hexane into a concentrated DCM solution. An ORTEP drawing and a table with selected bond lengths and angles are provided by Figure 23. The X-ray crystal structure confirms the anticipated absolute configuration *R* for P1 and *S* for P2 in the free ligand **134** (under the reasonable assumption that no selective epimerization takes place upon complex formation). Thus, electrophilic alkylation must take place from the *exo* side, whereas the nucleophilic attack of phenyl magnesium chloride at P1 occurs *endo*.[179]

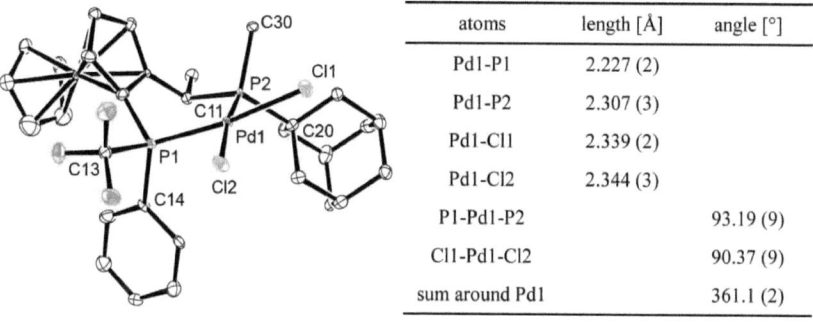

atoms	length [Å]	angle [°]
Pd1-P1	2.227 (2)	
Pd1-P2	2.307 (3)	
Pd1-Cl1	2.339 (2)	
Pd1-Cl2	2.344 (3)	
P1-Pd1-P2		93.19 (9)
Cl1-Pd1-Cl2		90.37 (9)
sum around Pd1		361.1 (2)

Figure 23: ORTEP representation of **138** and table with selected bond lengths and angles.

The distance between palladium and the trifluoromethylphosphino group P1 is considerably shorter than that to the bulky, electron-rich phosphine P2. This can be again explained with the higher s character of the lone pair orbital of P1 compared to P2 as well as with the higher steric crowd on P2. The bite angle of **134** is close to 90° and the sum of all angles is close to 360° indicating a nearly perfect square planar geometry.

3 P-Stereogenic Trifluoromethyl Phosphine Ligands

137 was used for the synthesis of the cationic rhodium complex as shown in Scheme 74. Complexes of this type are used as precursor in the rhodium-catalyzed hydrogenation of olefins. Unfortunately, the ^{31}P NMR spectrum was not resolved well enough to clearly identify the coupling constants but it seems that the coupling between P1 and rhodium is slightly larger than the one of P2 to the metal center. Again, this indicates the relatively high s character of the lone pair at P1 compared to that at P2.

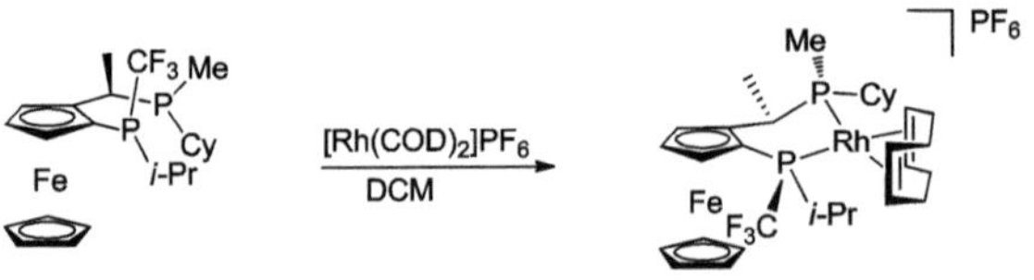

Scheme 74: Synthesis of a cationic rhodium complex of trifluoromethylphosphine **137**.

3.3 Intermolecular Substitution of Trifluoromethyl Groups

3.3.1 Synthesis of Trifluoromethylphosphines via Nucleophilic Substitution

In Section 3.2.1, we showed the intramolecular substitution of a trifluoromethyl group by a phosphido anion. Subsequently, we were interested to see if nucleophilic substitution at bis-(trifluoromethyl)phosphines is possible in a more general sense. If this is the case, the synthesis of a broad variety of trifluoromethylphosphines with different substituents should be possible by applying alkyl and aryl nucleophiles to RP(CF$_3$)$_2$ compounds. To explore the reactivity of these phosphines with organometallic reagents we treated bis(trifluoromethyl)-phosphine **120** with phenyl lithium as shown in Scheme 75.

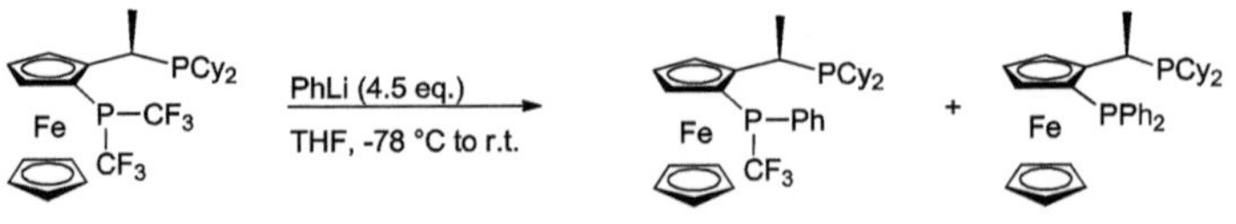

Scheme 75: Attempts for the intermolecular nucleophilic substitution of trifluoromethyl groups.

When one equivalent of phenyl lithium was added to a solution of **120** in tetrahydrofuran only minimal quantities of substitution products, however large amounts of unreacted starting material were observed by ^{31}P NMR spectroscopy. No separation of product **142** and starting material **120** was possible, thus the trifluoromethylphosphine **142** could not be isolated in pure form. Therefore, an excess of phenyl lithium was applied leading to full consumption of bis(trifluoromethyl)phosphine **120**. The formation of the desired products as mixture of stereoisomers as well as the presence of Josiphos, which is the product of a twofold CF_3 substitution, was confirmed by ^{31}P NMR spectroscopy. Purification by flash column chromatography yielded one diastereoisomer of (S_P)-**142** in pure form in 15% yield and an inseparable mixture of the other isomer (R_P)-**142** and Josiphos.

3.3.2 Synthesis of P-Stereogenic Trifluoromethyl Analogues of PPFA

Although we were pleased to find that the intermolecular substitution of trifluoromethyl groups is possible, the need for excess nucleophile and the unselective outcome of the reaction led us to modify the protocol. Instead of using the ferrocenyl substituted phosphine as the electrophile, we envisaged the use of lithiated Ugi's amine as the nucleophile in this transformation. The electrophilic counterpart in this reaction would then be and aryl or alkyl substituted bis(trifluoromethyl)phosphine. In the case of incomplete and/or unselective reaction, starting material (Ugi's amine) as well as unreacted phosphine should be easily separated from the products. As the synthesis of bis(trifluoromethyl)-2-naphthylphosphine (**143**) was known, we decided to use this phosphine as electrophile as shown in Scheme 76.

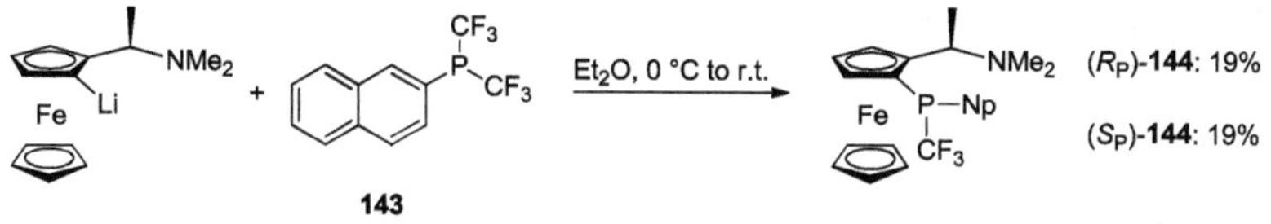

Scheme 76: Synthesis of trifluoromethylphosphine **144** by substitution with lithiated Ugi's amine.

Addition of lithiated Ugi's amine to a solution of **143** in diethyl ether at 0 °C afforded a dark brown inhomogeneous mixture, which was stirred for 60 hours. After aqueous work up and chromatographic purification both diastereoisomers of the product could be isolated in 19% each. The polarity of the two isomers differs in such a way as to allow easy separation by chromatography. The low yield is most likely explained by the low conversion (27% of Ugi's amine isolated) as well as with the formation of significant amounts of decomposition products. With these preliminary results in hand, we decided to use this methodology and optimize the conditions in view of an efficient synthesis of trifluoromethylphosphines. We planned to use bis(trifluoromethyl)phenylphosphine (**145**) as the electrophile in this reaction, because its precursor phenylphosphine, in contrast to naphthylphosphine, is commercially available. Moreover, the use of a phenyl substituted phosphine would allow the synthesis of a bidentate ligand very similar to Josiphos, with one phenyl group replaced by a trifluoromethyl

3 P-Stereogenic Trifluoromethyl Phosphine Ligands

group. The synthesis of both diastereoisomers of this diphosphine and the comparison to Josiphos should give a deeper insight into the electronic as well as steric properties of such trifluoromethylphosphines. Phenylphosphine **145** was synthesized according to the method used for the synthesis of the naphthyl analogue and the product was isolated in 60% yield (Scheme 77).[157]

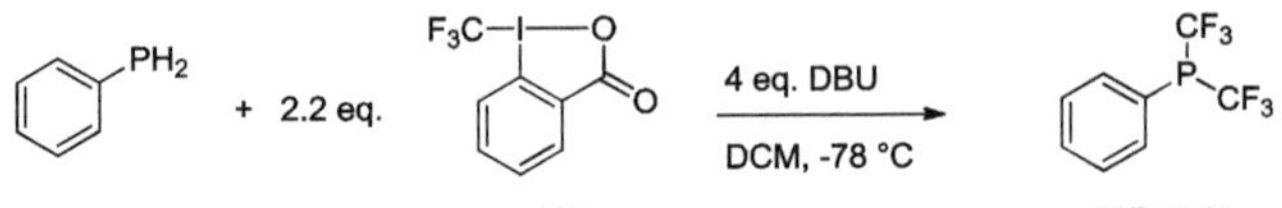

Scheme 77: Synthesis of bis(trifluoromethlyl)phenylphosphine (**145**) using Togni reagent **122**.

Since phenylphosphine and the trifluoromethylation agent **122** are rather expensive, we decided to apply the methodology of Caffyn[170] for the synthesis of **145** as shown in Scheme 78. Both, the phosphine precursor as well as the trifluoromethyl source are cheap and readily available and the desired product is obtained in two steps.

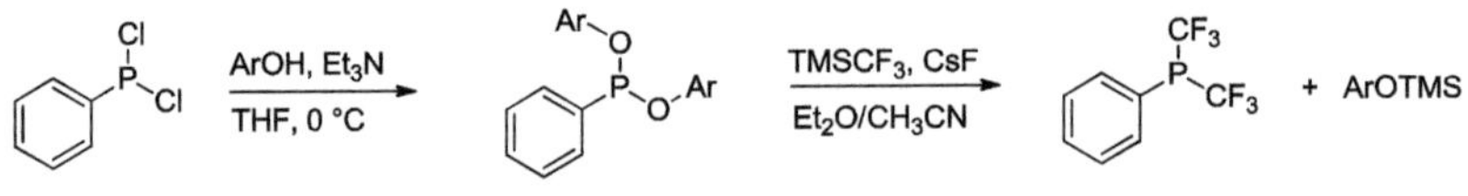

Scheme 78: Synthesis of bis(trifluoromethyl)phenylphosphine using the Caffyn's methodology.[170]

Firstly, dichlorophenylphosphine was transformed into the diarylphenylphosphonite **146** in quantitative yield. In a second step, **146** was converted to the bis(trifluoromethyl)phosphine **145** using Ruppert's reagent (TMSCF$_3$) and catalytic amounts of cesium fluoride. Using phenol in the first step of the synthesis led to PhOTMS as a byproduct, which was not separable from the product. However, when *para*-cyanophenol was used, filtration over silica yielded pure phosphine **145** in 60% yield over two steps.

Phosphine **145** was reacted with lithiated Ugi's amine thereby generating trifluoromethylphosphine **147**, as depicted in Scheme 79. Several reaction parameters such as reaction temperature during addition, concentration of the reactants as well as the time needed for the addition of the lithiated species were varied and some effects on the outcome of the reaction were observed.

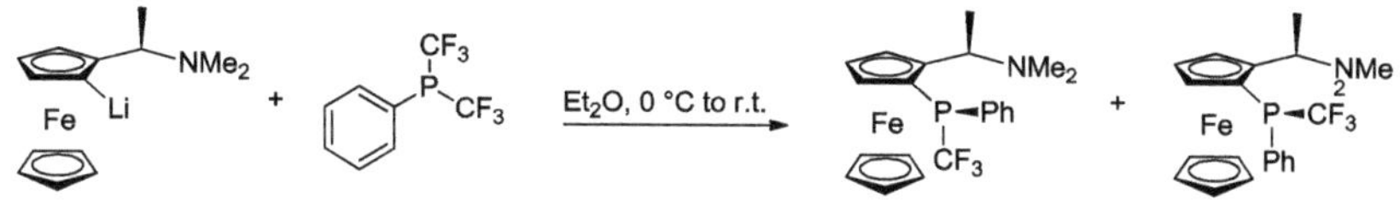

3 P-Stereogenic Trifluoromethyl Phosphine Ligands

Scheme 79: Synthesis of trifluoromethylphosphine X starting from lithiated Ugi's amine.

In a typical experiment, lithiated Ugi's amine was slowly added to a slight excess (1.2 eq.) of the bis(trifluoromethyl)phosphine **145**. The best result (49% yield) was obtained when the organolithium species was added drop wise at 0 °C and when relatively concentrated solutions (1M) were used. Ugi's amine as well as bis(trifluoromethyl)phosphine were recovered in most cases and significant amounts of unidentified decomposition products were also observed. As the reaction never reached full completion, we decided to apply an excess of the nucleophile instead of the electrophile, as we did with phenyl lithium. Indeed, when two equivalents of Ugi's amine were added to **145**, the two isomers of the product could be isolated in 36% and 45% yield, respectively. The ratio of the diastereomeric products was in all cases close to 1:1, indicating no or very modest diastereoselectivity in the substitution of the lithiated Ugi's amine.

In order to expand this methodology to alkyl substituted trifluoromethylphosphines, we decided to synthesize adamantylbis(trifluoromethyl)phosphine (**148**). The combination of a small electron-withdrawing and a bulky electron-rich substituent on the same P atom should give rise to unique steric and electronic properties. Since 1-adamantylphosphine was used for the synthesis of secondary phosphines (vide supra), we decided to use it as starting material for **148** using the electrophilic trifluoromethylation strategy (Scheme 80).

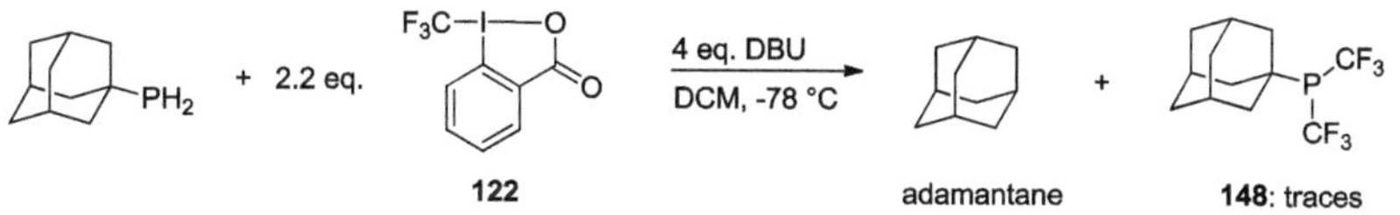

Scheme 80: Attempted synthesis of adamantylbis(trifluoromethyl)phosphine using Togni reagent **122**.

The typical conditions were applied, but only traces of the desired product were obtained and mainly adamantane was isolated from the reaction mixture. Since adamantyl radicals are known to be quite stable and a radical mechanism for the trifluoromethylation reaction with **122** is possible, we suggest that a radical pathway may lead to the cleavage of the adamantyl-phosphorus-bond. As **148** was obtained only in small amounts and was not separable from adamantane, we decided to apply the nucleophilic trifluoromethylation in this case as well.

3 P-Stereogenic Trifluoromethyl Phosphine Ligands

The synthetic strategy utilizes commercially available 1-bromoadamantane and is outlined in Scheme 81. For the preparation of the dichlorophosphine **149** we followed the protocol described by Koenig and co-workers.[180]

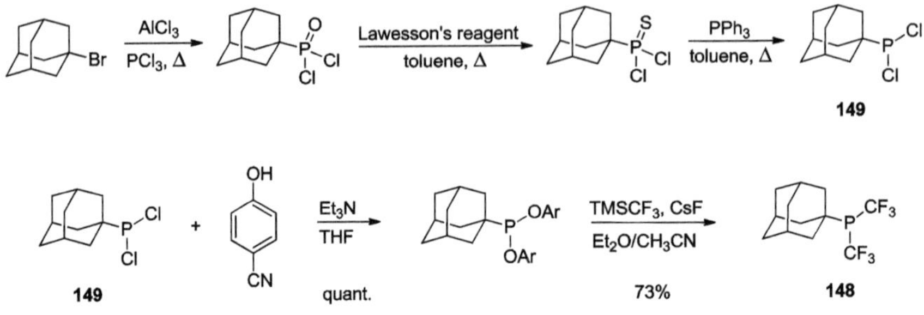

Scheme 81: Synthesis of 1-adamantylbis(trifluoromethyl)phosphine **148**.

149 was then converted to the desired bis(trifluoromethyl)phosphine **148** in two steps using the method described for the synthesis of the phenyl analogue. **148** was obtained as highly volatile clear oil in 73% yield over two steps.

Adamantylphosphine **148** was then reacted with lithiated Ugi's amine (Scheme 82) applying the previously optimized conditions. Interestingly, in this case only one isomer of the product ((S_P)-**150**) was formed, but in only 35% yield.

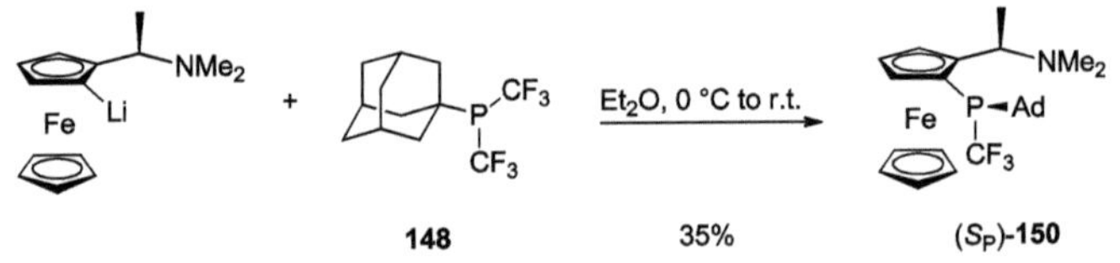

Scheme 82: Synthesis of P-stereogenic trifluoromethylphosphine **150**.

The steric difference between adamantyl and trifluoromethyl seems to be large enough that only one isomer is formed during the substitution reaction. The large steric bulk on the phosphine may then also be the reason for the low yield observed in this reaction. Significant amounts of Ugi's amine and byproducts were also observed. Therefore, we suggest that, although the lithiated species is used in a twofold excess, the reaction never reachs completion, due to steric hindrance. The (*S*)-configuration of the newly formed P-stereogenic phosphine was comfirmed by crystal structure analysis as shown in Figure 24.

3 P-Stereogenic Trifluoromethyl Phosphine Ligands

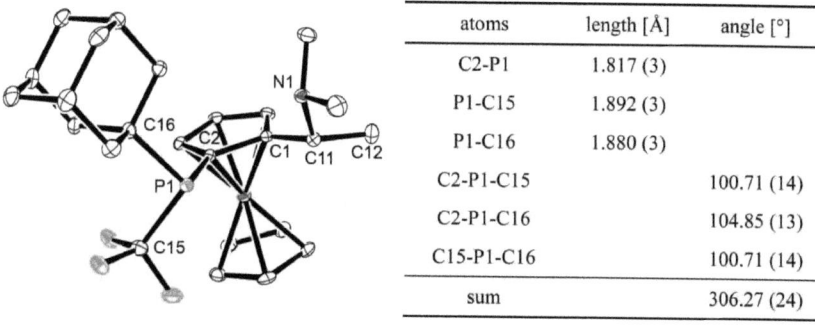

atoms	length [Å]	angle [°]
C2-P1	1.817 (3)	
P1-C15	1.892 (3)	
P1-C16	1.880 (3)	
C2-P1-C15		100.71 (14)
C2-P1-C16		104.85 (13)
C15-P1-C16		100.71 (14)
sum		306.27 (24)

Figure 24: ORTEP representation of **150** and table with selected bond lengths and angles. Only one of two independent molecules is shown. Hydrogen atoms are omitted for clarity and thermal ellipsoids are set to 30% probability.

From the crystal structure it is clear that the bulky adamantyl group assumes an exo orientation with respect to the ferrocenyl unit, thus minimizing steric repulsion with the unsubstituted ferrocene ring as well as the dimethylamino group. The angles around P1 are all larger than 100° and the sum of all these angles is more than 306°. The sum of these angles is a good measure of the hybridization at phosphorus and is influenced by the electronic as well as steric properties of the substituents. In general, small and electron-withdrawing substituents on the phosphine lead to a small total angle, which is also associated with a high degree of pyramidalization. In contrast, bulky as well as electron donating and aromatic groups give rise to larger angles as is also described in chapter 1. In the case of **150**, the bulkiness of the adamantyl group seems to dominate the electronic and steric influences of all substituents resulting in a value in the range of that of triphenylphosphine (306°) and PPFA (305°) and significantly larger than the one of bis(trifluoromethyl)phosphine **110** (298°).

3.3.3 Synthesis of P-Stereogenic Trifluoromethylphosphines of the Josiphos Type

In order to obtain bidentate phosphine ligands of the Josiphos type, the dimethylamino group in the trifluoromethylphoshine compounds **144/147/150** was substituted by dicyclohexylphosphine in acetic acid according to the known procedure (Table 15).

Table 15: Synthesis of P-stereogenic trifluoromethyl phosphines of the Josiphos-type.

entry	product	R	R'	yield [%]
1	(S_P)-**142**	Ph	CF_3	76
2	(R_P)-**142**	CF_3	Ph	83
3	(S_P)-**151**	2-Np	CF_3	82
4	(S_P)-**152**	Ad	CF_3	43
5	(R_P)-**152**	CF_3	Ad	15

The phenyl and naphthyl-substituted amine phosphines were converted into the corresponding diphosphines in good yields (76-83%) and without any epimerization of the stereogenic phosphorus center, in spite of the high temperatures applied. We suggest that the small size and highly electron-withdrawing character of the trifluoromethyl group increases the s character of the phosphorus lone pair, which results in this rather high stability towards pyramidal inversion, as is described in Chapter 1. In contrast, treating the adamantyl substituted compound **150** under the same conditions resulted in partial epimerization and the (S_P) and (R_P)-isomers were isolated in 43% and 15% yield, respectively. We discussed above the relatively low degree of pyramidalization around P1 in adamantylphosphine **150**. Generally, this is associated with a small contribution of the s orbital to the lone pair, which normally results in a higher rate of epimerization.

In the ^{31}P NMR and ^{19}F NMR spectra of (R_P)-**142** as well as (R_P)-**152** a through-lone-pair coupling of the phosphorus atom of the dicyclohexylphosphino group with the fluorine atoms of the trifluoromethyl group was observed. This implies a similar arrangement of both diphosphine compounds in which the CF_3 group is in close proximity to the lone pair of the phosphorus of the dicyclohexylphosphino group. In contrast, in the (S_P)-isomers of both structures **142** and **152** no such coupling was observed. (S_P)-**147** and (R_P)-**147** were also converted to the pyrazole substituted P,N-ligands (S_P)-**153** and (R_P)-**153** in low to moderate yield as depicted in Scheme 83.

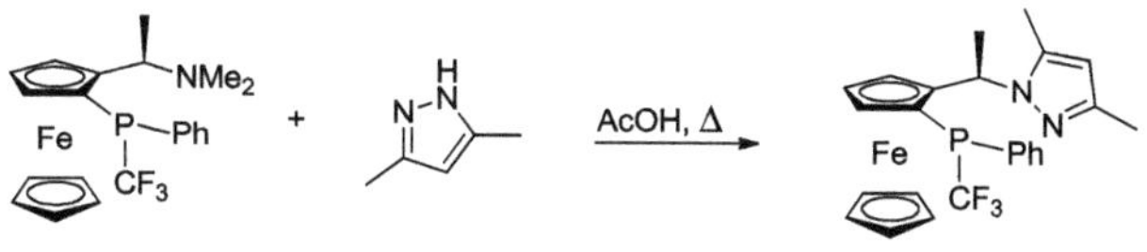

(R_P)-**147**
(S_P)-**147**

(R_P)-**153**: 70%
(S_P)-**153**: 37%

Scheme 83: Synthesis of pyrazole substituted ligands with a phosphorus and a nitrogen donor.

3.3.4 Transition-Metal Complexes of Bidentate Trifluoromethylphosphine Ligands

3.3.4.1 Palladium and Platinum Complexes

To assess the coordination behaviour of these new diphosphines toward late transition metals, **142**, **144** and **152** were reacted with [PtCl$_2$(COD)] or [PdCl$_2$(COD)] in dichloromethane to obtain the corresponding platinum and palladium complexes as shown in Table 16. In the platinum complexes **154** and **155**, the coupling between the ^{195}Pt nucleus and the coordinated phosphino groups were analyzed by ^{31}P NMR spectroscopy. For the trifluoromethylphosphino group of **154** and **155**, large coupling constants of $^1J_{P-Pt}$ = 3794 Hz and $^1J_{P-Pt}$ = 3747 Hz, respectively were observed. In both cases the coupling of platinum to the dicylohexylphosphino group was more than 400 Hz smaller.

Table 16: Synthesis of palladium and platinum complexes of trifluoromethylphosphine ligands.

entry	ligand	R	R'	M	product	Yield [%]
1	**144**	CF$_3$	Np	Pt	**154**	quant.
4	**152**	Ad	CF$_3$	Pt	**155**	76
2	(S_P)-**142**	Ph	CF$_3$	Pd	**156**	80
3	(R_P)-**142**	CF$_3$	Ph	Pd	**157**	70

3 P-Stereogenic Trifluoromethyl Phosphine Ligands

X-ray diffraction measurements of single crystals of **156** and **157** were performed and both structures analyzed. ORTEP representations are shown in Figure 25 and selected bond lengths and angles are listed in Table 17.

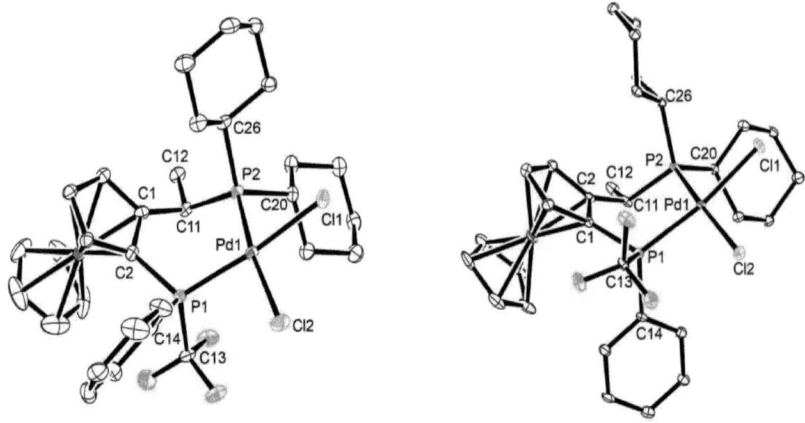

Figure 25: ORTEP representation of X-ray structures of palladium complexes **156** and **157**. Hydrogen atoms and co-crystallizing solvent molecules are omitted for clarity and thermal ellipsoids are set to 30% probability.

The influence of the P-stereogenicity on the coordination behavior of the ligands (S_P)-**142** and (R_P)-**142** seems to be rather small, since both complexes are structurally very similar as can be seen from the ORTEP representation of the structures as well as from the bond lengths and angles. **156** has a slightly longer Pd1-P1 bond than **157**, which conversely has the longer bond between the metal center and P2. The bite angle (P1-Pd1-P2) of both ligands (S_P)-**142** and (R_P)-**142** is around 95° and the corresponding palladium complexes slightly distorted square planar as shown by the sum of all angles around palladium, which is somewhat larger than 360° in both cases. In both complexes the six-membered chelate ring adopts a boat-like conformation.

Table 17: Selected bond lengths and angles of the palladium complexes **156** and **157**.

atoms	156		157	
	length [Å]	angle [°]	length [Å]	angle [°]
Pd1-P1	2.231 (1)		2.245 (1)	
Pd1-P2	2.291 (1)		2.273 (1)	
Pd1-Cl1	2.341 (1)		2.341 (1)	
Pd1-Cl2	2.351 (1)		2.343 (1)	
P1-Pd-P2		95.5 (5)		94.47 (4)
sum around Pd1		361 (1)		362(1)

156 and 157 also show great similarity to the doubly P-stereogenic palladium complex 138, shown in Figure 23.

We were able to grow single crystals of the dichloroplatinum complexes 154 and 155 suitable for X-ray diffraction by slow diffusion of hexane into a concentrated DCM solution of the metal complex. ORTEP representations of 154 and 155 are shown in Figure 26 and selected bond lengths and angles are listed in Table 18.

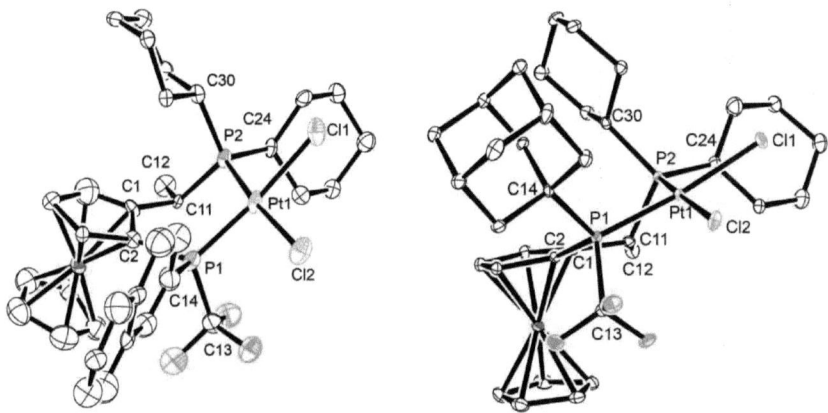

Figure 26: ORTEP representation of platinum complexes 154 and 155. Hydrogen atoms and co-crystallizing solvent molecules are omitted for clarity and thermal ellipsoids are set to 30% probability.

The most remarkable difference between 154 and 155 lies in the conformation of the six-membered chelate ring. Whereas 154 forms a boat conformation like the palladium complexes 156 and 157, the adamantyl-substituted complex 155 adopts a distorted half chair conformation, as seen clearly in the ORTEP representation. This conformational difference is also reflected in the angle between the ferrocene plane and the plane through the metal center and P1 and corresponds to the torsion angle C1-C2-P1-Pt1. In complex 155 the angle is close to 0°, whereas the naphthyl substituted complex has a torsion angle of 39°. In contrast, the torsion angle C2-C1-C11-P2 is very similar for both complexes, namely -68° for 254 and -62° for 255, respectively. This indicates that the conformational difference has mainly to do with the P-stereogenic phosphine P1. From a structural point of view, the dichloropalladium complex of the original Josiphos ligand lies in between these two "extremes". A torsion angle of 22° and a conformation of the six-membered chelate ring between a boat and half chair are observed in that case.

3 P-Stereogenic Trifluoromethyl Phosphine Ligands

Table 18: Selected bond lengths and angles of the platinum complexes **254** and **255**.

atoms	254 length [Å]	254 angle [°]	255 length [Å]	255 angle [°]
Pt1-P1	2.242 (1)		2.209 (8)	
Pt1-P2	2.267 (1)		2.267 (5)	
P1-Pt-P2		95.8 (4)		96.3 (2)
sum around Pt1		361.3 (1)		361.2 (5)
C1-C2-P1-Pt1		2.6 (4)		39 (2)
C2-C1-C11-P2		-62.1 (5)		-68 (2)

The short phosphorus-platinum bond P1-Pt1 in complex **254** (2.209 Å) compared to that in the adamantyl-substituted complex **255** (2.242 Å) is reasonable. In contrast, the bond from the metal center to P2 is exactly the same in both complexes.

3.3.4.2 Rhodium Complexes

Cationic rhodium diphosphine complexes are widely used as catalysts in hydrogenation of olefins and generally provide high activities and selectivities. In order to prepare such a metal precursor, trifluoromethylphosphine ligand (S_P)-**142** was reacted with [Rh(COD)$_2$]PF$_6$ to yield the corresponding rhodium complex **158** as depicted in Scheme 84.

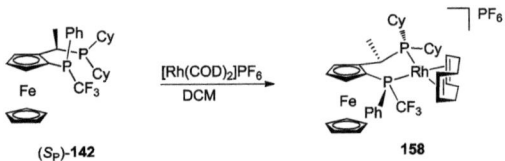

Scheme 84: Synthesis of a cationic rhodium complex **158**.

In the ^{31}P NMR spectrum the signal for P1 appears as a dqd due to coupling to the rhodium nucleus, the fluorine atoms of the CF$_3$ group and to P2. The cis coupling of the two phosphorus atoms is 33 Hz and the phosphorus-rhodium coupling for P1 and P2 is 159 Hz and 139 Hz, respectively.

Crystals suitable for X-ray diffraction were obtained and an ORTEP representation and a table with selected bond lengths and angles are shown in Figure 27. The six-membered chelate ring adopts a boat conformation as observed in the case of the palladium complex **156**.

3 P-Stereogenic Trifluoromethyl Phosphine Ligands

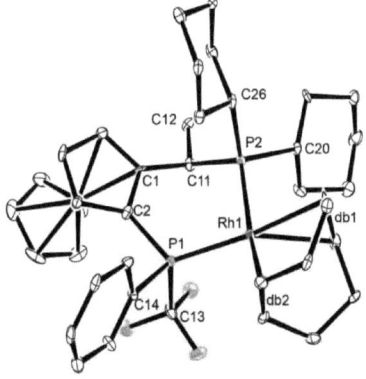

atoms	length [Å]	angle [°]
Rh1-P1	2.273 (2)	
Rh1-P2	2.353 (7)	
Rh1-db1	2.134	
Rh1-db1	2.130	
P1-Rh1-P2		92.15 (7)
C1-C2-P1-Rh1		32.3 (8)

Figure 27: ORTEP representation and selected bond lengths and angles of rhodium complex **158**. Hydrogen atoms are omitted for clarity and thermal ellipsoids are set to 30% probability.

The bond from the rhodium center to the trifluoromethylphosphino group P1 is significantly shorter than that to P2, again an indication for the high s character of the lone pair orbital of P1. Interestingly, the distance from the metal center to the midpoints of the two double bonds of the cyclooctadiene ligand (Rh1-db1 and Rh1-db2) is very similar. Therefore, it seems that in **158** the difference in bond lengths of the two phosphines to the metal is not directly reflected in the bond lengths to the olefin ligands as usually observed. A bite angle of 92° and a torsion angle C1-C2-P1-Rh1 of 32° are observed, which is very similar to that of the corresponding palladium complex **156**.

Carbonyl complexes are often used to evaluate the donor and acceptor abilities of certain ligands. Strong π accepting ligands decrease the Rh-CO π backbonding, which is reflected in a decrease of the CO bond length. This can be observed in crystal structures as well as in the infrared spectrum. For ligands trans to the carbonyl group this effect is much larger but it can also be seen in cis substituted complexes. In order to evaluate the π accepting behavior of trifluoromethyl substituted ligands, we synthesized a series of rhodium carbonyl complexes as shown in Scheme 85. As bidentate phosphine ligands we have chosen Josiphos, the trifluoromethylphosphines (S_P)-**142** and (R_P)-**142** as well as bistrifluoromethylated phosphine **120**. In the case of the trifluoromethylated phosphine ligands **120** and **142** only one of the two theoretically possible isomers were observed. Due to the fact that the chloride as well as the carbonyl group are rather small ligands we excluded steric reasons to be the source of this selectivity. According to the different electronic properties of the two phosphines we suggested the carbonyl group to be in the position trans to P2. P1, substituted with a phenyl group and the strongly electron-withdrawing CF_3 group should be the much stronger

3 P-Stereogenic Trifluoromethyl Phosphine Ligands

π accepting ligand than the alkyl substituted phosphine P2 and therefore adopt a cis position to the carbonyl group, which is also a strong π acceptor.

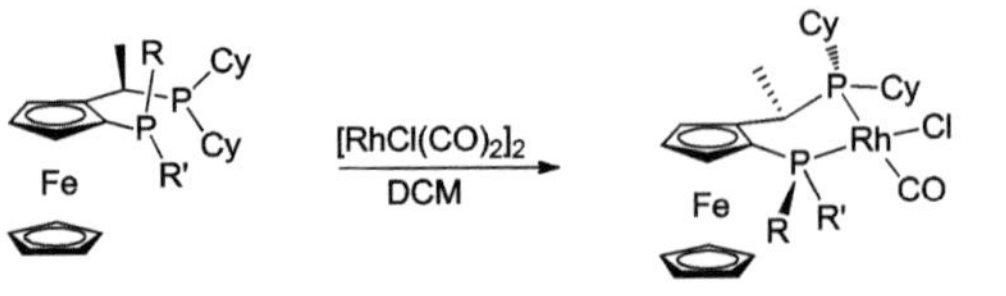

Scheme 85: Synthesis of rhodium carbonyl complexes with Josiphos ligands.

159: R = Ph, R' = Ph
160: R = Ph, R' = CF$_3$
161: R = CF$_3$, R' = Ph
162: R = CF$_3$, R' = CF$_3$

In contrast, when the original Josiphos was used as a ligand, an approximately 2:1 mixture of isomers (**159-A** and **159-B**) was observed. This emphasizes the influence of the strongly electron-withdrawing CF$_3$ goup on the electronic properties of these phosphines.

Spectroscopic data of these complexes such as ^{31}P NMR shifts and coupling constants as well as CO stretching frequencies are listed in Table 19. From the $^1J_{P-Rh}$ coupling constants one can clearly see that the phosphino group trans to the carbonyl group displays a much lower coupling (~120 Hz) to rhodium than that in cis position (159-188 Hz). In addition to the electronic differences between the phosphines such a difference in the phosphorus-metal coupling can also be caused by the CO ligand. The strong π acceptor properties of the carbonyl group lead normally to a weakening of the bond to the ligand trans to CO.

Table 19: Selected spectroscopic data for the complexes [RhCl(CO)(P,P)].

complex	R	R'	P1 δ [ppm]	P1 $^1J_{P-Rh}$ [Hz]	P2 δ [ppm]	P2 $^1J_{P-Rh}$ [Hz]	IR ν(CO) [cm^{-1}]
160	Ph	CF$_3$	50.4	175	58.2	123	2015
161	CF$_3$	Ph	54.6	175	52.6	122	2019
162	CF$_3$	CF$_3$	63.2	188	56.9	118	2029
159-A	Ph	Ph	34.5	160	47.5	124	1985
159-B	Ph	Ph	13.6	121	76.1	159	1985

A clear trend is observed in the carbonyl stretching frequencies obtained by IR spectroscopy. The isomeric mixture of the rhodium complex with Josiphos shows one broad band at 1985 cm^{-1}. The stretching frequency increases to 2015 and 2019 cm^{-1} for both stereoisomers of the trifluoromethyl complexes **160** and **161**, respectively. The bistrifluoromethylated phosphine complex **162** has a value of 2029 cm^{-1} for the CO stretching frequency. Thus, ascending electron-withdrawing properties of the ligands lead to a decrease of the Rh-CO π backbonding and therefore to an increase of ν(CO).

Single crystals of carbonyl complex **160** suitable for X-ray diffraction were obtained by slow diffusion of hexane into a concentrated ether solution. An ORTEP representation and a table with selected bond lengths and angles are shown in Figure 28.

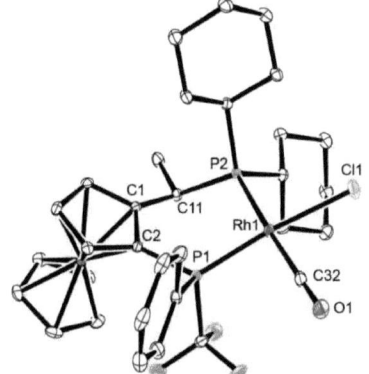

atoms	length [Å]	angle [°]
Rh1-P1	2.1899 (6)	
Rh1-P2	2.3710 (6)	
Rh1-Cl1	2.3878 (7)	
Rh1-C32	1.884 (3)	
C32-O1	1.127 (3)	
P1-Rh1-P2		94.36 (7)
sum around Rh1		359.9 (1)

Figure 28: ORTEP representation and selected bond lengths and angles of **160**. Hydrogen atoms are omitted for clarity and thermal ellipsoids are set to 30% probability.

The crystal structure of **160** confirms the suggestion that the carbonyl group is in cis position to the electron-withdrawing phosphine P1. A rather short bond from rhodium to the trifluoromethylphosphine P1 is observed, whereas the bond to P2 is much longer. This is caused by the different steric and electronic properties of the phosphines themselves as well as influenced by the different donor and acceptor properties of the ligands in the trans position (CO vs. Cl). The rhodium CO distance is very similar to those found in analogous diphosphine rhodium carbonyl complexes,[181-182] whereas the C-O triple bond has nearly the same bond length as in free carbon monoxide.[183] This is in the range of analogous systems but comparison is, in general, hampered by the relatively large standard deviations observed. The bite angle of the diphosphine is slightly larger than 90° and the complex perfectly square planar according to the sum of all angles around the rhodium center.

3.4 Some Mechanistic Aspects of the Formation of 1,2-Diphospholes

3.4.1 Cyclization with Bis(trifluoromethyl)phosphines as Substrates

In Section 3.2.1 we have shown the stereoselective synthesis of 1,2-diphospholes from the diphosphines **110** and **131** via intramolecular substitution of a trifluoromethyl group. Although the yield of this reaction, especially in the case of **131** was not very high and the formation of side products was detected, we did not observe the formation of another diastereoisomer of a 1,2-diphosphole. In principle the formation of four diastereoisomers is possible. Those are two cis isomers with the two P substituents (R, R') in exo or endo position to the ferrocene core **163** and **164** and two isomers having R and R' in a trans configuration **165** and **166** (Figure 29).

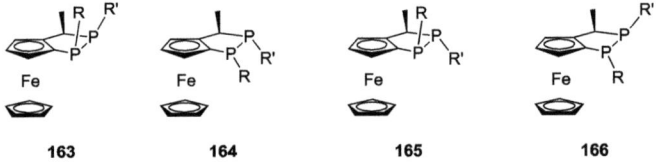

Figure 29: The four theoretically possible isomers of ferrocene based 1,2-diphospholes.

We suggested that steric repulsion between the three substituents on the five-membered diphosphole ring (Me, R and R') and the ferrocene core is the reason for the highly selective formation of **129** and **132**. In Figure 30 a schematic representation of the possible cyclization mechanism is shown on the basis of X-ray crystal structures of **110** and **129**.

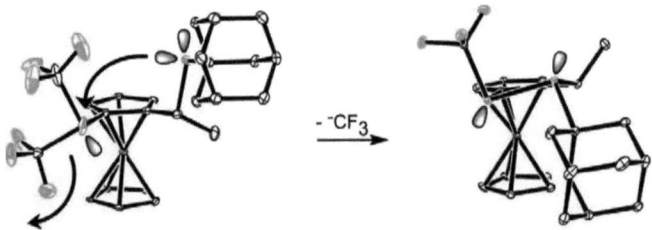

Figure 30: Schematic representation of the cyclization on the basis of X-ray structures of **110** and **129**.

Although the conformation observed in the crystal structure does not necessarily reflect exactly the situation present in solution, we clearly see from this picture that substitution of the endo trifluromethyl group by the adamantylphosphido group is possible and should give the 1,2-diphosphole **129**, displaying a trans arrangement of the two substituents on the phosphines.

3.4.2 Cyclization with Trifluoromethylphosphines as Substrates

Since only one of the two diastereotopic trifluoromethyl groups is substituted, we were interested to see the outcome of the analogous reaction but with monotrifluoromethyl phosphines instead the bistrifluoromethylated ones. One may ask several questions concerning this type of reaction:

- Are monotrifluoromethylphosphines reactive enough for CF_3 substitution?
- Do both diastereoisomers react in the same manner?
- What are the products? Diphospholes?
- How selective is the formation of these diphospholes?

In order to answer these questions, we synthesized the secondary phosphines (S_{P1})-**167** and (R_{P1})-**167** starting from the P-stereogenic amine-phosphines (S_P)-**147** and (R_P)-**147** as depicted in Scheme 86. The standard conditions were applied and the products isolated in good yields as a 12:1 mixture of diastereoisomers (P2).

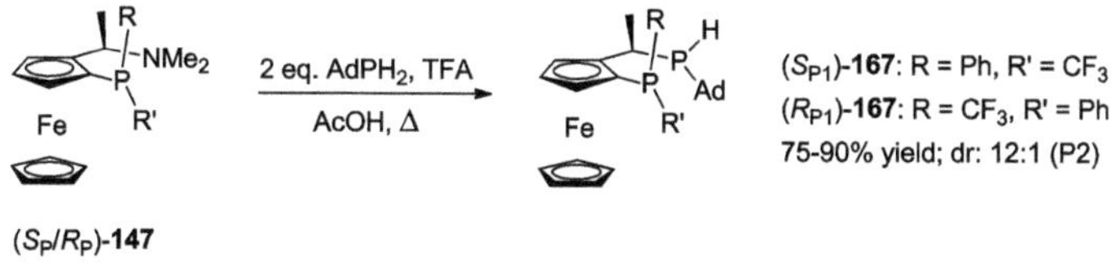

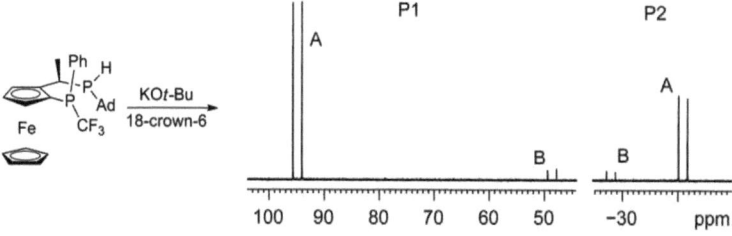

Scheme 86: Synthesis of P-stereogenic secondary phosphines (S_{P1})-**167** and (R_{P1})-**167**.

When (S_{P1})-**167** was reacted with potassium *tert*-butoxide and 18-crown-6 in tetrahydrofuran two 1,2-diphospholes were isolated as a diastereomeric mixture according to ^{31}P NMR spectroscopy. The two products (A and B) are easily identified as they appear as two doublets with a large $^1J_{P-P}$ coupling of around 200 Hz (Figure 31).

Figure 31: ^{31}P NMR spectrum of the crude product of the reaction of (S_{P1})-**167** with base.

Interestingly, when (R_{P1})-**167** was treated under the same conditions three isomers were observed as can be seen in Figure 32. In addition to the products denoted with A and B the signals of a new isomer C, which is also a 1,2-diphosphole derivative appeared in the ^{31}P NMR spectrum. Since no conditions were found to separate the isomers using column chromatography or crystallization, the isomeric mixtures were used in NMR analysis in order to ascertain the exact structures of the three isomers. 1-dimensional NMR experiments (^{1}H, ^{13}C, ^{31}P) as well as several different 2-D experiments (H-H-COSY, C-H-HMQC, C-H-HMBC and P-H-HMQC) were applied in order to assign all proton, carbon and phosphorus resonances to the structures A, B and C. Using H-H-NOESY methods the arrangement of the two substituents on the phosphines relative to each other as well as to other functional groups in the molecule such as the methyl group or the unsubstituted Cp-ring were examined.

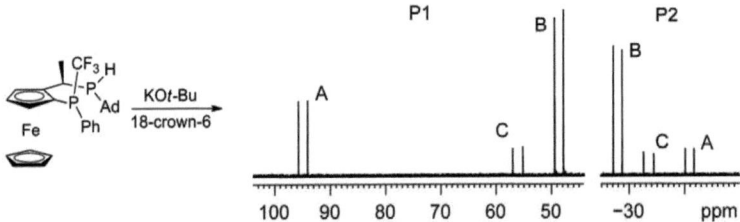

Figure 32: ^{31}P NMR spectrum of the crude product of the reaction of (R_{P1})-**167** with base.

Specific NOESY-contacts observed in the spectrum as well as contacts absent from the spectrum led us to the conclusions shown in Figure 33. Isomer A shows NOESY contacts from protons of the adamantyl group to the unsubstituted cyclopentadiene ring and contacts from the phenyl protons to the methyl group, whereas no contact between the phenyl substituents and the Cp-ring was found. These observations suggest a trans substituted 1,2-diphosphole structure having the adamantyl group in endo and the phenyl group in exo position similar to the structures described in 3.2.1. For isomer B, NOESY contacts between the protons of the phenyl groups and those of the Cp rings were observed. Together with the contacts absent between the methyl and the phenyl group as well as the adamantyl substituent and the Cp ring, a trans configuration with the adamantyl group exo and the phenyl group endo to the ferrocene core was deducted.

3 P-Stereogenic Trifluoromethyl Phosphine Ligands

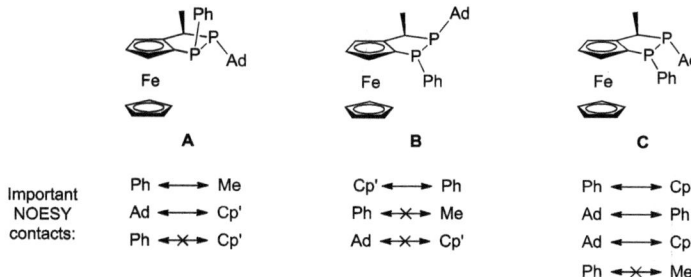

| Important NOESY contacts: | Ph ↔ Me
Ad ↔ Cp'
Ph ↮ Cp' | Cp' ↔ Ph
Ph ↮ Me
Ad ↮ Cp' | Ph ↔ Cp'
Ad ↔ Ph
Ad ↔ Cp'
Ph ↮ Me |

Figure 33: The three postulated structures A, B and C and important contacts observed by NOESY experiments.

For isomer C, NOESY contacts between protons of the adamantyl and the phenyl substituents as well as the protons of the cyclopentadienyl ring were found, implying a configuration in which all of these groups are in close proximity as is the case of a cis configuration in which both substituents on the phosphines are in an endo position.

As the stereoisomers A, B and C seem to be configurationally stable under reaction conditions and no interconversion from A to B or vice versa was observed, we suggest different mechanisms for the formation of each of them. The formation of the isomer A from the S_{P1}-isomer of **167** should occur according to a similar mechanism to the formation of **129** as shown in Scheme 87.

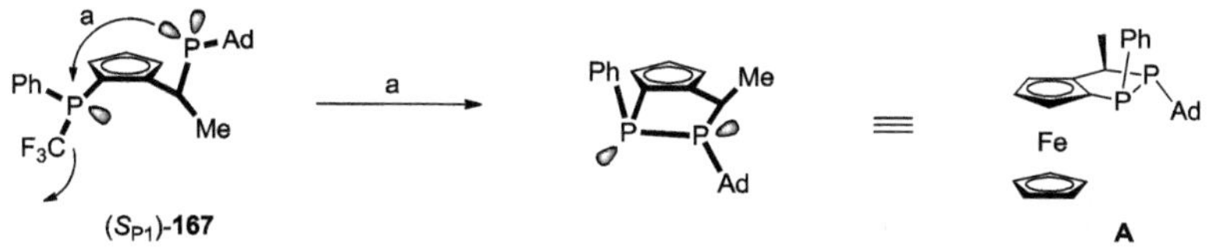

Scheme 87: Schematic representation of a possible mechanism for the formation of the stereoisomer A.

A related mechanism, with a different site of attack at P1 accounts for the formation of the isomers B and C from (R_{P1})-**167** (Scheme 88). As the nucleophilic attack can be by one of two diastereotopic lone pairs of the phosphido group, isomers B and C can be formed from the same intermediate, following either pathway b or pathway c, respectively. As the absolute configuration at P1 remains the same, the above described mechanisms do not explain the formation of B from the S_{P1}-isomer of **167** and the formation of A from (R_{P1})-**167**. An epimerization of the stereogenic center at P1 must occur during the course of the reaction.

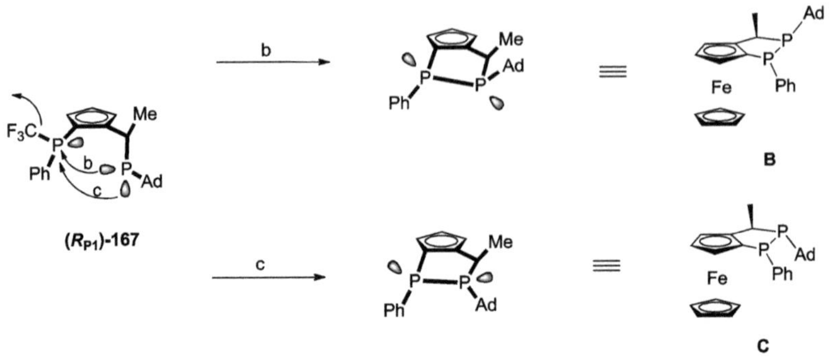

3 P-Stereogenic Trifluoromethyl Phosphine Ligands

Scheme 88: Schematic representation of possible mechanisms for the formation of the stereoisomers B and C.

3.4.2.1 Epimerization via a Pentacoordinated Intermediate

One possible mechanism for the epimerization of P1 involves a pentacoordinated phosphorus intermediate as shown in Scheme 89.

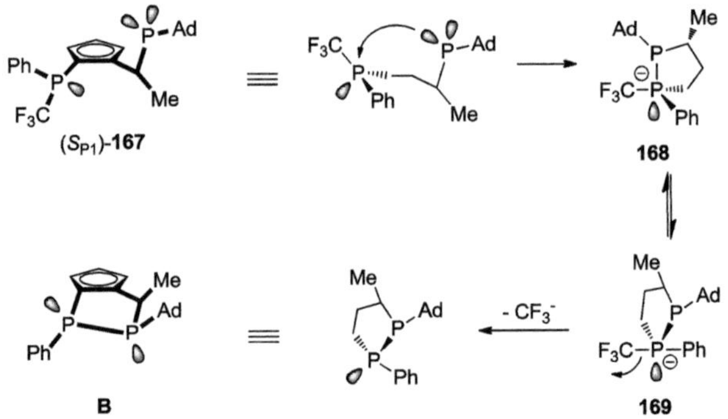

Scheme 89: Schematic representation of a possible mechanism for the formation of isomer B.

Attack of the phosphido group at the stereogenic phosphine forms the pentacoordinated structure **168** with the formal negative charge on P1. After stereopermutation into **169**, in which the trifluoromethyl group is in apical position, P-CF$_3$ bond cleavage can occur and the isomer B is released. Similar observations in which nucleophilic substitution at phosphorus compounds occurs with retention of configuration are reported and have been explained by such a mechanism.[184] However, in these cases phosphine oxides or phosphine boranes were used, structures that should stabilize a negatively charged phosphorus compound to a larger extent than in our case. However, the highly electron-withdrawing trifluoromethyl group may also stabilize such intermediates (**168** and **169**) and thus render this mechanism plausible. The reason that isomer C, in which P2 has opposite absolute configuration, was not observed in this case cannot be totally explained by this mechanism. Steric repulsion of the bulky adamantyl substituent on P2 may lead to preferential attack of one of the two diastereomeric lone pairs.

3.4.2.2 Epimerization via SET and Radical Recombination

Another reasonable explanation for the epimerization of P1 over the course of the reaction might involve a radical mechanism as depicted in Scheme 90. Single-electron transfer (SET) from the phosphido group to the P-stereogenic phosphine and subsequent loss of a trifluoromethyl anion yield the diradical **170**.

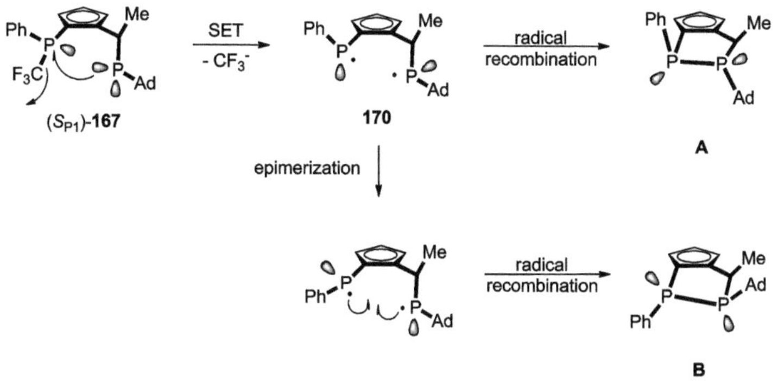

Scheme 90: Formation of stereoisomers A and B from (S_P)-**167** via a radical mechanism.

Epimerization of the stereocenter P1 of **170** followed by radical recombination leads to the formation of stereoisomer B. The addition of phosphine radicals to olefins[185-186] as well as single electron transfer from phosphides to alkyl halides[187-188] are known but no report of an SET from a phosphido group to a tertiary phosphino group is found in the literature. This mechanism cannot explain, why the formation of the isomer C was not observed when (S_P)-**167** was used as substrate. In principle, epimerization of P2 in the diradical compound could occur, which should provide isomer C after radical recombination.

3 P-Stereogenic Trifluoromethyl Phosphine Ligands

3.4.2.3 Epimerization via Nucleophilic Substitution with CF$_3$

The epimerization of the trifluoromethylphosphino group in **167** can in principle occur by a substitution of the CF$_3$ group through another CF$_3$ anion present in solution as shown in Scheme 91. If this process follows an S$_N$2-like mechanism, an equilibrium of (S_{P1})-**167** and (R_{P1})-**167** may be present in solution and both isomers of the 1,2-diphosphole product (A and B) are formed after cyclization. Similar to the mechanisms discussed above, the formation of B is easily explained but no clear reasons for the absence of stereoisomer C are provided by this mechanism.

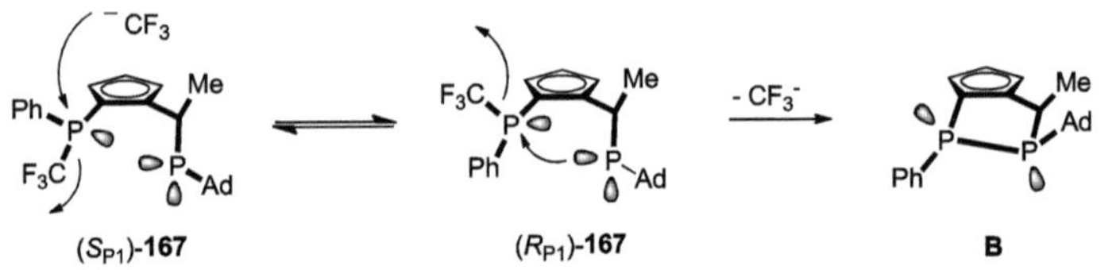

Scheme 91: Epimerization of the P-stereogenic phosphine via nucleophilic substitution with CF$_3$.

3.4.3 Cyclization in the Presence of Electrophiles

In order to verify if the presence of free trifluoromethyl anions in solution has an influence on the stereoselectivity of the cyclization of diphosphines to 1,2-diphosphole the reaction was carried out in the presence of an electrophile in order to remove the free CF$_3$ anion from the reaction mixture. We decided to use benzophenone as electrophile as this can efficiently trap trifluoromethyl anions (see Section 3.2.1). When (S_{P1})-**167** was mixed with potassium *tert*-butoxide in the presence of benzophenone, isomer A was the only product observed in the ^{31}P NMR spectrum as shown in Figure 34.

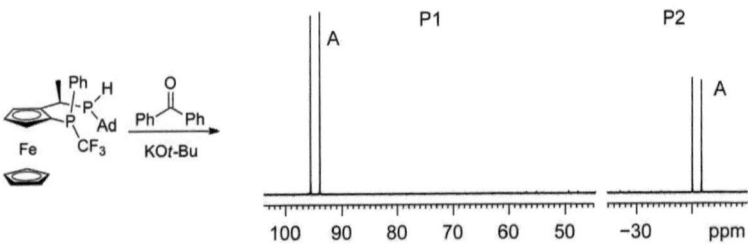

Figure 34: ^{31}P NMR spectrum of the reaction of (S_{P1})-**167** with base in the presence of benzophenone.

Thus, the electrophile seems to suppress the epimerization at P1 and therefore the formation of stereoisomer B. This observation would support a mechanism in which the free trifluoromethyl anion is the source of epimerization, since trapping of CF_3 should inhibit epimerization. In order to substantiate this observation, also the (R_{P1})-isomer of **167** was treated with base in the presence of benzophenone. Most surprisingly, the formation of all three isomers was observed in a different ratio than observed in the reaction carried out in the absence of benzophenone (Figure 35).

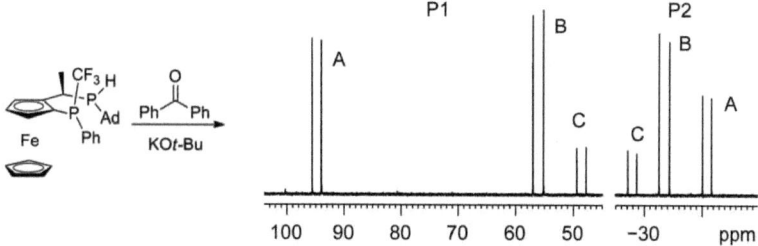

Figure 35: ^{31}P NMR spectrum of the reaction of (R_P)-**167** with base in the presence of benzophenone.

Therefore, it seems that the presence of the electrophile does not suppress the epimerization of the P-stereogenic center but somehow favors the formation of isomer A over isomers B and C. Of course, this observation does not support any of the mechanisms discussed above since the formation of a specific product and not the epimerization seems to be influenced by addition of benzophenone. Thus, we postulate that the mechanisms leading to the observed isomeric ratio may be significantly more complicated than those speculated above and are also strongly influenced by traces of reactive species such as CF_3 anions, benzophenone and associated byproducts formed over the course of the reaction.

3.5 Trifluoromethyl Phosphines in Asymmetric Catalysis

3.5.1 Rhodium-Catalyzed Hydrogenation of Olefins

Since the hydrogenation of dimethyl itaconate (DMI) was successful with the secondary phosphine and SPO ligands described in Chapter 2.7 we decided to test the trifluoromethyl substituted phosphines with this substrate as well. As we observed some influence of the solvent on the hydrogenation, a brief solvent screening was carried out with (S_P)-**142** as ligand as shown in Table 20. Methanol is one of the standard solvents for this type of hydrogenation but in our case poor results were observed. The reaction was not complete after 20 hours and the enantioselectivity was only 58%. In dichloromethane the hydrogenation was much faster and the *ee*s up to 88% when the reaction was carried out at 0 °C. Tetrahydrofuran gave only moderate selectivity, but the use of fluorinated solvents such as trifluoroethanol (TFE) or hexafluoroisopropanol gave high *ee*s of up to 95%. The reason for this remarkable difference between methanol and TFE is not clear but was observed for this substrate using a phosphine-phosphoramidite ligand as well.[189] We suggest that the solubility of the metal-substrate complex is enhanced or the higher acidity of the protic solvent has a positive effect on the reaction.

Table 20: Screening of solvents and conditions in the rhodium-catalyzed hydrogenation of DMI.

entry	solvent	M-precursor	T [°C]	conversion [%]	*ee* [%]
1	MeOH	[Rh(COD)$_2$]SbF$_6$	r.t.	95	58
2	DCM	[Rh(COD)$_2$]SbF$_6$	r.t.	100	81
3	DCM	[Rh(COD)$_2$]SbF$_6$	0	100	88
4	TFE	[Rh(COD)$_2$]SbF$_6$	r.t.	100	91
5	TFE	[Rh(COD)$_2$]SbF$_6$	0	100	89
6	THF	[Rh(COD)$_2$]SbF$_6$	r.t.	100	77
7	F$_6$-*i*-PrOH	[Rh(COD)$_2$]SbF$_6$	r.t.	100	95
8	TFE	[Rh(COD)$_2$]PF$_6$	r.t.	100	95

With the screening results in hand, we developed a set of standard conditions for testing the diphosphine ligands in the hydrogenation of DMI. Thus, we carried out the ligand screening in TFE at room temperature and with [Rh(COD)$_2$]PF$_6$ as the catalyst precursor. The results with all of the different P-stereogenic trifluoromethylphosphine ligands are listed in Table 21. Entries 1-5 concern diphosphines having one P-stereogenic phosphine and a dicyclohexylphosphino group, prepared using the intermolecular substitution strategy. The ligands with phenyl or 2-naphthyl substituents on P1 afforded enantioselectivities of 93-97%. Surprisingly, the absolute configuration at P1 turned out to have only a small influence on the enantiodiscrimination, since very similar enantioselectivities in the hydrogenation were observed for both configurations. In contrast, the ligands containing a bulky adamantyl group attached to the trifluoromethyl phosphines behaved differently. The (S_P)-isomer of the ligand provided the S-isomer of the product in 49% enantiomeric excess, but (R_P)-**152** afforded the R-product, although in only 15% *ee*. This demonstrates that the P-stereogenicity becomes important when the substituents on the phosphine are very different in size as is the case for adamantyl and CF$_3$. Thus, we suggest that in the case of the phenylsubstituted phosphines the size discrimination between the two substituents is too small to yield a pronounced difference in enantioselectivity.

Table 21: Screening of trifluoromethylphosphine ligands in the rhodium-catalyzed hydrogenation of DMI.

entry	ligand	config. of P1	conversion [%]	*ee* [%]	config. of the product
1	(S_P)-**142**	(*S*)	100	95	(*S*)
2	(R_P)-**142**	(*R*)	100	97	(*S*)
3	(S_P)-**151**	(*R*)	100	93	(*S*)
4	(S_P)-**152**	(*S*)	100	49	(*S*)
5	(R_P)-**152**	(*R*)	85	15	(*R*)
6	**133**	(*R*)	100	77	(*S*)
7	**134**	(*R*)	100	81	(*S*)
8	**135**	(*R*)	100	99	(*S*)
9	**136**	(*R*)	100	98	(*S*)
10	**137**	(*R*)	100	>99	(*S*)

3 P-Stereogenic Trifluoromethyl Phosphine Ligands

The last five entries in Table 21 concern the ligands with two P-stereogenic phosphines that were synthesized from the corresponding 1,2-diphospholes. In general, the cyclohexyl substituted ligands (entries 9 and 10) outperformed those with an adamantyl group (7, 8) and ligands with an isopropyl group on P1 (8, 10) gave better enantioselectivity than the phenyl substituted ones (7, 9). Therefore, the best ligand is the one with a substituent combination of isopropyl/trifluoromethyl at P1 and methyl/cyclohexyl at P2, providing more than 99% *ee* in the hydrogenation of DMI.

In order to determine the reactivities of these ligands they were again tested in the frame of a collaboration with Solvias AG.[125] The six best ligands were tested in the rhodium-catalyzed hydrogenation of DMI and MAA and the turnover frequencies (TOF) were determined. The results for DMI in methanol as well as trifluoroethanol (TFE) are shown in Table 22. The influence of the solvent is obvious since only two ligands showed high activity in methanol and the selectivities are, in general, lower than those observed in TFE. Only ligand **137** provided an enantiomeric excess of more than 99% in methanol. In TFE high turnover frequencies up to 6000 per hour were observed for all six ligands. The rhodium precursor was shown to have no influence since the same selectivities as with our system were observed.

Table 22: Hydrogenation of DMI in methanol and TFE with various trifluoromethylphosphine ligands.

entry	ligand	solvent	conversion [%]	TOF [1/h]	*ee* [%]
1	(S_P)-**142**	MeOH	90	99	74
2	(R_P)-**142**	MeOH	84	169	80
3	**134**	MeOH	76	185	65
4	**135**	MeOH	100	1000	93
5	**136**	MeOH	100	342	97
6	**137**	MeOH	100	1000	>99
7	(S_P)-**142**	TFE	100	3000	95
8	(R_P)-**142**	TFE	100	4000	97
9	**134**	TFE	100	5450	81
10	**135**	TFE	100	6000	99
11	**136**	TFE	100	5450	98
12	**137**	TFE	100	6000	>99

The same six ligands were also tested in the rhodium-catalyzed hydrogenation of MAA and the results are shown in Table 23. A remarkable solvent effect was observed in this case too but with better results for the hydrogenation in methanol than in TFE. The turnover frequencies observed in TFE were only slightly inferior to those obtained in MeOH but the enantioselectivities were quite low, therefore only the results obtained in methanol are shown in the table.

Table 23: Rh-catalyzed hydrogenation of MAA with various trifluoromethylphosphine ligands.

entry	ligand	solvent	conversion [%]	TOF [1/h]	ee [%]
1	(S_P)-**142**	MeOH	100	6000	88
2	(R_P)-**142**	MeOH	100	6000	98
3	**134**	MeOH	100	4000	88
4	**135**	MeOH	100	4000	91
5	**136**	MeOH	100	6000	97
6	**137**	MeOH	100	3000	>99

In general high TOF's of up to 6000 per hour and good to high enantioselecivities were observed. A more pronounced influence of the absolute stereochemistry of the P-stereogenic phosphine was observed compared to the hydrogenation of DMI, since the (R_P)-isomer of **142** provided 10% higher enantioselectivity than (S_P)-**142** and was the second best trifluoromethylphosphine ligand tested. The isopropyl substituted ligand **137** was again the best in terms of enantioselectivity, providing more than 99% ee.

3.5.2 Pd-Catalyzed Asymmetric Allylic Alkylation

The palladium catalyzed asymmetric allylic alkylation (AAA) is one of the most successful methods for the enantioselective formation of carbon-carbond bonds.[109-111] Josiphos as well as ferrocenyl ligands with a phosphine and a pyrazole donor are known to show high activity as well as selectivity in allylic substitution reactions.[12, 190] Therefore, we decided to test several of our trifluoromethyl substituted diphosphine ligands as well as the P,N-ligands (S_P)-**153** and (R_P)-**153** in the Pd-catalyzed AAA. In a typical reaction, 1,3-diphenylallyl acetate as the substrate and malonate as the nucleophile were used and the results are shown in Table 24.

Table 24: Pd-catalyzed allylic alkylation using trifluromethylphosphine ligands.

entry	ligand	config. of P1	conversion [%]	ee [%]	config. of the product
1	(S_P)-**142**	(S)	100	82	(S)
2	(R_P)-**142**	(R)	100	88	(S)
3	**135**	(R)	100	92	(S)
4	**136**	(R)	100	77	(S)
5	**137**	(R)	100	90	(S)
6	(S_P)-**153**	(S)	100	57	(R)
7	(R_P)-**153**	(R)	100	13	(S)

High activities were observed, since the reaction went to full conversion in less than two hours for all ligands tested. The reactions with both (S_P)-**142** and (R_P)-**142** showed that the absolute stereochemistry of P1 has some influence on the enantiodiscrimination since the enantiomeric excess observed for the products differed by approximately 6%. An even more pronounced effect of the P-stereogenicity was observed for the ligands containing a pyrazole moiety as the second donating group. (S_P)-**153** afforded the (R)-isomer of the product in 57% ee but the (R)-isomer provided the (S)-product in only 13% ee. The best trifluoromethyl phosphine ligands were the isopropyl substituted **135** and **137**, providing 92% and 90% ee, respectively, which is only slightly lower than the enantioselectivities observed with the parent Josiphos ligand (93% ee).[12]

3.6 Conclusion and Outlook

In this chapter we have shown that the CF_3 group of bis(trifluoromethyl)phosphines can act as a leaving group and thus may be substituted in intra- as well as intermolecular fashions. We were able to stereoselectively transform secondary phosphines of type **110** into ferrocenyl 1,2-diphospholes **129**. These diphospholes could then be converted into Josiphos-type ligands containing two P-stereogenic phosphines by sequential addition of methyltriflate and Grignard reagents. We showed that this transformation is highly regio- and stereoselective since the products were observed as single diastereoisomers. The electrophilic alkylation of P2 was limited to the use of methyl triflate so far, but various substituents were successfully introduced at P1 using different nucleophiles. Therefore, further investigations concerning the introduction of substituents on P2 are of interest in order to overcome these limitations. This strategy would then be a very useful method for the stereoselective synthesis of Josiphos-type ligands containing two P-stereogenic phosphines.

By intermolecular substitution of a CF_3 group on bis(trifluoromethyl)phosphines it is possible to synthesize both diastereoisomers of a PPFA-like structure containing a P-stereogenic trifluoromethyl substituted phosphine. These ligand precursors were then successfully transformed into bidentate ligands containing two phosphines or a phosphine and a pyrazole donor. The extension of this method to a broad variety of bistrifluoromethylated alkyl and aryl phosphines should give rise to a comprehensive library of P-stereogenic trifluoromethylphosphine ligands. This is currently under investigation in our research group.

Palladium, platinum as well as rhodium complexes of the P-stereogenic trifluoromethyl-phosphine ligands were synthesized and the special behaviour of these ligands compared to the parent Josiphos demonstrated. In general, the typical attributes of electron-poor phosphines such as short phosphorus-metal bonds as well as large phosphorus-metal coupling constants were observed in crystal structures and by ^{31}P NMR, respectively.

Application of the new ligands in asymmetric catalytic reactions such as Rh-catalyzed hydrogenation of olefins or Pd-catalyzed allylic alkylation showed these ligands to be highly useful in catalysis. The importance of P-stereogenicity on the enantiodiscrimination could be demonstrated in several examples. Ligand **137** provided exceptionally high selectivities of more than 99% *ee* in the hydrogenation of two different olefins and 90% *ee* in the asymmetric allylic alkylation was obtained. The high activities and moderate to very high enantioselectivities observed call for a broad screening of these ligands in other important transition-metal-catalyzed reactions, an ongoing project in our research group. In addition, these relatively electron poor ligands should be tested in Lewis-acid catalysis or applications were the stabilization of the metal center in a low oxidation state is required.

4 P-Stereogenic Tridentate Ligands of the Pigiphos-Type

4.1 Introduction[139, 191-192]

Compared to the large field of mono- and bidentate phosphine ligands known in the literature, the number of reports dealing with tridentate versions of phosphorus-based ligands is rather small. Thus, only a few examples of chiral tridentate phosphines are known and used as ligands in metal-catalyzed asymmetric reactions. Since chelation plays a major role in controlling a rigid geometry around the metal center, chiral tridentate phosphine ligands may generate a less dynamic catalyst, thus providing high selectivities in asymmetric catalysis. Depending on the geometry of the ligand as well as the nature of the metal, tripodal ligands can form tetrahedral, square planar or octahedral complexes upon coordination (Figure 36).

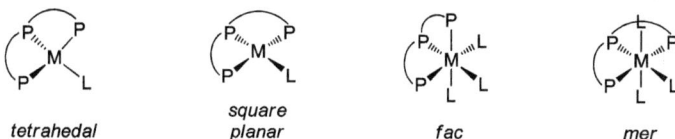

tetrahedal   square planar   fac   mer

Figure 36: Different coordination arrangements of tridentate phosphine ligands to certain metal centers.

In octahedral complexes, the tridentate ligand can coordinate to the metal in two different fashions, namely in a facial (*fac*) or in a meridional (*mer*) arrangement. In tetrahedral and square planar complexes, only one substrate molecule coordinates, thus providing a high degree of stereochemical control around the metal center. In contrast, with transition metals that are able to coordinate five or more ligands, three free coordination sites on the metal (denoted with L) should allow sufficient space for the coordination of two or more substrates.

Johnson and Imamoto[193] and Ward and Venanzi[40] reported the synthesis of triphosphine ligands containing two or three P-stereogenic phosphino groups as shown in Figure 37.

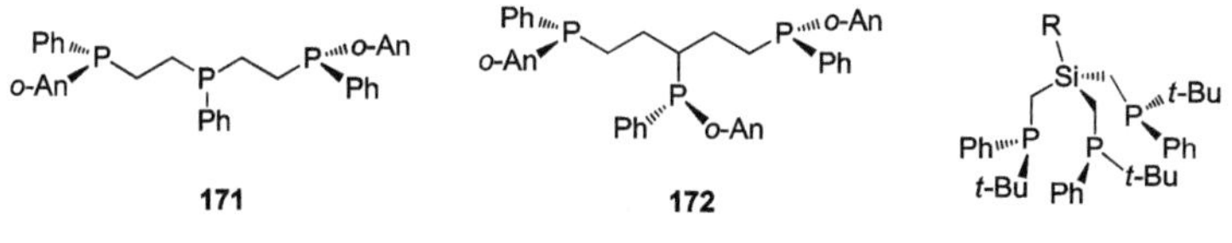

171   **172**

Figure 37: Selected examples of P-stereogenic triphosphine ligands as reported by Johnson/Imamoto[193] and Ward/Venanzi.[40]

171 and **172** were successfully applied in the rhodium-catalyzed hydrogenation of α-acetamido acrylic acids, providing moderate to good (47-88%) enantiomeric excess.

4 P-Stereogenic Pigiphos Ligands

4.1.1 Ferrocene-based Tridentate Phosphine Ligands

In 1995, a ferrocene-based tridentate phosphine ligand was developed in our group and named Pigiphos after the nickname of one of the authors.[127] The two-step synthesis starts from the commercially available chiral precursor Ugi's amine and is analogous to that of Josiphos and related ligands (Scheme 92).

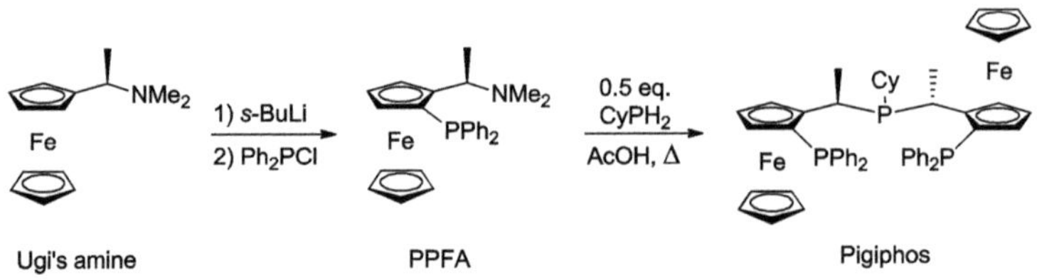

Scheme 92: Synthesis of the tridentate ferrocene-based phosphine ligand Pigiphos.[127]

Later, the yield of the Pigiphos ligand could be improved by using highly concentrated solutions and a stoichiometric amount of trifluoroacetic acid in the last step. Moreover, the scope of ligands was extended using different substituents on the phosphino groups as shown in Figure 38.[194-195] In addition, a tridentate ferrocenyl ligand, in which the ferrocene moieties were connected over the phosphino group attached to the Cp-ring, was reported. This tripodal phosphine was named Gipiphos, since it shows a "reversed" relation in connectivity between the two ferrocene units.

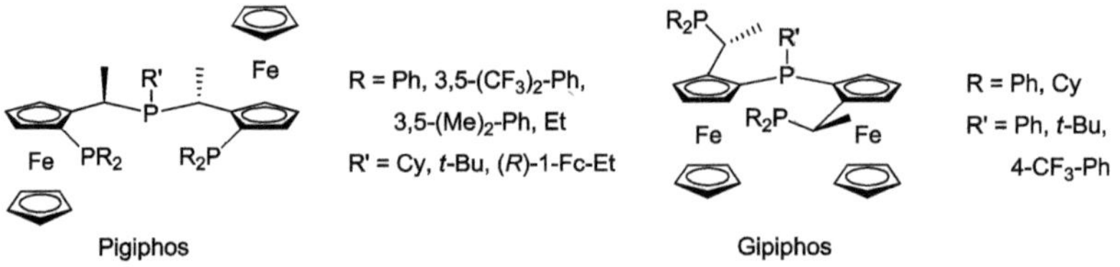

Figure 38: Pigiphos and Gipiphos ligands with different substituents on the phosphino groups.[194-195]

Since the substituents (cyclohexyl/t-butyl) on the phosphino group of the backbone are not C2-symmetric, the Pigiphos ligands display a pseudo C2-symmetry. The use of C2-symmetric ligands is advantageous since the number of metal-ligand-substrate intermediates and/or transition states is limited, therefore minimizing the number of reaction pathways. This can have a beneficial effect on the enantioselective outcome of the reaction compared to the situation involving lower-symmetry ligands. Although the concept of C2-symmetry has been substantiated by the appearance of several very successful C2-symmetric ligands over the last decades, there is no fundamental reason why nonsymmetrical ligands should be inferior in asymmetric catalysis.[196]

In 2002, the first tridentate phosphine ligand combining planar, phosphorus and carbon chirality was developed in our group.[197-198] The reaction sequence starts from PPFA and is depicted in Scheme 93. The diastereoisomers could be separated by column chromatography after borane protection and were converted to the triphosphine ligands by reduction of the phosphine oxide moiety followed by deboranation. Both P-stereoisomers of the triphosphine product **173** were isolated and the absolute configuration at the phosphorus centers was determined using X-ray crystallography.

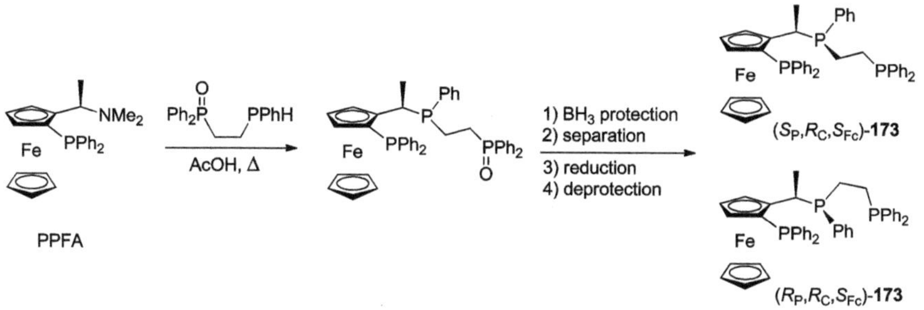

Scheme 93: Synthesis of the first tridentate ligand combining planar, phosphorus and carbon chirality.[197-198]

4.1.2 Transition-Metal Complexes of Tridentate Phosphine Ligands

The coordination behavior of these tridentate phosphines to transition metals was tested and for d^8-metals such as Pd(II), Pt(II) and Ni(II) as well as Rh(I) and Ir(I) the formation of square planar complexes was observed (Scheme 94).[127, 139, 199] Halides as well as coordinating solvents such as acetonitrile and tetrahydrofuran were used as the fourth ligand. The d^6-metal centers Rh(III) and Ir(III) formed octahedral complexes with a meridional arrangement of the tripodal phosphorus ligand upon coordination.[199-200] Interestingly, with Ru(II), four structurally different complexes were obtained.[140] The neutral five-coordinate complexes of type [RuCl$_2$(PPP)] ((PPP) stands for Pigiphos) adopt a *fac* square-pyramidal structure (**174**) with one of the PPh$_2$ group in an apical position or a *fac* trigonal-pyramidal arrangement (**175**) with one PPh$_2$ and one chloride in the axial positions. Electronic reasons were thought to cause these structural differences since electron-withdrawing aryl groups such as *p*-CF$_3$-Ph or 3,5-(CF$_3$)$_2$-Ph resulted in the formation of **174**, whereas Pigiphos and xylyl-

4 P-Stereogenic Pigiphos Ligands

substituted Pigiphos gave complex **175**. The dicationic complexes of type [Ru(PPP)(MeCN)$_3$]$^{2+}$ were obtained either as *fac* octahedral or *mer* octahedral isomers (**176**), depending on the reaction sequence used.

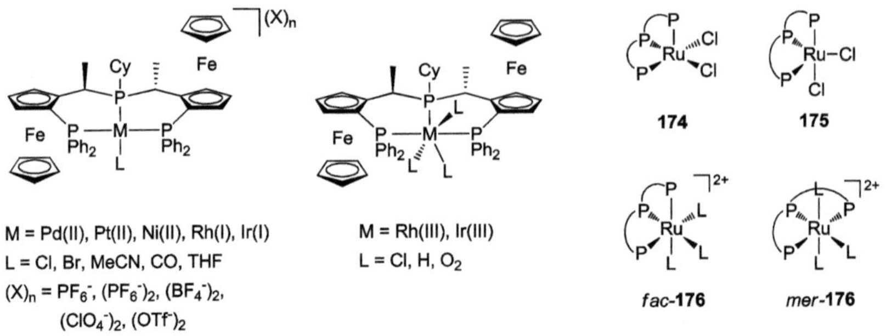

M = Pd(II), Pt(II), Ni(II), Rh(I), Ir(I)
L = Cl, Br, MeCN, CO, THF
(X)$_n$ = PF$_6^-$, (PF$_6^-$)$_2$, (BF$_4^-$)$_2$, (ClO$_4^-$)$_2$, (OTf$^-$)$_2$

M = Rh(III), Ir(III)
L = Cl, H, O$_2$

174   **175**

fac-**176**   *mer*-**176**

Scheme 94: Selected complexes of Pigiphos with d^8- and d^6-metals.

4.1.3 Transition-Metal Asymmetric Catalysis with Pigiphos Ligands

To date, Pigiphos has been applied as a ligand with a variety of transition metals and in several catalytic applications such as transfer hydrogenation, acetalization and Lewis acid catalysis. In the ruthenium-catalyzed transfer hydrogenation of acetophenone full conversions could be obtained when a cationic complex and high temperatures were applied. The best results were obtained with the *fac* complex **176**, providing moderate enantiomeric excess of 72%.[140] High yields were also obtained in the acetalization of benzaldehyde using Ni(II)- or Rh(III)-Pigiphos complexes but the cis/trans ratio of the products was always lower than 1:1.5.

4.1.3.1 Hydroamination of Activated Olefins

The first positive results concerning selectivity were obtained when the Ni(II)-Pigiphos system was used as a Lewis acid catalyst in the hydroamination of activated olefins.[142, 194-195, 201-202] It was shown that olefins containing an electron-withdrawing group (EWG) such as ethylcrotonate, crotonitrile or methacrylonitrile can be activated to allow a 1,4-addition upon coordination to the catalyst. Among the olefins tested, only methacrylonitrile gave a reasonable enantiomeric excess as can be seen in Table 25 (entries 1-3). Six- and five-membered cyclic amines were the best reagents tested and *ee*'s of over 90% were obtained when the reactions were performed at low temperatures (entries 4-6).

4 P-Stereogenic Pigiphos Ligands

Table 25: Hydroamination of activated olefins using a Ni-pigiphos catalyst.

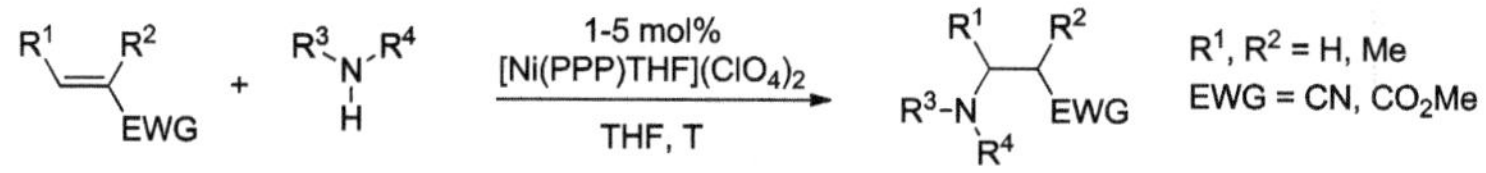

entry	olefin	amine	T [°C]	yield [%]	ee [%]
1	ethylcrotonate	morpholine	25	77	rac
2	crotononitrile	morpholine	25	99	3
3	methacrylonitrile	morpholine	25	99	69
4	methacrylonitrile	morpholine	-80	84	95
5	methacrylonitrile	piperidine	-80	69	92
6	methacrylonitrile	N-methylpiperazine	-80	68	96

By using ionic liquids as solvent system the activity of the Ni-Pigiphos catalyst could be enhanced and rendered a recycling of the catalyst possible.[142, 194] The catalytic experiments using methacrylonitrile and morpholine were repeated applying Pigiphos ligands with modified phosphino groups as well as Gipiphos and the P-stereogenic (S_P,R_C,S_{Fc})-**173** and (R_P,R_C,S_{Fc})-**173**. In nearly all cases yields of more than 90% were obtained but the enantioselectivities were far lower (<40% ee) than for the parent Pigiphos ligand. It was surprising and somewhat disappointing that the first ligand synthesized achieved the best results in the asymmetric hydroamination.

4.1.3.2 Hydrophosphination of Methacrylonitrile

Later, also secondary phosphines were found to be suited nucleophiles for a Michael type addition at methacrylonitrile activated by the Ni-Pigiphos system.[203-204] A strong influence of the solvent, the counterion as well as the temperature was encountered for this hydrophosphination. The best results were obtained when [Ni(H$_2$O)$_6$](ClO$_4$)$_2$ was used as catalyst precursor and the reactions were done in acetone at low temperatures as shown in Scheme 95. With sterically demanding phosphines such as diadamantylphosphine enantioselectivities up to 94% ee were achieved under these conditions.

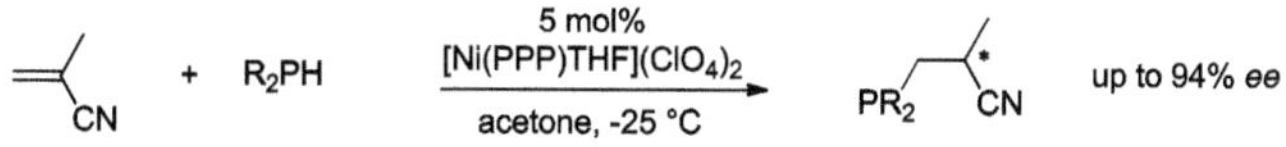

Scheme 95: Hydrophosphination of methacrylonitrile using a Ni-Pigiphos catalyst.

4 P-Stereogenic Pigiphos Ligands

Subsequently, small modifications such as the choice of solvent and the method of forming the active catalyst were undertaken and the same yields and enantioselectivities as described above could be obtained in toluene at room temperature.[139] In Figure 39 the proposed catalytic cycle for the Ni-Pigiphos catalyzed hydrophosphination of methacrylonitrile is shown.[204]

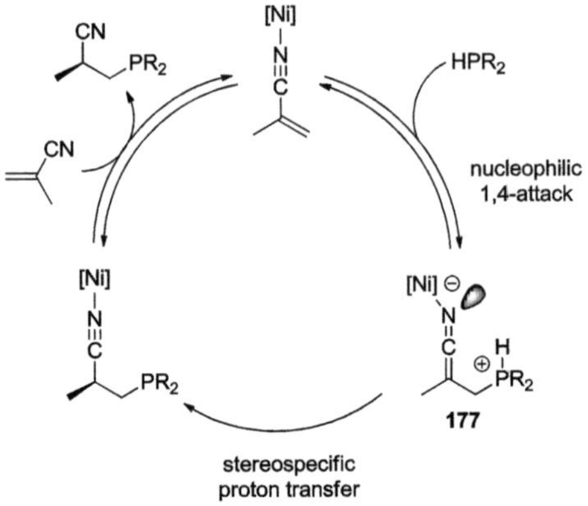

Figure 39: Proposed catalytic cycle for the Ni-Pigiphos catalyzed hydrophosphination of methacrylonitrile.[204]

The mechanism is based on structural, kinetic and computational studies and the fact that a large primary kinetic isotope effect was observed, which strongly suggest proton transfer to be the rate determining step. It was also postulated that the enantioselectivity originates from the formation of two energetically different diastereomers **177** that undergo a stereospecific proton transfer from the phosphonium unit to the prochiral carbon center. Even though the mechanistic investigations are based on phosphines as nucleophiles the catalytic cycle proposed is most likely general and therefore holds for the Ni-Pigiphos-catalyzed hydroamination. However, considering crystal structures of Ni-Pigiphos methacrylonitrile complexes, the prochiral carbon center of the substrate seems to be rather far apart from the chiral environment of the catalyst. Therefore, the high enantioselectivities observed in the hydrophosphination as well as in the hydroamination are still quite surprising.

4.1.3.3 Enantioselective Nazarov Cyclization

Irene Walz, a former member of our group, showed that the Ni-Pigiphos system is also active as a catalyst for the enantioselective Nazarov reaction, the cyclization of divinyl ketones to cyclopentenones, which is promoted by Brønstedt or Lewis acids. [205-206] The general reaction scheme and part of the results are shown in Table 26.

Table 26: Ni-Pigiphos catalyzed Nazarov cyclization of divinyl ketones.[205-206]

$$R^1\text{-CO-C(CO}_2\text{Et)=CHR}^2 \text{ (Ph)} \xrightarrow[\text{2) DCM, r.t.}]{\text{1) 10 mol-\% [Ni(H}_2\text{O)}_6\text{](ClO}_4\text{)}_2, \text{ 20 mol-\% Pigiphos, THF, r.t.}} \text{cyclopentenone}$$

entry	R^1	R^2	yield [%]	ee [%]
1	Me	2,4,6-trimethoxyphenyl	84	86
2	Ph	2,4,6-trimethoxyphenyl	85	87
3	Me	4-methoxyphenyl	32	71
4	Ph	4-methoxyphenyl	96	83

The best results were obtained when the catalyst was preformed in tetrahydrofuran and the reaction was run in weakly coordinating solvents such as dichloromethane. In solvents such as acetonitrile or tetrahydrofuran the cyclization did not proceed or only to a low extent, clearly indicating that coordinating solvents compete with the substrate for catalyst coordination. The use of suitable substituents at specific positions on the substrate was crucial to obtain high reactivity as well as stereoselectivity. In general, long reaction times up to 15 days were required, especially for the less activated substrates (entries 3-4). It was suggested that the product may inhibit the catalyst by coordination, therefore lowering the concentration of active catalyst, resulting in long reaction times. However, high yields and enantioselectivities of up to 88% *ee* were achieved with the dicationic Ni-Pigiphos system, indicating again its usefulness as chiral Lewis acid. Also in this case, the parent Pigiphos bearing "simple" diphenylphosphino groups attached to the Cp-rings turned out to be the best ligand with respect to both, activity as well as enantioselectivity.

4.2 Aim of the Project

The Ni-Pigiphos system has proven to be a suitable catalyst for the activation of functionalized olefins as well as β-ketoesters, thereby providing high yields and enantioselectivities in hydroamination and hydrophosphination as well as in the Nazarov cyclization. Other Ni-Pigiphos-catalyzed reactions such as 1,3-dipolar addition of nitrones to methacrylonitrile are under current investigation in our group. All of these different catalytic reactions have one feature in common, i.e. the fact that the parent Pigiphos ligand is the best suited ligand in terms of reactivity and selectivity. To date, all attempts to modify the Pigiphos system at different positions to improve catalyst performance failed.

This fact and our discovery of a stable ferrocenyl secondary phosphine (vide supra) led us to the idea of synthesizing Pigiphos-type ligands bearing two electronically and/or sterically different phosphino groups on the Cp-rings. This would lead to a total breakdown of the pseudo C2-symmetry and generate ligands with a P-stereogenic phosphino group on the backbone. The retrosynthetic strategy for such ligands can be seen in Scheme 96.

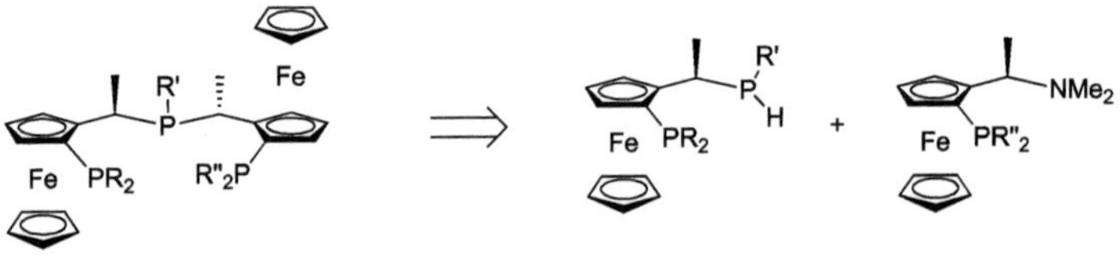

Scheme 96: Retrosynthetic strategy towards a Pigiphos ligand containing a P-stereogenic phosphino group.

Examination of the coordination behavior toward transition metals of the new ligands and their performance in asymmetric catalysis is another goal of this project. The use of phosphino groups bearing substituents of different size may lead to higher enantioselectivity. We were also interested to see if the use of strongly electron-withdrawing groups, such as trifluoromethyl, generates Ni-Pigiphos systems with a higher Lewis acidity, which may result in higher activity in the Nazarov reaction.

4.3 Pigiphos Ligands with a P-Stereogenic Phosphine on the Backbone

4.3.1 Ligand Synthesis

According to the strategy described above, the adamantyl-substituted secondary phosphine **89** was heated with the amine **109** in acetic acid for 12 hours, but the desired product was not formed. Significant amounts of starting material **89** was isolated from a mixture of decomposition products. Changing to a substrate containing silyl substituted arenes on the phosphine (**178**) in place of the trifluoromethyl groups did not have any positive influence on the outcome of the reaction. Also, with the slightly smaller cyclohexyl substituted secondary phosphine **81** as the nucleophile, no formation of the desired product was observed. The rather harsh conditions used in this step seemed not to be suitable for this kind of reaction or these substrates. Furthermore, the application of lower temperatures or different solvents also failed, most likely because of the low solubility of **89** in polar solvents. Since the substitution reaction is following an S_N1 mechanism polar solvents are required in order to stabilize the cationic intermediate.

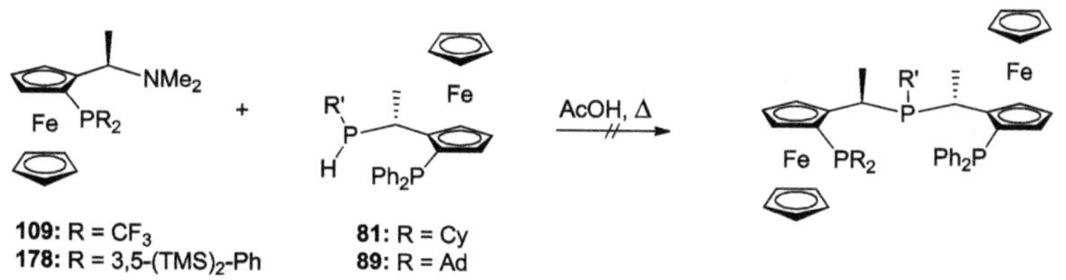

109: R = CF$_3$
178: R = 3,5-(TMS)$_2$-Ph
81: R = Cy
89: R = Ad

Scheme 97: Attempted synthesis of P-stereogenic Pigiphos ligands.

Therefore, we decided to change our strategy and use the bis(trifluoromethyl) substituted secondary phosphine **110** as the nucleophile. This is more soluble in polar solvents but has a lower diastereomeric purity (Scheme 98). As the electrophile we planned to use the acetate **100**, since acetic acid is the slightly better leaving group than dimethylamine. Fortunately, when **100** and **110** were reacted in acetic acid at room temperature for 20 hours, the desired product **179** was obtained in 91% yield as a mixture of two diastereoisomers. These stereoisomers arise from the P-stereogenity of the backbone phosphine and the ratio of 9:2 reflects quite well the dr observed for the P-stereogenic secondary phosphine **110**. Recrystallization of the diastereomeric mixture from ethyl acetate gave the major product as single isomer, but the minor isomer could not be isolated in diastereomerically pure form.

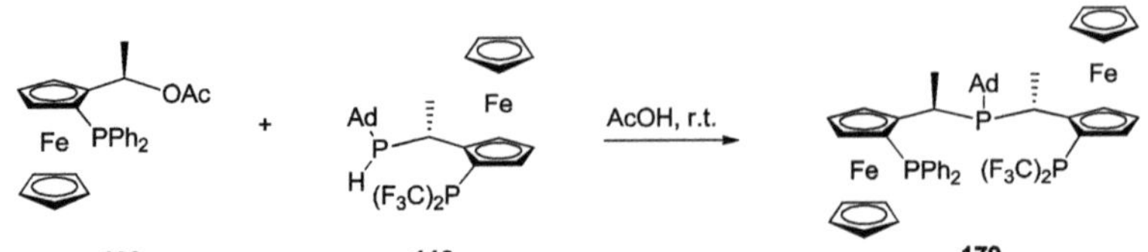

4 P-Stereogenic Pigiphos Ligands

Scheme 98: Synthesis of P-stereogenic Pigiphos ligand **179**.

With this new methodolgy in hand, we were able to synthesize three further P-stereogenic Pigiphos ligands with different substituents on the phosphino groups as shown in Figure 40. In **180** the phenyl groups on the phosphine were replaced by the bulkier and more electron-rich *t*-butyl substituents, whereas **181** shows a combination of a diphenylphosphino and a dicylcohexylphosphino group on the Cp-rings. Triphosphine **182** is very similar to **179**, since it also contains a bis(trifluoromethyl)phosphino and a diphenylphosphino group. The only difference to **179** is the opposite absolute configuration of one ferrocene unit. **179** was synthesized from the (R_C,S_{Fc})-precursors but **182** was prepared using (R_C,S_{Fc})-**110** and (S_C,R_{Fc})-**100**, therefore containing two ferrocenyl units with opposite planar and carbon stereochemistry.

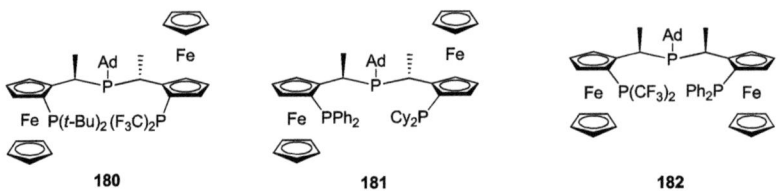

Figure 40: P-stereogenic Pigiphos ligands with different substituents on the phosphino groups.

We were able to grow single crystals of **179** and **182** suitable for X-ray diffraction by slow diffusion of methanol into a saturated DCM-solution. The ORTEP drawings can be seen in Figure 41 and Figure 42 and selected bond lengths and angles are listed in Table 27.

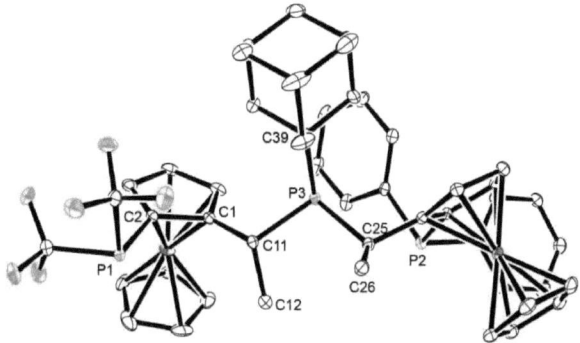

Figure 41: ORTEP representation of Pigiphos derivative **179** (inverted). Hydogen atoms are omitted for clarity and thermal ellipsoids are set to 30% probability.

From these structures it is obvious that the major diastereoisomer of both, **179** and **182** has an *S*-configuration at the backbone-phosphine (P3). Despite this, a clear difference in the arrangement of P3 in both isomers becomes evident from the ORTEP drawings. This is reflected by the torsion angles α and β, which are used to describe the relative arrangement of P3 with respect to the Cp-planes of both ferrocenyl units. α is the torsion angle between the bis(trifluoromethyl)phosphine substituted cyclopentadienyl ring and P3 and β that between the Fc-unit bearing the diphenylphosphino group and P3.

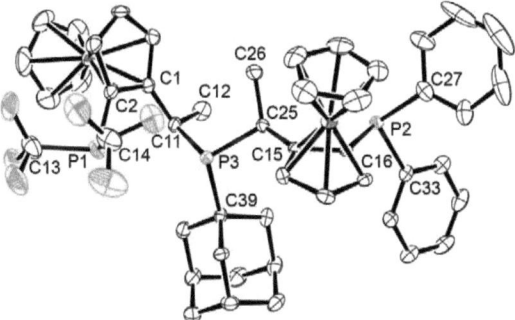

Figure 42: ORTEP representation of Pigiphos derivative **182**. Hydrogen atoms and co-crystallizing solvent molecules are omitted for clarity and thermal ellipsoids are set to 30% probability.

4 P-Stereogenic Pigiphos Ligands

In **179** P3 points away from P1, which is reflected in a torsion angle with an absolute value of more than 90° ($\alpha = -144.7°$), whereas in **182** P3 is bent towards P1, reflected by an angle less than 90° ($\alpha = -60.4°$). As a consequence, the methyl group (C12) is in **179** perpendicular to the Cp-plane but in **182** it lies nearly perfectly in the plane. For the torsion angle β the situation is reversed, since P3 is bent towards P2 in **179** and points away from P2 in **182**. The change in algebraic sign of β in **179** and **182** is caused by the opposite absolute configuration of one of the ferrocenyl units. The bond lengths of P3 to all three substituents are very similar in both isomers **179** and **182** as well as comparable to the situation observed for secondary phosphine **110**. The pyramidalization angles around P3, which are the sums of all three angles caused by the three substituents of P3, are nearly equal in both structures. However, these angles are considerably larger than that in Pigiphos, which contains the sterically less demanding cyclohexyl group.

Table 27: Selected bond lengths and angles of the Pigiphos derivatives **179** and **182**.

atoms	179 length [Å]	179 angle [°]	182 length [Å]	182 angle [°]
P3-C11	1.8907 (18)		1.889 (4)	
P3-C25	1.8918 (18)		1.865 (4)	
P3-C39	1.8855 (18)		1.892 (4)	
α (C2-C1-C11-P3)		-144.7 (2)		-60.4 (5)
β (C16-C15-C25-P3)		-70.6 (2)		147.3 (3)
pyramid. P3		315.6 (1)		313.4 (3)

4.3.2 Transition-Metal Complexes of P-Stereogenic Pigiphos

In order to gain more detailed information about the coordination properties of these asymmetric Pigiphos ligands, transition-metal complexes were synthesized. **179** reacted smoothly with Pd(II) and Ni(II) precursors to afford the corresponding square planar complexes **183** and **184** shown in Scheme 99.

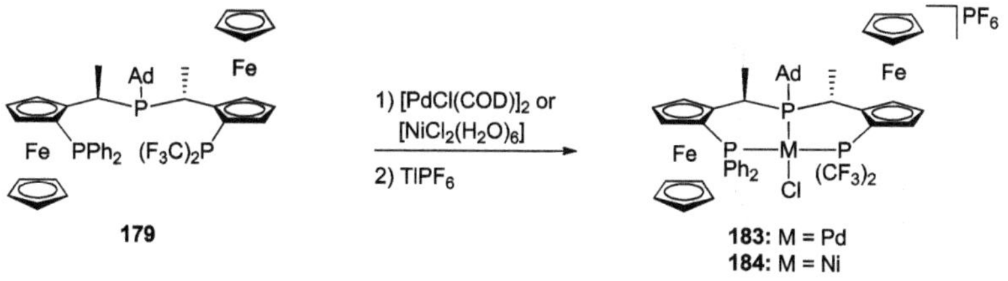

Scheme 99: Synthesis of Pd(II) and Ni(II) complexes with the P-stereogenic Pigiphos derivative **179**.

The ^{31}P NMR spectra of the nickel as well as the palladium complex showed broad signals indicating a dynamic behavior of the square planar complexes. Only the signal of the diphenylphosphino group was sharp enough to determine the coupling constant to the other phosphino groups. Large J-couplings to the phosphino group in trans position of $^2J_{P-P} = 450$ Hz for **183** and $^2J_{P-P} = 370$ Hz for **184** was obtained, respectively. A similar behavior was obtained for the Ni(II) complex of the dicyclohexyl substituted ligand **181**, shown in Scheme 100.

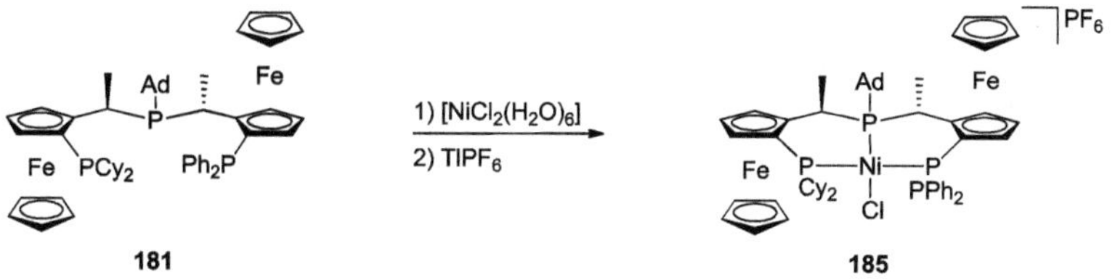

Scheme 100: Synthesis of Ni-Pigiphos complex **185**.

At room temperature, sharp signals of one Ni-Pigiphos species were observed in the ^{1}H NMR spectra, but only broad, nearly undetectable signals in the ^{31}P NMR spectrum. Cooling to -70 °C showed the signals of two isomers in the region of the PPh$_2$ and the PCy$_2$ groups as can be seen in Figure 43.

4 P-Stereogenic Pigiphos Ligands

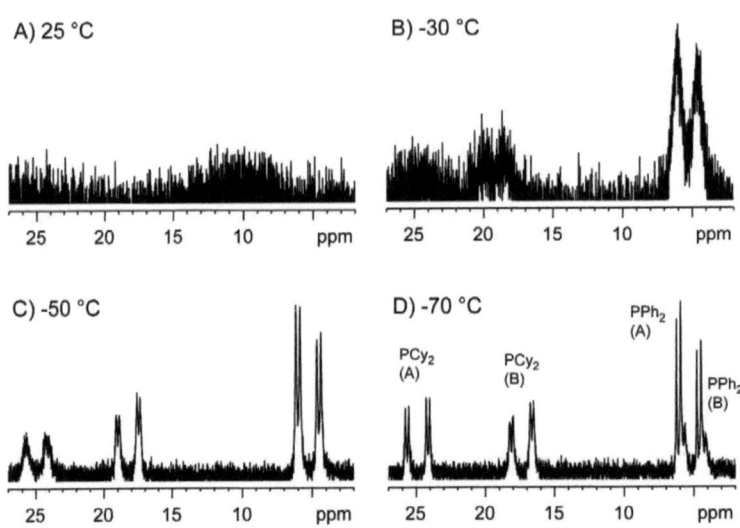

Figure 43: ^{31}P NMR spectra of the Ni-Pigiphos complex **185** at different temperatures.

A similar behavior of Pigiphos transition-metal complexes has been observed previously and can be attributed to the existence of two isomers, differing in the conformation of the two six-membered rings formed by the metal center and the three coordinating phosphino groups.[139, 199] It was suggested that the complex is either in the double boat conformation **186** or the boat-half-chair conformation **187** shown in Figure 44. We could not find any evidence that **185** also exists in these two proposed conformations nor could we determine which of the isomers observed in the ^{31}P NMR spectrum belongs to which conformation. However, the fact that ^{31}P NMR spectroscopically only one signal was observed for the backbone phosphorus (P3) indicates that the main difference between the two isomers involves primarily the trans phosphino groups P1 and P2.

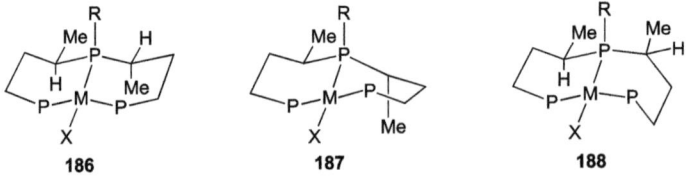

Figure 44: Double boat and boat-half-chair conformation of Pigiphos transition-metal complexes.[139]

Crystals suitable for X-ray diffraction of the palladium complex **183** were obtained and an ORTEP representation is shown in Figure 45. Two crystallographically independent molecules of **183** are contained in the asymmetric unit but due to their conformational similarity only one of them (**183**-A) is shown. However, they show significant differences in bond lengths and angles, therefore the geometric parameters are listed separately in Table 28, labeled as **183**-A and **183**-B, respectively.

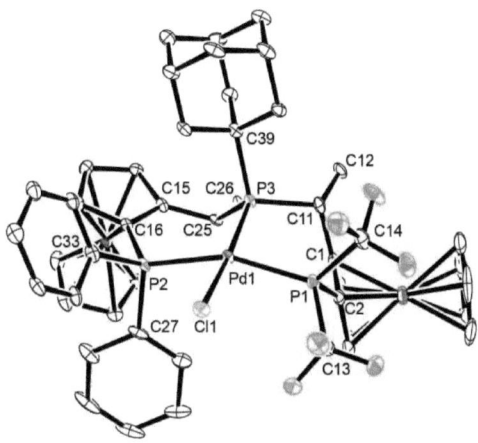

Figure 45: ORTEP representation of the Pd(II) Pigiphos complex **183**. Only one of the two crystallographically independent molecules observed in the asymmetric unit is shown (**183**-A). Hydrogen atoms as well as PF_6 counterions and co-crystallizing solvent molecules are omitted for clarity. Thermal ellipsoids are set to 30% probability.

As can be seen in Figure 45, the two six-membered chelate rings both adopt a boat conformation with the one on the left slightly more distorted then the other. Interestingly, this double boat conformation (a schematic drawing (**188**) can be found in Figure 44) is different from the double boat structure **186**, which is normally observed in Pigiphos transition metal complexes. Whereas in **186** the ferrocene part on the right is pointing towards the substituent on P3, in **183** (**188**) it is bent towards one of the phenyl groups of P2. As a consequence the methyl group of the right subunit in **183** is in an equatorial position, lying nearly exactly in the plane α, defined by the three phosphorus atoms. In conformations **186** and **187** this methyl group is axial, and is therefore in a position perpendicular to this plane. These differences observed for **183** may be caused by the smaller substituents on P1 (CF_3) or the larger substituent on P3 (Ad) compared to the parent Pigiphos ligand (Ph on P1, Cy on P3).

4 P-Stereogenic Pigiphos Ligands

Since the bond lengths and angles observed in the two conformers differ significantly a coherent discussion of them is rather difficult. Whereas in **183**-A the bond from the metal center to P1 is shorter than that to P2, in **183**-B the opposite situation is observed. Pd1-P1 and Pd1-P2 are comparable to those observed in the parent Pd-Pigiphos complex (~2.33 Å) but Pd1-P3 is significantly longer (2.255 Å).[127] Inspection of the crystal structure as well as the bond angles listed in Table 28 indicate the distorted square planar geometry around the metal center, typical for Pigiphos transition-metal complexes. The non-ideal square planar geometry is also reflected in the deviation of the metal center and the chloride ligand from the plane α. Pd1 lies ~0.4 Å and the chloride more than 1 Å above this plane, which is considerably more as observed in the parent Pd-Pigiphos system (0.18 Å and 0.47 Å, respectively). Therefore, the deviation from the square planar geometry in **183** is much more pronounced than in the original palladium Pigiphos complex, most likely because of the different boat-boat conformation.

Table 28: Selected bond lengths and angles of the two conformers **183**-A and **183**-B. α defines the plane through P1, P2 and P3.

atoms	length [Å]		atoms	angle [°]	
	183-A	**183**-B		**183**-A	**183**-B
Pd1-P1	2.3062 (18)	2.3442 (18)	P1-Pd1-P3	94.42 (6)	93.94 (6)
Pd1-P2	2.3304 (17)	2.3217 (16)	P2-Pd1-P3	92.80 (6)	94.19 (6)
Pd1-P3	2.3053 (16)	2.3144 (16)	P1-Pd1-P2	157.42 (6)	154.56 (6)
Pd1-Cl	2.3318 (16)	2.3392 (16)	P3-Pd1-Cl	174.42 (6)	172.76 (6)
Pd1-α	0.396	0.446	P1-Pd1-Cl	88.56 (6)	91.30 (6)
Cl1-α	1.003	1.131	P2-Pd1-Cl	86.16 (6)	83.12 (6)

In order to obtain complexes suitable for catalytic application, we envisaged the preparation of Ni complexes with a free coordination site. Unfortunately, the synthesis of such a dicationic Ni complex with the bis(trifluoromethyl)phosphino ligand **179** failed. Two strategies towards the desired complex were tested as shown in Scheme 101.

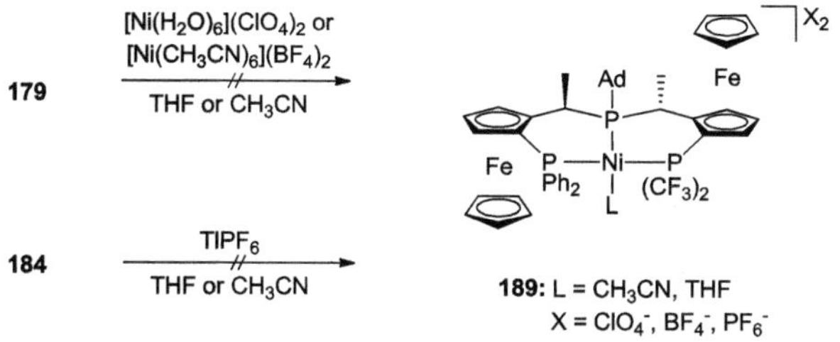

Scheme 101: Attempted syntheses of dicationic Ni complexes **189**.

Neither direct synthesis using tetrafluoroborate or perchlorate Ni precursors nor chloride abstraction from the nickel chloro-complex **184** in coordinating solvents afforded the desired complexes **189**. It seems that the relatively electron-poor phosphino group P1 or the tridentate geometry of **179** is not suited to form such complexes.

4.4 Pigiphos Ligands with P-Stereogenic Phosphines on the Cp-Ring

4.4.1 Ligand Synthesis

In addition to the Pigiphos ligands having the P-stereogenic phosphorus on the backbone, we also envisaged the syntheses of tridentate ligands bearing P-stereogenic phosphino groups on the Cp-rings. Since the preparation of their precursors is known and straightforward, the synthesis of a library of P-stereogenic Pigiphos ligands should be a simple task. Applying the method reported by Chen et al.[98], three different P-stereogenic Pigiphos ligands were prepared as depicted in Scheme 102. Lithiated Ugi's amine was reacted with a dichlorophosphine followed by the addition of an organolithium or organomagnesium reagent to afford the P-stereogenic phosphino amines **190-192**. The amines bearing a naphthyl- or an anisylphosphino group were obtained in diastereomerically pure form and the adamantyl substituted phosphine as a 3:1 mixture, which could be purified by recrystallization. In order to circumvent epimerization of the stereogenic phosphorus centers, the amino group was not replaced directly with the phosphine by applying the usual method of heating the mixture in acetic acid. Thus, **190** and **191** were first transformed into the acetoxy compounds **193** and

4 P-Stereogenic Pigiphos Ligands

194 by reacting with acetic anhydride over 72 hours. The compounds obtained were then reacted with the primary phosphine in acetic acid at room temperature to afford the desired tridentate ligands (**195/196**). Some epimerization of the P-stereogenic phosphine was observed in the case of the naphthyl-substituted triphosphine but recrystallization yielded the diastereomerically pure product. Since the phosphines bearing an adamantyl group turned out to be very stable to epimerization, Pigiphos ligand **197** was synthesized directly from the amine applying the conventional conditions.

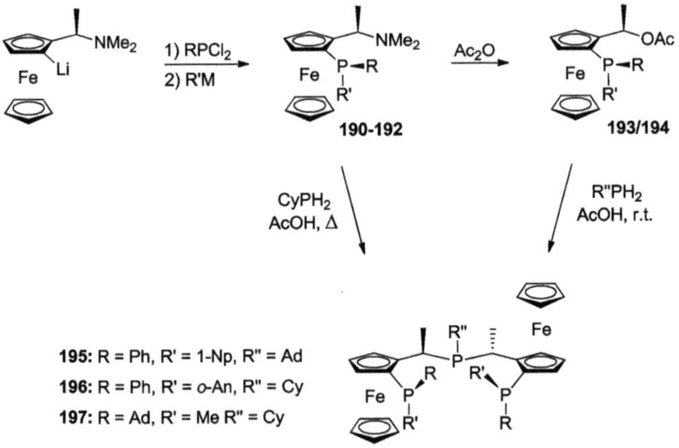

195: R = Ph, R' = 1-Np, R" = Ad
196: R = Ph, R' = o-An, R" = Cy
197: R = Ad, R' = Me R" = Cy

Scheme 102: Synthesis of Pigiphos ligands **195-197** bearing two P-stereogenic phosphino groups.[98]

In contrast to the case of the anisyl- and naphthyl-substituted compounds, in which the absolute configuration of the phosphorus center was known, the phosphines bearing an adamantyl group were new. Although we were convinced that the stereochemistry was the same as the reported compounds, it had to be verified. Therefore, the phosphino amine **192** was converted into the Pt-Josiphos complex **198** in two steps as shown in Scheme 103.

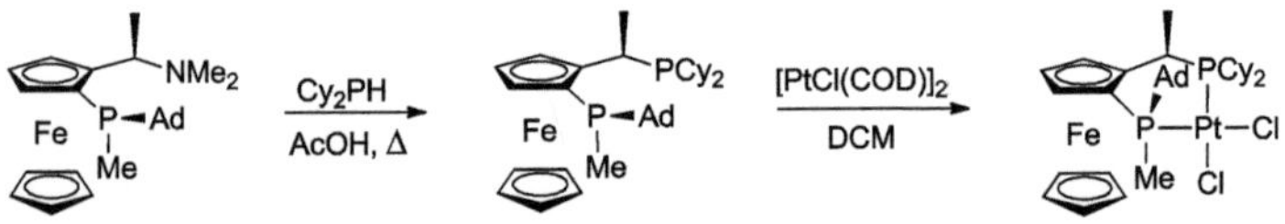

Scheme 103: Synthesis of P-stereogenic platinum complex **198** for the determination of the absolute configuration.

4 P-Stereogenic Pigiphos Ligands

The X-ray structure analysis of the platinum complex **198** showed that the P-stereogenic phosphine adopts an *S*-configuration (Figure 46). Under the assumption that no full epimerization occurs in the two steps towards **198**, the amine **192** and the Pigiphos ligand **197** should also have *S*-configured phosphino groups.

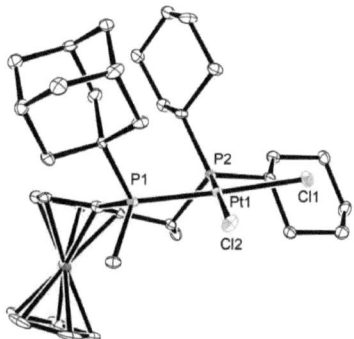

Figure 46: ORTEP representation of platinum complex **198**. Hydrogen atoms are omitted for clarity and thermal ellipsoids are set to 30% probability.

Since the tridentate ligand **179** containing a bis(trifluoromethyl)phosphino group did not form dicationic nickel complexes we decided to prepare Pigiphos ligands containing two P-stereogenic trifluoromethylphosphino groups. This ligand may have the ability to effectively coordinate to the metal center while increasing the Lewis acidity of the metal due to the electron-poor phosphines. The synthesis starts from the P-stereogenic trifluoromethyl-phosphines (*S*$_P$)-**147** and (*R*$_P$)-**147** described in Chapter 2 and is shown in Scheme 104. The tridentate ligands **199** and **200** were obtained without any epimerization of the stereogenic phosphorus by heating the monophosphines in acetic acid in the presence of 0.5 equivalents of cyclohexylphosphine.

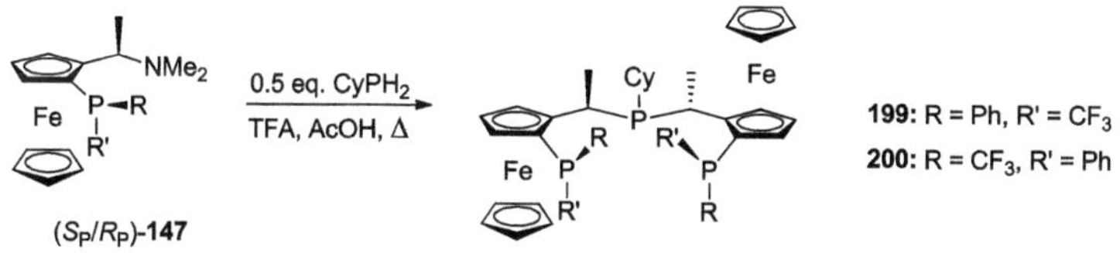

Scheme 104: Synthesis of Pigiphos ligands containing two P-stereogenic trifluoromethylphosphino groups.

4.4.2 Nickel(II) Complexes of P-Stereogenic Pigiphos

According to the standard procedure, Ni(II) chloro-complexes **201-203** were prepared as shown in Scheme 105. Similar to the Ni-Pigiphos complexes described in Section 4.3.2 broad resonances were observed in the ^{31}P NMR spectrum of these complexes. Nevertheless, a large J-coupling of $^2J_{P-P}$ = 295 Hz between P1 and P2 was detected in **203**.

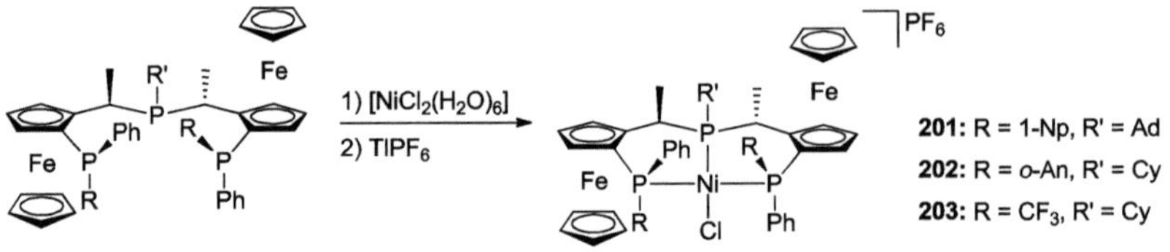

Scheme 105: Synthesis of Ni(II) complexes **201-203** with P-stereogenic phosphine ligands.

The same ligand was also tested in the synthesis of a dicationic nickel(II) complex to see if the trifluoromethylphosphine-substituted Pigiphos is better suited to form such complexes than the ligand bearing a bis(trifluoromethyl)phosphino group (**179**). Unfortunately, this ligand also failed to form the desired dicationic nickel complex upon chloride abstraction.

Since the transition-metal Pigiphos complexes we prepared showed a variety of different colors depending on the metal as well as the ligand used we analyzed them using UV-Vis spectroscopy. In Table 29 the color of the specific complex as well as the wavelength, λ, of its maximum absorption is listed.

Table 29: Wavelenghts of maximum UV-absorption and colors of different transition-metal Pigiphos complexes.

complex	metal	observed color	λ [nm]
202	Ni(II)	orange	470
183	Pd(II)	red	510
203	Ni(II)	grape	525
201	Ni(II)	purple	550
185	Ni(II)	royal blue	570
184	Ni(II)	navy blue	610

The observed colors correlate very well with the wave lengths of the visible light that is absorbed by the complexes. For example, **184** absorbs orange light, which is in the range of 610 nm. As a consequence the transmitted light is of complementary color, which is in this case blue, exactly what is observed. The absorption is caused by an electronic transition in the complexes, most likely d-d-transitions (from a filled d-orbital of the metal into an empty or half-filled one) or charge-transfer transitions (electronic transition from metal d-orbitals into empty ligand orbitals). The energy gap between two such orbitals is dependent on the nature of the metal center, the geometry of the metal-ligand-complex as well as on the electronic nature of the ligands. Thus, the different electronic as well as steric properties of the Pigiphos ligands described above obviously generate metal complexes with very different absorption properties as observed by the varying colors.

4.5 Conclusion and Outlook

In this chapter we demonstrated the synthesis of two different types of Pigiphos ligands containing one or two P-stereogenic phosphino groups by applying two different strategies. The synthesis of triphosphines of type **179**, having the stereogenic P atom in the backbone involves two ferrocene moieties with different phosphino groups on the Cp-rings. The secondary phosphine **110** reacts in an S_N1 type fashion with the acetoxy substituted ferrocene compound **100** affording the P-stereogenic triphosphine **179**. In general, good yields and a diastereomeric mixture were obtained but isolation of the major isomer in pure form was possible. The absolute configuration of the stereogenic phosphorus center could be determined by X-ray crystallography. The second type of P-stereogenic Pigiphos ligands, having the stereogenic phosphino groups on the Cp-rings, makes use of the methodology introducing P-stereogenicity into PPFA-type compounds as developed by Chen et al. The monophosphines thus obtained were then reacted with a primary phosphine to afford the P-stereogenic triphosphines of type **195** in moderate to good yields.

Palladium and nickel complexes of the new Pigiphos ligands were prepared and spectroscopically and crystallographically analyzed. It was observed, using NMR spectroscopy, that **185** exists as a mixture of isomers in solution, most likely with differing conformations of the two six-membered chelate rings formed by the metal center and the three phosphine donors. X-ray crystal structure of palladium complex **183** showed the unusual double-boat geometry around the palladium center. The synthesis of dicationic nickel complexes with a coordinating solvent as the fourth ligand failed for the trifluoromethyl substituted phosphines **179** and **199**. Therefore, we conclude, these types of ligands are electronically or sterically not suited for the stabilization of such metal centers. Due to this fact, these ligands were not applied in the Ni-Pigiphos-catalyzed Nazarov reaction as was intended.

The preparation and analyses of complexes of these new tridentate ligands with different transition metals such as Pt, Ir, Rh, Cu etc. should be the subject of further investigations. Additionally, the application of these P-stereogenic ligands in asymmetric catalysis is another important task for the future. Over and above the known catalytic applications, hydroamination, hydrophosphination, Nazarov cyclization, new reactions should also be applied to test these ligands.

5 General Conclusions and Outlook

This thesis reports the synthesis of P-stereogenic bidentate and tridentate ferrocenyl phosphines, their transition-metal complexes and their application as ligands in asymmetric catalysis. We demonstrated that secondary phosphines of the type **81/110** are stable against air and moisture and exist as a diastereomeric mixture with a specific dr. In addition, it was shown that such phosphines are, on one hand, able to form stable transition-metal complexes (**112/116**) and can provide in part high enantioselectivities in asymmetric catalysis. On the other hand, these compounds can be reacted under certain conditions with suited metal precursors to generate new metal-ligand systems (**115**). This special reactivity of the secondary phosphine moiety makes these compounds to versatile precursors for the synthesis of a variety of different phosphine ligands such as SPOs (**97/111**), Pigiphos-like triphosphines (**179**) or 1,2-diphospholes (**129**).

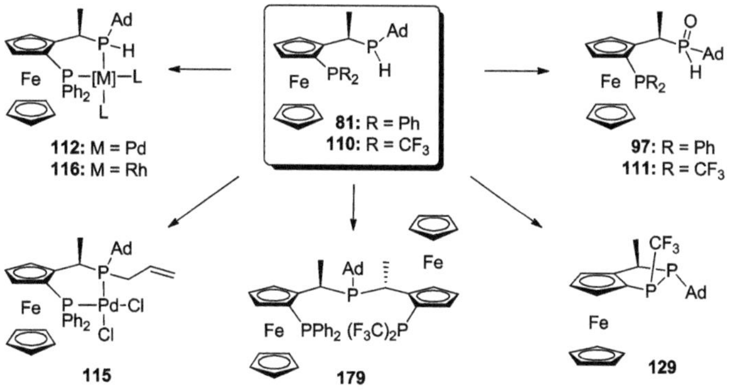

Scheme 106: Secondary phosphines **81/110** as versatile precursors for the synthesis of P-stereogenic ligands and complexes.

While the bis(trifluoromethyl)phosphino-substituted compound **110** was easily transformed into the SPO **111**, no selective oxidation methods were found for the synthesis of SPOs bearing an electron-rich tertiary phosphino group such as **97**. Since these SPO-P-ligands showed their potential in asymmetric catalysis, their selective and efficient synthesis still remains an important target for the future. A laborious screening of a broad variety of different oxidation agents as well as reaction conditions may lead to some success. On the other hand, in order to circumvent the problems of unselective oxidation, one may synthesize bidentate ferrocenyl ligands consisting of two P-stereogenic SPO-functionalities. Such ligands could form, after coordination and loss of a proton, a second chelate ring through hydrogen

5 General Conclusions and Outlook

bond stabilization (Scheme 107). Similar, SPO-variations of Pigiphos ligands would provide complexes with two trans-SPO groups and a tertiary phosphine donor. Such transition-metal SPO complexes might serve as catalysts in various asymmetric reactions.

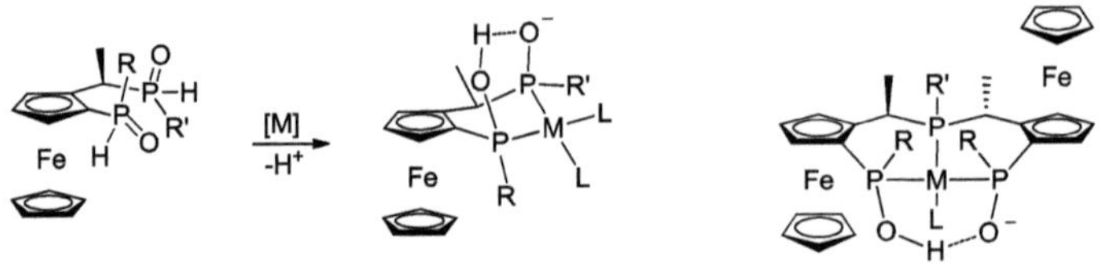

Scheme 107: Ferrocene-based bidentate and tridentate complexes containing two P-stereogenic SPO-functionalities.

In Chapter 3, it was shown that the CF_3 group of bis(trifluoromethyl)phosphines can act as a leaving group and thus may be substituted in intra- as well as intermolecular fashions. This finding opens a new strategy for the stereoselective synthesis of Josiphos-like ligands containing a P-stereogenic trifluoromethylphosphino group. The catalytic potential of these ligands was demonstrated and high activities and enantioselectivities up to >99% ee were obtained in the rhodium-catalyzed hydrogenation of olefins. These results clearly disclose that the small and electron-withdrawing substituent CF_3 does not need to be detrimental to both catalytic activity and stereoselectivity. To further establish this special ligand class, a broad screening in different transition-metal-catalyzed reactions should be undertaken. Especially Lewis-acid catalysis or applications where the stabilization of the metal center in a low oxidation state is required, may disclose catalytic reactions, for which these relatively electron-poor phosphines are very well suited.

Moreover, Josiphos-type ligands bearing the trifluoromethylphosphino group on the side chain should be synthesized and their structural, spectroscopic and catalytic properties be compared with those of Josiphos and the ligands reported herein (Scheme 108).

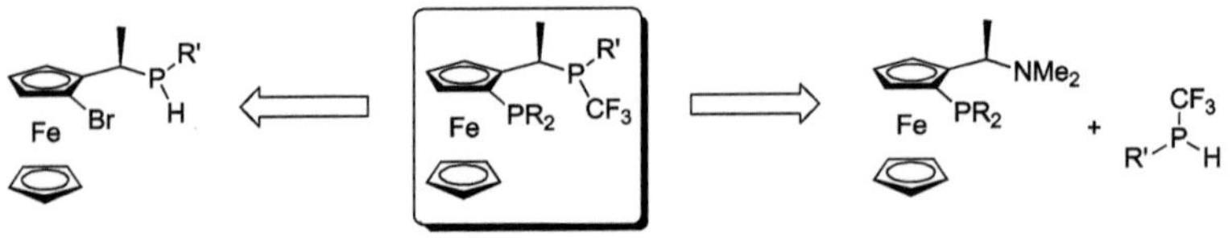

Scheme 108: Strategies for the synthesis of Josiphos-type ligands with the trifluoromethylphosphino group on the side chain.

5 General Conclusions and Outlook

One possible, synthetic approach towards such compounds would use the trifluoromethylation of a ferrocenyl secondary phosphine as the key step. A different synthetic strategy may implement the substitution of the amino functionality in a PPFA-like precursor with a secondary trifluoromethylphosphine. The synthesis of such CF_3-Josiphos ligands is under further investigation in our research group.

In order to examine its potential in more detail, the introduction of the P-stereogenic trifluoromethylphosphino group into the core of ligand systems other than Josiphos or Pigiphos could be the subject of further investigation. One may think about variations of the BoPhoz- or Taniaphos-type ligands, which should be easily achieved by applying our synthetic strategy to known precursors (Figure 47). P-Stereogenic trifluoromethylphosphines based on the binaphthyl-skeleton such as CF_3-MOP or CF_3-BINAP would also present highly interesting ligands, but their preparation displays most likely a high synthetic challenge.

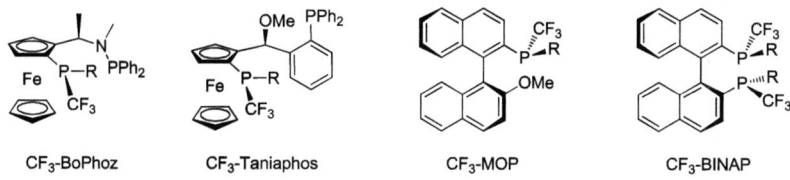

Figure 47: Different ligand-families containing a P-stereogenic trifluoromethylphosphino group.

Although, ferrocene-based trifluoromethylphosphines could be successfully synthesized using the CF_3 substituent as a leaving group on phosphines, a more general applicability of this methodology has yet to be demonstrated. The major drawbacks of this strategy is the use of an excess of the nucleophile and the relatively low yield of product accompanied by high amounts of side-products and unreacted starting material. These problems are most likely caused by the inferior leaving group-character of the CF_3 group (compared to Cl) as well as undesired side reactions of the free, relatively reactive, CF_3 anion. A suited electrophile able to trap this reactive CF_3 species without interfering with the desired reaction pathway may offer a solution to these drawbacks. If conditions could be found that allow a clean and selective synthesis of the desired trifluoromethylphosphine products, the methodology of using CF_3 as a leaving group could be used as a general strategy for the synthesis of P-stereogenic trifluoromethylphosphine ligands. In addition, the examination of the leaving-group character of the CF_3 group not only in P-CF_3 compounds, but in general, could be an interesting target for the future.

5 General Conclusions and Outlook

In Chapter 4, we disclosed the synthesis of two different families of P-stereogenic Pigiphos ligands. The corresponding palladium(II) and nickel(II) complexes were synthesized and the coordination behavior of the new ligands to these metals analyzed. However, the original goal, to use trifluoromethyl substituted Ni-Pigiphos complexes in the Nazarov cyclization was not achieved, since no dicationic nickel complex could be synthesized with these special ligands. The synthetic strategies we applied to prepare these complexes seemed not to be suited for this ligand-metal precursor combination. Therefore, new methods for the preparation of such dicationic complexes with electron-withdrawing phosphine ligands should be developed. To ascertain if the electron-poor tridentate ligands may increase the Lewis-acidity of these catalysts and thus increase their activity in catalytic reactions such as the Nazarov cyclization could be an important target for future studies.

6 Experimental Part

6.1 General Remarks

6.1.1 Techniques

All reactions with air- or moisture-sensitive materials were carried out under argon using standard Schlenk techniques or in a glove box (MBraun MB 150B-G and Lab Master 130) under an atmosphere of nitrogen. Glassware was preheated at 140 °C in an oven or dried under vacuum with a Bunsen burner and set then under argon. The solvents used for synthetic and recrystallization purposes were of "puriss p.a." quality (Fluka AG, Riedel-de-Häen, J.T. Bakker or Merck). Anhydrous solvents, when needed, were freshly distilled under argon from Na/benzophenone (toluene, THF, Et_2O), Na/benzophenone/diglyme (n-pentane), Na/benzophenone/tetraglyme (n-hexane), Na/diethyl phthalate (EtOH) or CaH_2 (MeOH, CH_2Cl_2, MeCN). For FC and TLC, technical grade solvents were generally used. Deuterated solvents were purchased from Cambridge Isotope Laboratories or Armar Chemicals ($CDCl_3$). For sensitive compounds the deuterated solvents were purified by bulb-to-bulb distillation from CaH_2 (CD_2Cl_2), degassed by 3 freeze-pump-taw cycles and stored under argon in a Young-Schlenk.

6.1.2 Chemicals

(R)-Ugi amine was kindly provided by Solvias AG (Basel) as the tartrate salt and was obtained as enantiomerically pure free amine following a modified procedure by Ugi.[124] Commercially available chemicals were purchased from ABCR, Acros AG, Aldrich, Fluka AG or Pressure Chemical Co. and used without further purification. The following reagents and metal complexes were synthesized according to reported procedures: 1-trifluoromethyl-1,2-benziodoxol-3-(1H)-one (**122**),[171] (R_C,S_{Fc})-1-(diphenylphosphino)-2-[1-(dimethylamino)-ethyl]ferrocene,[207] (R_C,S_{Fc})-1-(diphenylphosphino)-2-(1-acetoxyethyl)ferrocene (**100**),[91] (R_C,S_{Fc})-1-[bis(trifluoromethyl)phosphino]-2-[1-(dimethylamino)ethyl]ferrocene (**109**),[171] [$PdCl_2(COD)$],[108] [$PtCl_2(COD)$],[108] [$PdCl(C_3H_5)]_2$,[209] [$Rh(COD)_2$]PF_6.[210]

6.1.3 Analytical Techniques and Instruments

Thin layer chromatography (TLC): *Merck Silica gel 60 F_{254}* visualized by fluorescence quenching at 254 nm. In addition, TLC plates were typically stained using vanillin (15 g vanillin, 2.5 mL H_2SO_4 conc., 250 mL EtOH). **Flash column chromatography (FC):** chromatographic purification of products was performed on *Fluka Silica Gel 60* (230-400

6 Experimental Part

mesh) or *MP Alumina N, Akt. I* (*MP Biomedicals GmbH*) using the given solvent ratios and a forced flow of eluent at 0.1-0.2 bar pressure. **Optical rotations** were measured at 589 nm (Na/Hal) and room temperature (22 °C) on a Perkin Elmer 341 polarimeter using a 10 cm cell at a concentration of 1 g substance per 100 mL (c = 1.0) in the given solvent. **NMR spectra** were recorded on *Bruker* DPX-500, DPX-400, DPX-300, DPX-250 or AC-200 spectrometers operating at the given spectrometer frequency. The samples were measured as solutions in the given solvent at room temperature (if not indicated differently) and in non-spinning mode. The chemical shifts (δ) are expressed in part per millions (ppm) relative to TMS as an external standard for ^{1}H- and ^{13}C-NMR spectra and are calibrated against the solvent residual peak.[211] For ^{19}F-NMR spectra, CFCl$_3$ and for ^{31}P-NMR spectra H$_3$PO$_4$ (85%), respectively, were used as external standards. Coupling constants J are given in Hertz (Hz) as absolute values. The multiplicity of the signals is abbreviated as follows: s = singlet, d = doublet, t = triplet, q = quartet, pent. = pentet, hept. = heptet, b = broad, m = multiplet. Other abbreviations used: Ad = adamantyl, Cp = cyclopentadienyl, Cp' = unsubstituted Cp of the ferrocene, Cy = cyclohexyl, Np = naphthyl, An = *o*-anisyl, Pz = pyrazole. **Infrared (IR)** spectra were recorded on a *Perkin-Elmer BX II* using ATR FT-IR technology and are reported as absorption maxima in cm^{-1} (w = weak, m = medium, s = strong) for the range between 4000 and 1300 cm^{-1}. **High-resolution mass spectra (HRMS)** were measured by the MS-Service of the "Laboratorium für organische Chemie der ETHZ". **Elemental Analysis (EA)** were carried out by the microelemental analysis service of the *"Laboratorium für organische Chemie der ETHZ"* on a *LECO CHN-900* analyzer. The content of the specified element is expressed in mass percent (%). **High Pressure Liquid Chromatography (HPLC)** was run on either a *Hewlett-Packard 1050 Series* or an *Agilent 1100 Series* with detection at five different wave lengths (210, 215, 220, 230, 254 nm) using the specified column (*Daicel Chiralcel* OJ, OD-H), flow rate of the solvents (mL/min), ratio of *n*-hexane/*i*-PrOH and sample injection volume (µl; sample concentration approximately 1-5 mg/mL). Retention times are given in minutes (min). **Crystallography:** Intensity data of single crystals glued to a glass capillary were collected at the given temperature (usually 100 K) on Bruker SMART 1k or Bruker SMART APEX platforms with CCD detector and graphite monochromated Mo-K$_\alpha$-radiation (λ = 0.71073 Å). The program SMART served for data collection; integration was performed with the software SAINT.[212] The structures were solved by direct or Patterson methods using the program SHELXS-97.[213] The refinement and all further calculations were carried out using SHELXL-97.[214] If nothing else mentioned all non-hydrogen atoms were refined anisotropically using weighted full-matrix least-squares on F^2. The hydrogen atoms were included in calculated positions and treated as riding atoms using SHELXL default parameters. In the end absorption correction was applied (SADABS)[215] and weights were optimized in the final refinement cycles. The absolute configuration of chiral compounds was determined on the basis of the Flack parameter.[216-217] The standard uncertainties (s.u.) are rounded according to the "Note for Authors" of the *Acta Crystallographica*.[218] Associated crystallographic data are given in Appendix A.

6.2 Secondary Phosphines and SPO

Adamantyldichlorophosphine oxide.[180,219] **Method A**: In a 250 mL Schlenk balloon under argon atmosphere aluminium trichloride (5 g, 37.5 mmol, 1 eq.) was suspended in 20 mL of dichloromethane. Then trichlorophosphine (3.3 mL, 37.5 mmol, 1 eq.) was added, the mixture stirred for 15 min and then cooled to 0 °C. Then a solution of 1-bromoadamantane (10 g, 46.9 mmol, 1.25 eq.) in dichloromethane was added dropwise yielding a yellow solution, which was then allowed to warm up to room temperature and stirred for 2 h. No change in aspect could be observed until another 10 mL of trichlorophosphine was added to give a white precipitate. The mixture was then poured on concentrated HCl (50 mL) and ice and the organic phase separated. The water phase was extracted with chloroform and the combined organic phases were washed with water and brine. Drying over Na_2SO_4 and removal of the solvent gave a mixture of product and a side product that was most probably di-1-adamantylphosphonic chloride. FC (silica; $CHCl_3$:cyclohexane 1:1 – CH_2Cl_2:MeOH 20:1) gave 4.69 g (40%) of adamantylphosphonic acid dichloride in the second fraction.

Method B: In a 100 mL Schlenk balloon under argon atmosphere at 0 °C, aluminium trichloride (3.8 g, 28.4 mmol, 1.2 eq.) was suspended in 10 mL of PCl_3. Then 1-bromoadamantane (5.1 g, 23.7 mmol, 1 eq.) was slowly added and the brown suspension first warmed up to room temperature and then heated to 70 °C. After 30 min, 10 mL of PCl_3 were added to dilute the mixture and the suspension heated for 3 h. Then the mixure was cooled down to room temperature, the solid material filtered off and washed with benzene. The filter cake was then poured on a mixture of crushed ice and concentrated HCl, stirred for ten minutes and extracted with chloroform. The organic phases were washed with water and brine and dried over $MgSO_4$. Removal of the solvent under reduced pressure gave 4.98 g (83%) of the product. ^{1}**H NMR** (300 MHz, C_6D_6): δ 1.38-1.50 (m, 6 H, CH_2); 1.77 (bs, 3 H, CH); 1.94 (d, J = 7.5 Hz, 6 H, CH_2); $^{13}C\{^1H\}$ **NMR** (75.5 MHz, C_6D_6): δ 27.44 (d,

6 Experimental Part

$^3J_{\text{C-P}}$ = 15.2 Hz, C*H*); 34.83 (d, $^2J_{\text{C-P}}$ = 4.2 Hz, C*H*$_2$); 35.54 (d, $^4J_{\text{C-P}}$ = 2.8 Hz, C*H*$_2$); 48.63 (d, $^1J_{\text{C-P}}$ = 89, quat. *C*); ^{31}**P{^{1}H} NMR** (121 MHz, C$_6$D$_6$): δ 62.23 (s); **EA** calcd for C$_{10}$H$_{15}$Cl$_2$OP (253.11): C 47.45, H 5.97, Cl 28.01, O 6.32, P 12.24; found: C 47.34, H 5.93, Cl 27.82, O 6.57, P 12.13; **HRMS** (EI) calcd (*m/z*) for C$_{10}$H$_{15}$Cl$_2$OP: 252.0238 ([M]$^+$); found: 252.0235 ([M]$^+$); 135.1169 ([M-P(O)Cl$_2$]$^+$).

Adamantylphosphine.[180, 219] In a 100 mL Schlenk flask under argon atmosphere LiAlH$_4$ (1.16 g, 30.58 mmol, 1.8 eq.) was suspended in 50 mL of dry diethyl ether and the mixture cooled down to 0 °C. Under an Ar counter flow 1-adamantyldichlorophosphine oxide (4.3 g, 16.98 mmol, 1 eq.) was then added portion by portion. An exothermic reaction and formation of hydrogen gas was observed. The mixture was then allowed to warm up to room temperature and stirred for 1 h. To circumvent oxidation of the product the work up was also done under Ar atmosphere and with degassed solvents. Hydrochloric acid (35%, 5-10 mL) and water were added and the mixture stirred until formation of H$_2$ ceased. The ether phase was separated and the water phase extracted with diethyl ether (2x). The combined organic phases were washed with water, dried over Na$_2$SO$_4$ and the solvent removed under reduced pressure to give 1-adamantylphosphine as a colourless oil. Yield: 1.96 g (68%). ^{1}**H NMR** (300 MHz, C$_6$D$_6$): δ 1.55 (bs, 6 H, C*H*$_2$); 1.69 (bs, 6 H, C*H*$_2$); 1.77 (bs, 3 H, C*H*); 2.68 (d, $^1J_{\text{H-P}}$ = 188 Hz, 2 H, P*H*$_2$); ^{13}**C{^{1}H} NMR** (75.5 MHz, C$_6$D$_6$): δ 29.02 (d, $^3J_{\text{C-P}}$ = 7.62 Hz, C*H*); 36.25 (d, $^2J_{\text{C-P}}$ = 0.83 Hz, C*H*$_2$); 44.74 (d, $^4J_{\text{C-P}}$ = 8.6 Hz, C*H*$_2$); ^{31}**P{^{1}H} NMR** (121 MHz, C$_6$D$_6$): δ -82.77 (s, P*H*$_2$); ^{31}**P NMR** (121 MHz, C$_6$D$_6$): δ -82.77 (t, $^1J_{\text{P-H}}$ = 187 Hz, P*H*$_2$).

1-Adamantyldichlorophosphine sulphide.[180] To a stirred suspension of Lawesson's reagent (14.45 g, 35.72 mmol, 1 eq.) in toluene (100 mL) was added 1-adamantyldichlorophosphine oxide (9.04 g, 35.72 mmol, 1 eq.). The mixture was heated to 110 °C, and stirred for 20 h at this temperature. The suspension was filtered and then concentrated in vacuo. The residue was suspended in diethyl ether and dissolved Lawesson's reagent precipitated by addition of pentane. The suspension was again filtered and concentrated in vacuo. This was reapeated several times to get the product contaminated with about 10% side products and Lawesson's reagent. The product was used for the next step without further purification. Yield: 6.5 g (68%). ^{1}**H NMR** (300 MHz,

CDCl$_3$): δ 1.72-1.83 (m, 6 H, *Ad*); 2.13 (dd, *J* = 8.4, 2.5 Hz, 6 H, *Ad*); 2.22 (bs, 3 H, *Ad*); ^{31}P{^{1}H} NMR (121 MHz, CDCl$_3$): δ 115.72 (s).

1-Adamantyldichlorophosphine (149).[180] 1-Adamantyldichlorophosphine sulfide (6.48 g, 24.07 mmol, 1 eq.) and triphenylphosphine (6.31 g, 24.07 mmol, 1 eq.) were dissolved in degassed toluene under an argon atmosphere. The mixture was refluxed in the dark for four h. The solution was cooled down to room temperature and the solvent removed under reduced pressure. The residue was purified by sublimation (60 °C @ 0.1 mbar) to yield the product as white crystals. Yield: 3.8 g (66%). ^{1}H NMR (300 MHz, CDCl$_3$): δ 1.63-1.89 (m, 12 H, *Ad*); 2.14 (bs, 3 H, *Ad*); ^{31}P{^{1}H} NMR (121 MHz, CDCl$_3$): δ 191.66 (s).

2-Naphthyl trifluoromethanesulfonate.[220] To a solution of 2-naphthol (5 g, 34.68 mmol, 1 eq.) in 150 mL of anhydrous dichloromethane was added pyridine (5.6 mL, 69.36 mmol, 2 eq.) and the solution cooled down to 0 °C. Then trifluoromethanesulfonic anhydride (6.9 mL, 41.62 mmol, 1.2 eq.) was added and the now pink-orange solution warmed up to room temperature. After 30 min hydrochloric acid (10%, 40 mL), was added and the organic phase washed with saturated aqueous NaHCO$_3$ and brine. Drying over Na$_2$SO$_4$, and removal of the solvent under reduced pressure yielded an orange oil, which was purified by FC (silica, cyclohexane) to give the title compound as a colourless oil. Yield: 8.81 g (92%); ^{1}H NMR (300 MHz, CDCl$_3$): δ 7.39 (dd, *J* = 9.0, 2.4 Hz, 1 H); 7.55-7.62 (m, 2 H); 7.77 (d, *J* = 1.8 Hz, 1 H); 7.86-7.94 (m, 3 H); ^{13}C{^{1}H} NMR (50.32 MHz, CDCl$_3$): δ 118.8 (q, J^1_{C-F} = 321 Hz, *CF$_3$*); 119.3, 119.6, 127.2, 127.6 127.9, 128.1 130.7, 132.4, 133.4, 147.1; ^{19}F{^{1}H} NMR (188 MHz, CDCl$_3$): δ -72.75 (s).

Naphthyl-2-diethylphosphonate.[143,221] **Method A:** Into a 250 mL Schlenk flask 2-naphthyl trifluoromethanesulfonate (8.6 g, 31.13 mmol, 1 eq.), palladium acetate (699 mg, 3.11 mmol, 0.1 eq.), 1,4-bis-(diphenylphosphino)butane (1.32 g, 3.11 mmol, 0.1 eq.) and sodium formate (466 mg, 6.85 mmol, 0.22 eq.) were weighed and the flask set under argon atmosphere. Then anhydrous dimethylsulfoxide (50 mL), diethylphosphite (12.1 mL, 93.4 mmol, 3 eq.) and *N,N*-

6 Experimental Part

diisopropylethylamine (30.8 mL, 186.8 mmol, 6 eq.) were added, the mixture heated to 95 °C and stirred overnight. The solvent was then removed under reduced pressure and the dark oil diluted with diethyl ether and water. The aqueous phases was extracted two times with diethyl ether and the combined organic phase washed with hydrochloric acid (10%, 2x), water, saturated aqueous $NaHCO_3$ (2x), water and brine. Drying over Na_2SO_4, and removal of the solvent under reduced pressure yielded a dark oil, which was purified by FC (silica, cyclohexane:EtOAc 1:1-0:1) to give 6.99 g (85%) of the title compound as a slightly yellow oil.

Method B: Into a 250 mL Schlenk flask 2-naphthylbromide (5 g, 24 mmol, 1 eq.), palladium acetate (108 mg, 0.48 mmol, 0.02 eq.) and triphenylphosphine (380 mg, 1.45 mmol, 0.06 eq.) were weighed and the flask set under argon atmosphere. Then, anhydrous ethanol (80 mL), diethylphosphite (3.7 mL, 28 mmol, 1.2 eq.) and N,N-diisopropylethylamine (6 mL, 36.22 mmol, 1.5 eq.) were added, the mixture heated up to 80 °C and stirred overnight. The mixture was then cooled down to room temperature, diluted with ethyl acetate and the organic phase washed with hydrochloric acid (1M), saturated aqueous $NaHCO_3$ and brine. Drying over Na_2SO_4, and removal of the solvent under reduced pressure yielded a yellow oil, which was purified by FC (Silica, cyclohexane:EtOAc 3:1-0:1) to give 6.36 g (quant.) of the title compound as a slightly yellow oil. **^{1}H NMR** (300 MHz, $CDCl_3$): δ 1.33 (t, J = 7.1 Hz, 6 H, OCH$_2$CH_3); 4.06-4.23 (m, 4 H, OCH_2CH$_3$); 7.52-7.61 (m, 2H, ArH); 7.24-7.79 (m, 1H, ArH); 7.85-7.94 (m, 3H, ArH); 8.43 (d, J_{P-H} = 15.6 Hz, 1 H, ArH); **^{13}C{^{1}H} NMR** (75.47 MHz, $CDCl_3$): δ 16.2 (d, J_{C-P} = 6.3 Hz); 62.0 (d, J_{C-P} = 5.4 Hz); 125.4 (d, J_{C-P} = 188 Hz); 126.3 (d, J_{C-P} = 9.8 Hz); 126.7 (d, J_{C-P} = 1.3 Hz); 127.6 (d, J_{C-P} = 1.0 Hz); 128.1 (s); 128.2 (d, J_{C-P} = 14.3 Hz); 128.7 (s); 132.2 (d, J_{C-P} = 16.6 Hz); 133.8 (d, J_{C-P} = 10.2 Hz); 134.8 (d, J_{C-P} = 2.7 Hz); **^{31}P{^{1}H} NMR** (121.49 MHz, $CDCl_3$): δ 19.2.

2-Naphthylphosphine.[143] To a suspension of lithium aluminium hydride (2.94 g, 77.48 mmol, 3.25 eq.) in tetrahydrofurane (80 mL) was added trimethylsilyl chloride (9.5 mL, 73.4 mmol, 3.1 eq.) at -80 °C and the mixture allowed to warm up to room temperature. After stirring 30 min at ambient temperature the suspension was cooled to -45 °C in a dry ice/acetonitrile bath and naphthyl-2-diethylphosphonate (6.3 g, 23.84 mmol, 1 eq.) in tetrahydrofuran (15 mL) was added. The reaction mixture was allowed to warm up to room temperature and stirred overnight before the solvent was removed under reduced pressure. The residue was then suspended in degassed diethyl ether (80 mL) and the excess of lithium aluminium hydride quenched with degassed water (20 mL). Then degassed aqueous NaOH (10%, 20 mL) and degassed water (40 mL) were added and the organic phases separated. The aqueous phase was extracted with diethyl ether (2 x 40 mL) and the combined organic phase dried over MgSO$_4$. Evaporation of the solvent under reduced pressure gave the title compound as a white crystalline solid that was handled and stored in a glove box due to its air sensitivity. Yield: 3.36 g (88%); ^{1}**H NMR** (300 MHz, CD$_2$Cl$_2$): δ 4.14 (d, $^1J_{P-H}$=202 Hz, 2 H, P*H*$_2$); 7.46-7.50 (m, 3 H, Ar*H*); 7.75-7.79 (m, 3 H, Ar*H*); 8.02 (d, J_{P-H} = 8.5 Hz, 1 H, Ar*H*); ^{13}**C{^{1}H} NMR** (62.90 MHz, CD$_2$Cl$_2$): 126.2 (d, J_{C-P} = 7.8 Hz); 127.4 (s); 127.7 (s); 127.7 (d, J_{C-P} = 5.1 Hz); 131.6 (d, J_{C-P} = 10.8 Hz); 132.8 (s); 133.3 (d, J_{C-P} = 7.9 Hz); 134.3 (d, J_{C-P} = 20.7 Hz); ^{31}**P{^{1}H} NMR** (101.25 MHz, CD$_2$Cl$_2$): δ -122.98 (s); ^{31}**P NMR** (101.25 MHz, CD$_2$Cl$_2$): δ -122.98 (tdd, J_{P-H} = 202, 8.0, 6.0 Hz).

(*R*$_C$,*S*$_{Fc}$)-1-[Di-(*tert*-butyl)phosphino]-2-[1-(dimethylamino)ethyl]-ferrocene. At -78 °C 1.5 M *t*-BuLi in pentane (6.6 mL, 8.56 mmol, 1.1 eq.) was added dropwise to a solution of (*R*)-[1-(dimethylamino)ethyl]ferrocene (2 g, 7.78 mmol) in Et$_2$O (60 mL). The orange solution was slowly warmed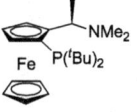
to room temperature and stirred at ambient temperature for 1 h. The red mixture thus obtained was cooled to -78 °C and a solution of di-*tert*-butylchlorophosphine (1.75 mL, 9.33 mmol, 1.2 eq.) in Et$_2$O (10 mL) was added. The resulting mixture was allowed to warm up to room temperature over night and was then heated to 45 °C and refluxed for 4 h. Then, a saturated aqueous solution of NaHCO$_3$ was added, the organic layer separated and the aqueous phase extracted with diethyl ether (2 x 100 mL). The combined organic phases were washed with water and brine, dried over MgSO$_4$ and the solvent evaporated in vacuo. The crude dark red oil was purified by FC (silica, pentane:EtOAc:Et$_3$N 150:10:1-50:5:1). Yield: 2.13 g (68%), deep red oil. ^{1}**H NMR** (250 MHz, C$_6$D$_6$): δ 1.09 (d, $^3J_{H-P}$ = 10.9 Hz, 9 H, C*Me*$_3$); 1.26 (d, ^3J = 6.9 Hz, 3 H, CH*Me*); 1.52 (d, $^3J_{H-P}$ = 11.9 Hz, 9 H, C*Me*$_3$) 2.18 (s, 6 H, N*Me*$_2$); 3.90-3.99 (m, 1 H, C*H*Me); 4.02 (s, 5 H, Cp'); 4.10-4.16 (m, 2 H, Cp); 4.21 (s, 1 H, Cp); ^{13}**C{^{1}H} NMR** (75.5 MHz, C$_6$D$_6$): δ 11.44 (s, CH*Me*); 30.43 (d, $^2J_{C-P}$ = 14.2 Hz, C*Me*$_3$); 30.84 (d, $^2J_{C-P}$ = 15.8 Hz, C*Me*$_3$); 31.46 (d, $^1J_{C-P}$ = 21.3 Hz, *C*Me$_3$); 32.95 (d, $^1J_{C-P}$ = 23.5 Hz, *C*Me$_3$); 40.03 (s, N*Me*$_2$); 56.00 (d, $^3J_{C-P}$ = 13.1 Hz, *C*HMe); 67.69 (d, J_{C-P} = 4.6 Hz, Cp); 68.38 (s, Cp); 70.07 (s, Cp'); 72.28 (d, J_{C-P} = 6.4 Hz, Cp); 78.49 (d, J_{C-P} = 39.4 Hz, Cp); 99.89 (d, J_{C-P} = 27.7 Hz, Cp); ^{31}**P{^{1}H} NMR** (101 MHz, C$_6$D$_6$): δ 13.54 (s, *P*^{t}Bu$_2$).

(S_P,R_C,S_{Fc})-1-[(1-Naphthyl)phenylphosphino]-2-[(1-dimethylamino)-ethyl]ferrocene (190).[98] At -78 °C a 1.3 M solution of s-BuLi (10 mL, 12.87 mmol, 1.1 eq.) was added dropwise to a solution of (R)-[1-(dimethylamino)ethyl]ferrocene (3.00 g, 11.7 mmol, 1 eq.) in Et$_2$O (30 mL). The orange solution was slowly warmed to room temperature and stirred at ambient temperature for 1 h. The red mixture obtained was cooled again to -78 °C and dichlorophenylphosphine (1.8 mL, 12.87 mmol, 1.1 eq.) was then added in one portion. The resulting yellowish brown mixture was allowed to warm up to room temperature and stirred for 1 h. Then a solution of 1-naphthyllithium [prepared from 1-bromonaphthalene (2.2 mL, 15.80 mmol, 1.35 eq.) and 1.9 M t-BuLi (17 mL, 31.59 mmol, 2.7 eq.) in Et$_2$O (30 mL) at -78 °C] was added slowly via teflon canula. The mixture was allowed to warm up to room temperature over night. Then diethyl ether and saturated aqueous NaHCO$_3$ solution were added, the organic layer was separated and the aqueous phase extracted with diethyl ether. The combined organic phases were dried over MgSO$_4$ and the solvent evaporated in vacuo. The crude dark oil was purified by FC (silica, pentane:EtOAc:Et$_3$N 90:6:4) to yield the product as a single diastereoisomer in the first coloured fraction. Yield: 4.718 g (82%), orange powder. **^{1}H NMR** (300 MHz, CDCl$_3$): δ 1.39 (d, ^3J = 6.7 Hz, 3 H, CH*Me*); 1.97 (bs, 6 H, N*Me*$_2$); 3.62 (s, 5 H, *Cp'*); 4.06 (bs, 1 H, *Cp*); 4.20-4.28 (m, 1 H, C*H*Me); 4.30 (bs, 1 H, *Cp*); 4.31 (bs, 1 H, *Cp*); 7.12-7.29 (m, 5 H, P*NpPh*); 7.40-7.62 (m, 3 H, P*NpPh*); 7.62-7.72 (m, 1 H, P*NpPh*); 7.91 (t, J = 8.1 Hz, 2 H, P*NpPh*); 9.31-9.48 (m, 1 H, P*NpPh*); **^{31}P{^{1}H} NMR** (121 MHz, CDCl$_3$): δ -40.16 (s, *PNpPh*); **EA** calcd for C$_{30}$H$_{30}$FeNP (491.39): C 73.33, H 6.27, N 2.85, P 6.30; found: C 73.06, H 6.27, N 2.86, P 6.44; **HRMS** (EI) calcd (m/z) for C$_{30}$H$_{30}$FeNP: 491.1460 ([M]$^+$); found: 491.14659 ([M]$^+$).

(S_P,R_C,S_{Fc})-1-[(2-Methoxyphenyl)phenylphosphino]-2-[(1-dimethyl-amino)ethyl]ferrocene (191).[98] At -78 °C a 1.3 M solution of s-BuLi (3.3 mL, 4.26 mmol, 1.1 eq.) was added dropwise to a solution of (R)-[1-(dimethylamino)ethyl]ferrocene (997 mg, 3.88 mmol, 1 eq.) in Et$_2$O (15 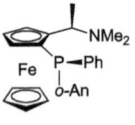 mL). The orange solution was slowly warmed to room temperature and stirred at ambient temperature for 1 h. The red mixture obtained was cooled to -78 °C and dichlorophenylphosphine (0.59 mL, 4.26 mmol, 1.1 eq.) was then added in one portion. The resulting yellowish brown mixture was allowed to warm up to room temperature and stirred for 1 h at this temperature. Then, a solution of 2-methoxyphenyllithium [prepared from 1-bromoanisole

(0.6 mL, 4.57 mmol, 1.2 eq.) and 1.9 M *t*-BuLi (5.7 mL, 9.12 mmol, 2.4 eq.) in Et$_2$O (20 mL) at -78 °C] was added slowly via teflon canula. The mixture was allowed to warm up to room temperature over night. Then, diethyl ether and saturated aqueous NaHCO$_3$ were added, the organic layer was separated and the aqueous phase extracted with diethyl ether. The combined organic phases were dried over MgSO$_4$ and the solvent evaporated in vacuo. The crude dark oil was purified by FC (silica, cyclohexane:EtOAc:Et$_3$N 85:10:5) to yield the product as a single diastereoisomer in the first coloured fraction. Yield: 1.67 g (91%), orange crystalline solid. ^{1}H NMR (300 MHz, CDCl$_3$): δ 1.31 (d, ^3J = 6.7 Hz, 3 H, CH*Me*); 1.82 (bs, 6 H, N*Me*$_2$); 3.95 (s, 3 H, O*Me*); 3.98 (bs, 1 H, *Cp*); 4.03 (s, 5 H, *Cp'*); 4.14 (dq, *J* = 6.5, 2.8 Hz, 1 H, C*H*Me); 4.28 (bs, 1 H, *Cp*); 4.39 (bs, 1 H, *Cp*); 6.90 (t, *J* = 7.4 Hz, 1 H, P*AnPh*); 6.97 (dd, *J* = 8.1, 4,6 Hz, 1 H, P*AnPh*); 7.13-7.26 (m, 6 H, P*AnPh*); 7.36 (t, *J* = 7.1 Hz, 1 H, P*AnPh*); ^{13}C{^{1}H} NMR (62.9 MHz, CDCl$_3$): δ 10.21 (s, CH*Me*); 39.17 (s, N*Me*$_2$); 55.40 (s, O*Me*); 56.84 (d, $^3J_{C-P}$ = 7.7 Hz, *C*HMe); 68.17 (s, *Cp*); 69.36 (d, J_{C-P} = 4.1 Hz, *Cp*); 69.56 (s, *Cp'*); 71.90 (d, J_{C-P} = 4.8 Hz, *Cp*); 76.20 (d, $^1J_{C-P}$ = 10.9 Hz, *Cp*); 97.13 (d, $^2J_{C-P}$ = 22.6 Hz, *Cp*); 110.21 (d, J_{C-P} = 2.5 Hz, P*AnPh*); 120.48 (d, J_{C-P} = 2.1 Hz, P*AnPh*); 126.52 (d, $^1J_{C-P}$ = 13.3 Hz, P*AnPh*); 126.68 (s, P*AnPh*); 127.04 (d, J_{C-P} = 6.4 Hz, P*AnPh*); 130.47 (s, P*AnPh*); 132.48 (d, J_{C-P} = 19.3 Hz, P*AnPh*); 137.60 (d, J_{C-P} = 4.0 Hz, P*AnPh*); 141.50 (d, $^1J_{C-P}$ = 8.8 Hz, P*AnPh*); 161.92 (d, J_{C-P} = 16.8 Hz, P*AnPh*); ^{31}P{^{1}H} NMR (121 MHz, CDCl$_3$): δ -40.38 (s, P*AnPh*); **EA** calcd for C$_{27}$H$_{30}$FeNOP (471.35): 68.80, H 6.41, N 2.97, O 3.39, P 6.57; found: C 68.42, H 6.43, N 2.91, O 3.84, P 6.76.

(R_P,R_C,S_{Fc})-1-[(1-Adamantyl)methylphosphino]-2-[(1-dimethylamino)-ethyl]ferrocene (192). At -78 °C 1.3 M *s*-BuLi (688 µL, 894 µmol, 1.15 eq.) was added dropwise to a solution of (*R*)-[1-(dimethylamino)-ethyl]ferrocene (200 mg, 778 µmol, 1 eq.) in Et$_2$O (3 mL). The orange 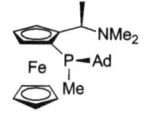 solution was slowly warmed to room temperature and stirred at ambient temperature for 1 h. The red mixture thus obtained was cooled to 0 °C and was then added over 15 min to a solution of 1-adamantyldichlorophosphine (149) (230 mg, 972 µmol, 1.25 eq.) in Et$_2$O (3 mL) at 0 °C. The resulting yellowish brown mixture was allowed to warm up to room temperature and stirred for 4 h. Then a 3 M solution of methylmagnesium chloride (363 µL, 1089 µmol, 1.4 eq.) was added and the mixture stirred over night. The mixture was then diluted with diethyl ether and saturated aqueous NaHCO$_3$, the organic layer separated and the aqueous phase extracted with diethyl ether. The combined organic phases were dried over MgSO$_4$ and

6 Experimental Part

the solvent evaporated in vacuo. The crude dark oil was purified by FC (silica, pentane:EtOAc:Et$_3$N 100:5:2) to yield the product as a 3:1-mixture of diastereoisomers in the first coloured fraction. The major isomer could be isolated after recrystallization from hot ethanol. Yield: 230 mg (68%), brown crystalline solid. **^{1}H NMR** (300 MHz, CDCl$_3$): δ 1.25-1.34 (m, 6 H, CH*Me*, P*Me*); 1.51-1.72 (m, 12 H, *Ad*); 1.86 (bs, 3 H, *Ad*); 2.10 (bs, 6 H, N*Me*$_2$); 3.90-3.98 (m, 1 H, C*H*Me); 4.08 (s, 5 H, *Cp'*); 4.15 (bs, 1 H, *Cp*); 4.27 (bs, 1 H, *Cp*); 4.34 (bs, 1 H, *Cp*); **^{13}C{^{1}H} NMR** (62.9 MHz, CDCl$_3$): δ 5.55 (d, $^1J_{C-P}$ = 16.2 Hz, P*Me*); 8.56 (s, CH*Me*); 29.17 (d, J_{C-P} = 8.1 Hz, *Ad*); 32.76 (d, $^1J_{C-P}$ = 17.0 Hz, *Ad*); 37.37 (d, J_{C-P} = 0.7 Hz, *Ad*); 39.22 (s, N*Me*$_2$); 39.38 (d, J_{C-P} = 11.0 Hz, *Ad*); 55.94 (d, $^3J_{C-P}$ = 8.9 Hz, *C*HMe); 67.95 (s, *Cp*); 68.87 (d, J_{C-P} = 3.7 Hz, *Cp*); 69.60 (s, *Cp'*); 70.18 (d, J_{C-P} = 6.1 Hz, *Cp*); 76.76 (d, $^1J_{C-P}$ = 23.2 Hz, *Cp*); 98.69 (d, $^2J_{C-P}$ = 25.8 Hz, *Cp*); **^{31}P{^{1}H} NMR** (121 MHz, CDCl$_3$): δ -29.74 (s, *P*AdMe); **EA** calcd for C$_{25}$H$_{36}$FeNP (437.39): C 68.65, H 8.30, N 3.20, P 7.08; found: C 68.45, H 8.21, N 3.14, P 7.11; **HRMS** (EI) calcd (*m/z*) for C$_{25}$H$_{36}$FeNP: 437.1930 ([M]$^+$); found: 437.1935 ([M]$^+$); 422.1696 ([M-Me]$^+$); 392.1351 ([M-NMe$_2$]$^+$); 302.0750 ([M-Ad]$^+$).

Minor isomer: ^{31}P{^{1}H} NMR (121 MHz, CDCl$_3$): δ -7.92 (s, *P*AdMe).

(R_C,S_{Fc})-1-[Di-(*tert*-butyl)phosphino]-2-[1-acetoxyethyl]ferrocene.

(R_C,S_{Fc})-1-[Di-(*tert*-butyl)phosphino]-2-[1-(dimethylamino)ethyl]ferrocene (250 mg, 623 µmol) was suspended in 2 mL of acetic anhydride. The mixture was heated up to 90 °C and the resulting red solution stirred for 14 h at this temperature. The solvent was removed at 70 °C under vacuum and the residue dried under reduced pressure for several h. The product was used in the next step without further purification. Yield: 259 mg (100%), dark red oil. **^{1}H NMR** (300 MHz, CDCl$_3$): δ 0.90 (d, $^3J_{H-P}$ = 11.2 Hz, 9 H, C*Me$_3$*); 1.46 (d, $^3J_{H-P}$ = 12.2 Hz, 9 H, C*Me$_3$*); 1.70 (d, ^3J = 6.5 Hz, 3 H, CH*Me*); 1.97 (s, 3 H, O*Ac*); 4.18 (s, 5 H, *Cp'*); 4.33 (s, 1 H, *Cp*); 4.48 (s, 1 H, *Cp*) 4.56 (s, 1 H, *Cp*); 6.23 (qd, ^3J = 6.5 Hz, J_{H-P} = 3.1 Hz, 1 H, C*H*Me); **^{13}C{^{1}H} NMR** (75.5 MHz, CD$_2$Cl$_2$): δ 20.18 (s, CH*Me*); 20.18 (s, O*Ac*); 30.02 (d, $^2J_{C-P}$ = 14.1 Hz, C*Me$_3$*); 30.43 (d, $^2J_{C-P}$ = 15.0 Hz, C*Me$_3$*); 31.47 (d, $^1J_{C-P}$ = 19.4 Hz, *C*Me$_3$); 32.71 (d, $^1J_{C-P}$ = 21.6 Hz, *C*Me$_3$); 40.03 (s, N*Me*$_2$); 67.59 (d, J_{C-P} = 4.8 Hz, *Cp*); 68.67 (d, $^3J_{C-P}$ = 13.8 Hz, *C*HMe); 69.71 (s, *Cp*); 70.28 (s, *Cp'*); 73.27 (d, J_{C-P} = 6.0 Hz, *Cp*); 78.58 (d, J_{C-P} = 39.2 Hz, *Cp*); 94.42 (d, J_{C-P} = 28.6 Hz, *Cp*); 169.52 (s, O*Ac*); **^{31}P{^{1}H} NMR** (121 MHz, CDCl$_3$): δ 13.64 (s, *P*^{t}Bu$_2$).

6 Experimental Part

**(S_P,R_C,S_{Fc})-1-[(1-Naphthyl)phenylphosphino]-2-[1-acetoxyethyl]-
ferrocene (193).**[98] (S_P,R_C,S_{Fc})-1-[(1-Naphthyl)phenylphosphino]-2-[(1-
dimethylamino)ethyl]ferrocene (**190**) (2.199 g, 5.93 mmol) was dissolved in
22 mL of acetic anhydride. The resulting red solution was stirred for 5 d at

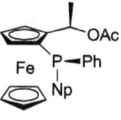

room temperature. The solvent was removed under reduced pressure to yield the product as a single diastereoisomer. Yield: 2.23 g (98%), orange powder. **^{1}H NMR** (300 MHz, CDCl$_3$): δ 1.29 (s, 3 H, O*Ac*); 1.69 (d, *J* = 6.4 Hz, 3 H, CH*Me*); 3.73 (s, 5 H, *Cp*'); 3.95 (s, 1 H, *Cp*); 4.36 (s, 1 H, *Cp*); 4.59 (s, 1 H, *Cp*); 6.30 (qd, *J* = 6.4, 2.9 Hz, 1 H, C*H*Me); 7.06-7.32 (m, 5 H, P*NpPh*); 7.36-7.48 (m, 2 H, P*NpPh*); 7.55 (t, *J* = 7.3 Hz, 1 H, P*NpPh*); 7.66 (t, *J* = 7.4 Hz, 1 H, P*NpPh*); 7.86-7.94 (m, 2 H, P*NpPh*); 9.3 (t, *J* = 7.2 Hz, 1 H, P*NpPh*); **^{13}C{^{1}H} NMR** (75.5 MHz, CDCl$_3$): δ 18.53 (s, CH*Me*); 20.18 (s, O*Ac*); 68.75 (d, $^3J_{C-P}$ = 10.2 Hz, CHMe); 69.33 (d, J_{C-P} = 4.1 Hz, *Cp*); 69.56 (s, *Cp*'); 69.72 (s, *Cp*); 72.17 (d, J_{C-P} = 4.3 Hz, *Cp*); 77.17 (d, J_{C-P} = 11.3 Hz, *Cp*); 92.24 (d, J_{C-P} = 26.1 Hz, *Cp*); 125.17 (d, J_{C-P} = 2.3 Hz, P*NpPh*); 126.11 (s P*NpPh*); 126.12 (d, J_{C-P} = 5.2 Hz, P*NpPh*); 127.00 (d, J_{C-P} = 32.4 Hz, P*NpPh*); 127.22 (s, P*NpPh*); 127.89 (d, J_{C-P}= 5.8 Hz, P*NpPh*); 128.56 (d, J_{C-P} = 1.9 Hz, P*NpPh*); 129.77 (d, J_{C-P} = 0.9 Hz, P*NpPh*); 132.43 (d, J_{C-P} = 18.6 Hz, P*NpPh*); 133.53 (d, J_{C-P} = 5.4 Hz, P*NpPh*); 134.99 (d, J_{C-P} = 11.9 Hz, P*NpPh*); 136.88 (d, J_{C-P} = 25.7 Hz, P*NpPh*); 140.84 (d, J_{C-P} = 10.3 Hz, P*NpPh*); 169.98 (s, O*Ac*); **^{31}P{^{1}H} NMR** (121 MHz, CDCl$_3$): δ -41.14 (s, *P*PhNp); **EA** calcd for C$_{30}$H$_{27}$FeO$_2$P (506.36): C 71.16, H 5.37, P 6.12; found: C 70.98, H 5.49, P 6.20; **HRMS** (MALDI) calcd (*m/z*) for C$_{30}$H$_{28}$FeO$_2$P: 507.1171 ([MH]$^+$); found: 507.1176 ([MH]$^+$); 467.0963 ([M-OAc]$^+$).

**(S_P,R_C,S_{Fc})-1-[(2-Methoxyphenyl)phenylphosphino]-2-[1-acetoxyethyl]-
ferrocene (194).**[98] (S_P,R_C,S_{Fc})-1-[(2-methoxyphenyl)phenylphosphino]-2-
[(1-dimethylamino)ethyl]ferrocene (**191**) (1.89 g, 4.02 mmol) was dissolved

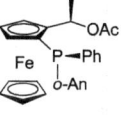

in 7 mL of acetic anhydride. The resulting red solution was stirred for 70 h at room temperature. The solvent was removed under reduced pressure and the residue diluted with water and diethyl ether. The organic phase was separated and dried over MgSO$_4$. After evaporation of the solvent the product was obtained as a single diastereoisomer. Yield: 1.49 g (76%), yellow powder. **^{1}H NMR** (300 MHz, CDCl$_3$): δ 1.21 (s, 3 H, O*Ac*); 1.66 (d, ^3J = 6.4 Hz, 3 H, CH*Me*); 3.93 (s, 4 H, O*Me*, *Cp*); 4.10 (s, 5 H, *Cp*'); 4.37 (t, *J* = 2.2 Hz, 1 H, *Cp*); 4.57 (bs, 1 H, *Cp*); 6.18 (qd, *J* = 6.2, 2.7 Hz, 1 H, C*H*Me); 6.89 (t, *J* = 7.4 Hz, 1 H, P*AnPh*); 6.98 (dd, *J* = 8.1, 4.9 Hz, 1 H, P*AnPh*); 7.09-7.23 (m, 6 H, P*AnPh*); 7.37 (t,

6 Experimental Part

$J = 7.7$ Hz, 1 H, P*AnPh*); ^{13}C{^{1}H} NMR (62.9 MHz, CDCl$_3$): δ 18.51(s, O*Ac*); 10.05 (s, CH*Me*); 55.44 (s, O*Me*); 68.70 (d, $^{3}J_{C-P}$ = 9.7 Hz, C*H*Me); 69.22 (d, J_{C-P} = 3.8Hz, *Cp*); 69.43 (s, *Cp*); 69.79 (s, *Cp'*); 72.52 (d, J_{C-P} = 4.2 Hz, *Cp*); 76.82 (d, $^{1}J_{C-P}$ = 12.0 Hz, *Cp*); 91.68 (d, $^{2}J_{C-P}$ = 25.2 Hz, *Cp*); 110.52 (d, J_{C-P} = 2.7 Hz, P*AnPh*); 120.67 (d, J_{C-P} = 2.6 Hz, P*AnPh*); 124.84 (d, $^{1}J_{C-P}$ = 12.7 Hz, P*AnPh*); 127.37 (s, P*AnPh*); 127.69 (d, J_{C-P} = 5.9 Hz, P*AnPh*); 130.95 (s, P*AnPh*); 132.76 (d, J_{C-P} = 19.4 Hz, P*AnPh*); 137.45 (d, J_{C-P} = 4.4 Hz, P*AnPh*); 140.59 (d, $^{1}J_{C-P}$ = 11.5 Hz, P*AnPh*); 162.04 (d, J_{C-P} = 16.8 Hz, P*AnPh*); 169,75 (s, O*Ac*); ^{31}P{^{1}H} NMR (121 MHz, CDCl$_3$): δ -41.14 (s, *P*PhNp); **EA** calcd for C$_{27}$H$_{27}$FeO$_3$P (486.32): C 66.68, H 5.60 O 9.87, P 6.37; found: C 65.93, H 5.68, O 10.34, P 6.37; **HRMS** (MALDI) calcd (*m/z*) for C$_{27}$H$_{27}$FeO$_3$P: 486.1042 ([M]$^{+}$); found: 486.1035 ([MH]$^{+}$).

(*S*$_C$,*R*$_{Fc}$)-1-(Diphenylphosphino)-2-[1-(dimethylamino)ethyl]ferrocene P,N-bis-borane complex (107). To a solution of (*R*$_C$,*S*$_{Fc}$)-1-(diphenylphosphino)-2-[1-(dimethylamino)ethyl]ferrocene (1.00 g, 2.27 mmol, 1 eq.) in 5 mL of tetrahydrofuran was added a 1.6 M solution of borane dimethyl sulfide adduct (13.5 mL, 22.67 mmol, 10 eq.) and the mixture stirred for 1 h. The mixture was then diluted with 20 mL of dichloromethane and 20 mL of water and the phases separated. The organic phase was washed with brine, dried over MgSO$_4$ and the solvent removed under reduced pressure. Yield: 1.03 g (97%), orange solid. ^{1}**H NMR** (300 MHz, CDCl$_3$): δ 0.80-2.20 (bs, 6 H, BH$_3$); 1.87 (d, J = 6.9 Hz, 3 H, CH*Me*); 1.98 (s, 3 H, N*Me*); 1.99 (s, 3 H, N*Me*); 4.05 (s, 5 H, *Cp'*); 4.32 (s, 1 H, *Cp*); 4.62-4.69 (m, 1 H, C*H*Me); 4.65 (s, 1 H, *Cp*); 4.85 (s, 1 H, *Cp*); 7.33-7.52 (m, 6 H, P*Ph*$_2$); 7.64-7.71 (m, 2 H, P*Ph*$_2$); 7.81-7.87 (m, 2 H, P*Ph*$_2$); ^{31}**P{^{1}H} NMR** (121 MHz, CDCl$_3$): δ 10.00 (bs, *P*Ph$_2$); ^{11}**B{^{1}H} NMR** (96.29 MHz, CDCl$_3$): -37.02 (bs, PPh$_2$*B*H$_3$); -10.69 (bs, NMe$_2$*B*H$_3$).

(*R*$_C$,*S*$_{Fc}$)-1-[Diphenylphosphino]-2-[1-(cyclohexylphosphino)ethyl]-ferrocene (81). To a suspension of (*R*$_C$,*S*$_{Fc}$)-1-(diphenylphosphino)-2-[1-(dimethylamino)ethyl]ferrocene (3.00 g, 6.80 mmol, 1 eq.) in 3 mL of degassed acetic acid was added trifluoroacetic acid (0.5 mL, 6.80 mmol, 1 eq.) giving a yellow solid. Then, cyclohexylphosphine (2.7 mL, 20.40 mmol, 3 eq.) was added, the mixture heated up to 90 °C and the red solution stirred at this temperature for 15 h. Then the solvent was removed under reduced pressure and the residue purified by FC (silica, cyclohexane:EtOAc 100:0-20:1). Yield: 2.70 g (75%), orange foamy solid, 2.2:1 mixture of isomers. Major isomer: **¹H NMR** (300 MHz, CDCl₃): δ 0.75-1.26 (m, 6 H, PH*Cy*); 1.47-1.69 (m, 8 H, CH*Me*, PH*Cy*); 2.54 (dd, $^1J_{P-H}$ = 213 Hz, ^2J = 5.9 Hz, 1 H, P*H*Cy); 3.37-3.48 (m, 1 H, C*H*Me); 3.93 (s, 1 H, *Cp*); 3.97 (s, 5 H, *Cp'*); 4.32-4.34 (m, 2 H, *Cp*); 7.18-7.31 (m, 5 H, P*Ph₂*); 7.34-7.41 (m, 3 H, P*Ph₂*); 7.59-7.65 (m, 2 H, P*Ph₂*); **³¹P{¹H} NMR** (121 MHz, CDCl₃): δ -18.33 (bs, *PHCy*); -25.98 (bs, *PPh₂*); **³¹P NMR** (121 MHz, CDCl₃): δ -17.93 (d, $^1J_{P-H}$ = 208 Hz *PHCy*); -25.97 (bs, *PPh₂*); **HRMS** (EI) calcd (*m/z*) for C₃₀H₃₃FeP₂: 511.1402 ([M-H]⁺); found: 511.1401 ([M-H]⁺); 447.0529 ([M-Cp]⁺); 429.0623 ([M-Cy]⁺).

Minor isomer: **³¹P{¹H} NMR** (121 MHz, CDCl₃): δ -16.23 (d, J_{P-P} = 7.1 Hz, *PHCy*); -25.89 (d, J_{P-P} = 7.1 Hz, *PPh₂*); **³¹P NMR** (121 MHz, CDCl₃): δ -15.37 (d, $^1J_{P-H}$ = 212 Hz *PHCy*).

(R_C,S_Fc)-2-[1-(Phenylphosphino)ethyl]-1-[diphenylphosphino]ferrocene (87). To a solution of (R_C,S_Fc)-1-(diphenylphosphino)-2-[1-(dimethylamino)-ethyl]ferrocene (2.00 g, 4.53 mmol, 1 eq.) in 7 mL of degassed acetic acid was added trifluoroacetic acid (0.35 mL, 4.53 mmol, 1 eq.) giving a yellow precipitate. Phenylphosphine (1.48 mL, 13.60 mmol, 3 eq.) was then added, the mixture heated up to 90 °C and the red solution stirred at this temperature for 15 h. Then the solvent was removed under reduced pressure and the residue purified by FC (silica, cyclohexane:EtOAc 100:0-10:1) Yield: 2.12 g (89%), orange, foamy solid, 2.1:1 mixture of isomers. Major isomer: **¹H NMR** (300 MHz, CDCl₃): δ 1.42 (dd, $^3J_{P-H}$ = 14.9 Hz, ^3J = 7.0 Hz, 3 H, CH*Me*); 3.52-3.60 (m, 1 H, C*H*Me); 3.84 (d, $^1J_{P-H}$ = 215 Hz, 1 H, P*H*Cy); 3.93 (s, 5 H, *Cp'*); 4.01 (s, 1 H, *Cp*); 4.35-4.36 (m, 2 H, *Cp*); 7.12-7.42 (m, 13 H, P*Ph₂*); 7.61-7.69 (m, 2 H, P*Ph₂*); **³¹P{¹H} NMR** (121 MHz, CDCl₃): δ -25.91 (d, J_{P-P} = 5.2 Hz, *PPh₂*); -29.62 (d, J_{P-P} = 5.2 Hz, *PHPh*); **³¹P NMR** (121 MHz, CDCl₃): δ -28.63 (dm, $^1J_{P-H}$ = 218 Hz *PHPh*); **HRMS** (EI) calcd (*m/z*) for C₃₀H₂₈FeP₂: 506.1011 ([M]⁺); found: 506.1006 ([M]⁺); 397.0803 ([M-PHPh]⁺).

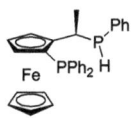

Minor isomer: **³¹P{¹H} NMR** (121 MHz, CDCl₃): δ -13.90 (d, J_{P-P} = 11.9 Hz, *PHPh*); -29.62 (d, J_{P-P} = 11.8 Hz, *PPh₂*); **³¹P NMR** (121 MHz, CDCl₃): δ -13.04 (d, $^1J_{P-H}$ = 215 Hz *PHCy*).

6 Experimental Part

(R_C,S_{Fc})-1-[Diphenylphosphino]-2-{1-[(2-naphthyl)phosphino]ethyl}-ferrocene (88). To a solution of (R_C,S_{Fc})-1-(diphenylphosphino)-2-[1-(dimethylamino)ethyl]ferrocene (1.50 g, 3.40 mmol, 1 eq.) in 3 mL of degassed acetic acid was added trifluoroacetic acid (0.26 mL, 3.40 mmol, 1 eq.) giving a yellow precipitate. 2-Naphthylphosphine (1.20 g, 13.60 mmol, 2.2 eq.) was then added, the mixture heated to 90 °C and the red solution stirred at this temperature for 15 h. Then, the solvent was removed under reduced pressure and the residue purified by FC (silica, cyclohexane:EtOAc 100:0-10:1). Yield: 1.29 g (68%), orange foamy solid, 2.1:1 mixture of isomers. Major isomer: **^{1}H NMR** (300 MHz, CDCl$_3$): δ 1.42 (dd, $^3J_{P-H}$ = 14.9 Hz, ^3J = 7.0 Hz, 3 H, CH*Me*); 3.52-3.60 (m, 1 H, C*H*Me); 3.84 (d, $^1J_{P-H}$ = 215 Hz, 1 H, P*H*Cy); 3.93 (s, 5 H, Cp'); 4.01 (s, 1 H, Cp); 4.35-4.36 (m, 2 H, Cp); 7.12-7.42 (m, 13 H, P*Ph$_2$*); 7.61-7.69 (m, 2 H, P*Ph$_2$*); **^{31}P{^{1}H} NMR** (121 MHz, CDCl$_3$): δ -25.91 (d, J_{P-P} = 5.2 Hz, P*Ph$_2$*); -29.62 (d, J_{P-P} = 5.2 Hz, *P*HPh); **^{31}P NMR** (121 MHz, CDCl$_3$): δ -28.63 (dm, $^1J_{P-H}$ = 218 Hz *P*HPh); **HRMS** (EI) calcd (*m/z*) for C$_{30}$H$_{28}$FeP$_2$: 506.1011 ([M]$^+$); found: 506.1006 ([M]$^+$); 397.0803 ([M-PHPh]$^+$).

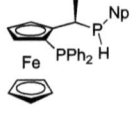

Minor isomer: **^{31}P{^{1}H} NMR** (121 MHz, CDCl$_3$): δ -13.90 (d, J_{P-P} = 11.9 Hz, *P*HPh); -29.62 (d, J_{P-P} = 11.8 Hz, P*Ph$_2$*); **^{31}P NMR** (121 MHz, CDCl$_3$): δ -13.04 (d, $^1J_{P-H}$ = 215 Hz P*H*Cy).

(R_C,S_{Fc})-1-[Diphenylphosphino]-2-{1-[(1-adamantyl)phosphino]ethyl}-ferrocene (89). To a suspension of (R_C,S_{Fc})-1-(diphenylphosphino)-2-[1-(dimethylamino)ethyl]ferrocene (2.03 g, 4.59 mmol, 1 eq.) in 3 mL of degassed acetic acid was added trifluoroacetic acid (0.35 mL, 4.59 mmol, 1 eq.) giving a yellow solid. 1-Adamantylphosphine (1.70 g, 10.11 mmol, 2.2 eq.) was then added, the mixture heated to 90 °C and the red solution stirred at this temperature for 15 h. The yellow precipitate, which was formed overnight, was filtered off and washed with MeOH to give the title compound as a 10:1-mixture of diastereoisomers. Yield: 2.33 g (90%), yellow solid, crystals suitable for X-ray were obtained by recrystallization from boiling hexane. **^{1}H NMR** (300 MHz, C$_6$D$_6$): δ 1.86-1.52 (m, 18 H, PH*Ad*, CH*Me*); 2.81 (d, $^1J_{H-P}$ = 210 Hz, 1 H, P*H*Ad); 3.89-3.82 (m, 1 H, C*H*Me); 3.91 (s, 5 H, Cp'); 3.97 (s, 1 H, Cp); 4.14 (s, 1 H, Cp); 4.48 (s, 1 H, Cp); 7.10-6.98 (m, 6 H, P*Ph$_2$*); 7.48-7.40 (m, 2 H, P*Ph$_2$*); 7.71-7.68 (m, 2 H, P*Ph$_2$*); **^{13}C{^{1}H} NMR** (62.9 MHz, C$_6$D$_6$): δ 21.73 (t, J_{C-P} = 7.5 Hz, PCH*Me*); 22.13 (d, J_{C-P} = 30.2 Hz, PC*H*Me); 29.96 (d, J_{C-P} = 7.6 Hz, *Ad*); 32.26 (d, J_{C-P} = 14.5 Hz, *Ad*); 36.80 (s,

Ad); 42.30 (d, J_{C-P} = 9.4 Hz, Ad); 68.87 (t, J_{C-P}= 4.4 Hz, Cp); 69.39 (s, Cp'); 70.09 (d, J_{C-P} = 4.4 Hz, Cp); 73.18 (d, J_{C-P} = 8.8 Hz, Cp); 101.80 (dd, J_{C-P} = 27.7, 9.4 Hz, Cp); 128.07-127.66 (m, PPh_2); 129.01 (s, PPh_2); 133.32 (d, J_{C-P} = 18.9 Hz, PPh_2); 135.66 (d, J_{C-P} = 22.7 Hz, PPh_2); 138.56 (d, J_{C-P} = 9.4 Hz, PPh_2); 140.73 (d, J_{C-P} = 9.4 Hz, PPh_2); ^{31}P{^{1}H} NMR (121 MHz, C_6D_6): δ 3.38 (d, J_{P-P} = 11.8 Hz, PHAd); -25.92 (d, J_{P-P} = 10.4 Hz, PPh_2); ^{31}P NMR (121 MHz, C_6D_6): δ 3.33 (d, $^1J_{P-H}$ = 204 Hz PHAd). **EA** calcd for $C_{34}H_{38}P_2Fe$ (564.47): C 72.35, H 6.79, P 10.97; found: C 72.44, H 6.78, P 10.96; **HRMS** (MALDI) calcd (m/z) for $C_{34}H_{38}FeP_2$: 563.1715 ([M-H]$^+$); found: 563.1725 ([M-H]$^+$); 443.1 ([M-FeCp]$^+$); 397.0 ([M-PHAd]$^+$).

Minor isomer: ^{31}P{^{1}H} NMR (121 MHz, $CDCl_3$): δ -25.00 (d, J_{P-P} = 12.0 Hz, PPh_2); 10.75 (d, J_{P-P} = 12.1 Hz, PHPh); ^{31}P NMR (121 MHz, $CDCl_3$): δ 10.76 (d, $^1J_{P-H}$ = 199.4 Hz PHCy).

(S_C,R_{Fc})-1-{Bis[3,5-bis(trifluoromethyl)phenyl]phosphino}-2-{1-[(1-adamantyl)phosphino]ethyl}ferrocene (90). To a suspension of (S_C,R_{Fc})-2-[1-(dimethylamino)ethyl]-1-{bis[3,5-bis(trifluoromethyl)-phenyl]phosphino}ferrocene (500 mg, 0.70 mmol, 1 eq.) in 1 mL of degassed acetic acid was added trifluoroacetic acid (54 µL, 0.70 mmol, 1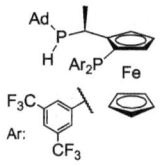
eq.). 1-Adamantylphosphine (236 mg, 1.4 mmol, 2 eq.) was added, the mixture heated to 90 °C and the red solution stirred at this temperature for 15 h. The mixture was then cooled down to room temperature, diluted with dichloromethane and saturated aqueous $NaHCO_3$, the organic phase washed with brine, dried over $MgSO_4$ and the solvent removed under reduced pressure to give a oily foam. FC (silica, pentane:EtOAc 100:0-50:2) gave the product as a mixture of two diastereoisomers (5:1). Yield: 503 mg (86%), orange crystalline solid. Major isomer: ^{1}**H NMR** (300 MHz, C_6D_6): δ 1.48-1.73 (m, 15 H, Ad, CH*Me*); 1.78 (b s, 3 H, Ad); 2.44 (d, J_{P-H} = 202 Hz, 1 H, P*H*); 3.48-3.59 (m, 2 H, C*H*Me, Cp); 3.75 (s, 5 H, Cp'); 3.99 (s, 1 H, Cp); 4.21 (s, 1 H, Cp); 7.69 (d, J_{P-H} = 11.7 Hz, 2 H, PAr_2); 7.88 (d, J = 5.4 Hz, 2 H, PAr_2); 8.07 (t, J = 6.9 Hz, 2 H, PAr_2); ^{19}**F NMR** (188 MHz, C_6D_6): δ -62.51 (s, CF_3); -62.67 (s, CF_3'); ^{31}**P{^{1}H} NMR** (121 MHz, C_6D_6): δ -22.72 (s, PAr_2); 5.32 (s, PHAd); ^{31}**P NMR** (121 MHz, C_6D_6): δ 5.32 (d, $^1J_{P-H}$ = 203 Hz, PHAd); **EA** calcd for: $C_{38}H_{34}F_{12}FeP_2$ (836.46): C 54.57, H 4.10, F 27.26, P 7.41; found: C 54.82, H 4.35, F 27.19, P 7.54; **HRMS (EI)** calcd (m/z) for $C_{38}H_{34}F_{12}FeP_2$: 836.1294; found: 836.1280 ([M]); 817.1324 ([M]-F); 701.01 ([M]-Ad).

6 Experimental Part

Minor isomer: **^{19}F NMR** (188 MHz, C_6D_6): δ -62.49 (s, CF_3); -62.62 (s, CF_3'); **^{31}P{^{1}H} NMR** (121 MHz, C_6D_6): δ -21.64 (d, J_{P-P} = 16.6 Hz, PAr_2); 14.45 (d, J_{P-P} = 16.6 Hz, $PHAd$); **^{31}P NMR** (121 MHz, C_6D_6): δ 14.45 (d, J_{P-H} = 193.6 Hz, $PHAd$).

(S_C,R_{Fc})-1-(Dicyclohexylphosphino)-2-{1-[(1-adamantyl)phosphino]-ethyl}ferrocene (91). To a solution of (S_C,R_{Fc})-1-(dicyclohexylphosphino)-2-[1-(dimethylamino)ethyl]ferrocene (500 mg, 1.10 mmol, 1 eq.) in 1 mL of degassed acetic acid was added trifluoroacetic acid (82 μL, 1.10 mmol, 1 eq.). 1-Adamantylphosphine (371 mg, 2.21 mmol, 2 eq.) was then added, the mixture heated to 90 °C and the red solution stirred at this temperature for 12 h. The solvent was then removed under reduced pressure at 60 °C, and the residue diluted with diethyl ether and saturated aqueous $NaHCO_3$. The organic phase were separated and dried over $MgSO_4$ and the solvent was removed under reduced pressure. FC (silica, pentane:Et_2O 1:0-20:1) gave the product as a mixture of two diastereoisomers (3:1). Yield: 448 mg (70%), orange foam. Major isomer: **^{1}H NMR** (300 MHz, C_6D_6): δ 1.14-2.43 (m, 40 H, Ad, Cy,$CHMe$); 3.27 (d, $^1J_{H-P}$ = 207 Hz, 1 H, PH); 3.56-3.66 (m, 1 H, $CHMe$); 4.00 (bs, 5 H, Cp); 4.09 (bs, 1 H, Cp); 4.14 (s, 1 H, Cp'); 4.28 (s, 1 H, Cp); **^{31}P{^{1}H} NMR** (121 MHz, C_6D_6): δ -17.24 (s, PCy_2); -2.49 (s, $PHAd$); **^{31}P NMR** (121 MHz, C_6D_6): δ -17.24 (bs, PCy_2); -2.47 (d, $^1J_{P-H}$ = 210 Hz, $PHAd$); **EA** calcd for $C_{34}H_{50}FeP_2$ (576.56): C 70.83, H 8.74, P 10.74; found: C 70.81, H 8.70, P 10.91; **HRMS** (EI) calcd (m/z) for $C_{34}H_{49}FeP_2$ ([M-H]$^+$): 575.2654; found: 575.2656 ([M-H]$^+$); 492.1793 ([M-HCy]$^+$); 441.1551 ([M-Ad]$^+$).

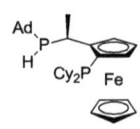

Minor Isomer: **^{31}P{^{1}H} NMR** (121 MHz, C_6D_6): δ -16.16 (d, J_{P-P} = 19.3 Hz, PCy_2); 6.32 (d, J_{P-P} = 20.2 Hz, $PHAd$); **^{31}P NMR** (121 MHz, C_6D_6): δ -17.24 (bs, PCy_2); 6.38 (d, $^1J_{P-H}$ = 203 Hz, $PHAd$).

(S_C,R_{Fc})-1-Bromo-2-{1-[(1-adamantyl)phosphino]ethyl}ferrocene (92). To a solution of (R_C,S_{Fc})-1-bromo-[2-(dimethylamino)ethyl]ferrocene (500 mg, 1.49 mmol, 1 eq.) in 2 mL of degassed acetic acid was added trifluoroacetic acid (110 μL, 1.49 mmol, 1 eq.). 1-Adamantylphosphine (375 mg, 2.23 mmol, 1.5 eq.) was then added, the mixture heated to 120 °C and the red solution stirred at this temperature for 13 h. After cooling down to room temperature, the

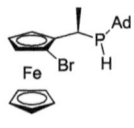

mixture was diluted with diethyl ether and saturated aqueous NaHCO$_3$, the organic phase separated, dried over MgSO$_4$ and the solvent removed under reduced pressure. FC (silica, pentane:Et$_2$O 1:0-10:1) gave the product as a mixture of two diastereoisomers (5:1). Yield: 660 mg (97%), orange crystalline solid. Major isomer: ^{1}**H NMR** (300 MHz, CD$_2$Cl$_2$): δ 1.53 (dd, $^3J_{H-P}$ = 15.2 Hz, ^3J = 7.0 Hz, 3 H, CH*Me*); 1.62-1.91 (m, 12 H, *Ad*); 1.95 (b s, 3 H, *Ad*); 2.78 (d, J_{H-P} = 209 Hz, 1 H, P*H*); 3.19-3.29 (m, 1 H, C*H*Me); 4.01 (s, 1 H, *Cp*); 4.06 (s, 1 H, *Cp*); 4.17 (s, 5 H, *Cp'*); 4.39 (s, 1 H, *Cp*); ^{31}**P{^{1}H} NMR** (121 MHz, CD$_2$Cl$_2$): δ -1.14 (s, *P*HAd); ^{31}**P NMR** (121 MHz, CD$_2$Cl$_2$): δ -1.12 (d, $^1J_{P-H}$ = 213 Hz, *P*HAd).

Minor Isomer: ^{31}**P{^{1}H} NMR** (121 MHz, CD$_2$Cl$_2$): δ 9.49 (s, *P*HAd); ^{31}**P{^{1}H} NMR** (121 MHz, CD$_2$Cl$_2$): δ 9.48 (d, $^1J_{P-H}$ = 195 Hz, *P*HAd).

(R_C,S_p)-1-[Bis(trifluoromethyl)phosphino]-2-{1-[(1-adamantyl)-phosphino]ethyl}ferrocene (110). To a solution of (R_C,S_p)-1-[bis(trifluoromethyl)phosphino]-2-[1-(dimethylamino)ethyl]ferrocene (2.00 g, 4.70 mmol, 1 eq.) in 4 mL of degassed acetic acid was added trifluoroacetic acid (350 µL, 4.70 mmol, 1 eq.). 1-Adamantylphosphine (1.58 g, 9.41 mmol, 2 eq.) was then added, the mixture heated to 95 °C and the red solution stirred at this temperature for 21 h. After cooling down to room temperature, the mixture was diluted with diethyl ether and saturated aqueous NaHCO$_3$, the organic phase separated, dried over MgSO$_4$ and the solvent removed under reduced pressure. FC (silica, pentane:Et$_2$O 1:0-20:1) gave the product as a mixture of two diastereoisomers (6:1). Yield: 2.49 g (96%), orange crystalline solid. Major isomer: ^{1}**H NMR** (300 MHz, C$_6$D$_6$): δ 1.45 (dd, $^3J_{H-P}$ = 14.8 Hz, ^3J = 6.8 Hz, 3 H, CH*Me*); 1.60 (bs, 6 H, *Ad*); 1.82 (bs, 9 H, *Ad*); 3.02 (d, J_{H-P} = 204 Hz, 1 H, P*H*); 3.23-3.26 (m, 1 H, C*H*Me); 4.01 (s, 5 H, *Cp'*); 4.08 (t, ^3J = 2.6 Hz, 1 H, *Cp*); 4.18 (s, 1 H, *Cp*); 4.35 (s, 1 H, *Cp*); ^{13}**C{^{1}H} NMR** (62.9 MHz, C$_6$D$_6$): δ 21.62 (dd, J_{C-P} = 15.8, 12.1 Hz, PC*H*Me); 22.01 (d, J_{C-P} = 29.3 Hz, PCH*Me*); 28.84 (d, J_{C-P} = 7.4 Hz, *Ad*); 32.72 (d, J_{C-P} = 14.9 Hz, *Ad*); 36.63 (s, *Ad*); 42.48 (d, J_{C-P} = 10.3 Hz, *Ad*); 57.04 (d, J_{C-P} = 6.8 Hz, *Cp*); 70.02 (s, *Cp'*); 70.28 (s, *Cp*); 71.43 (d, J_{C-P} = 4.4 Hz, *Cp*); 72.23 (d, J_{C-P} = 2.3 Hz, *Cp*); 103.80 (dd, J_{C-P} = 35.11, 8.6 Hz, *Cp*); ^{19}**F NMR** (282 MHz, C$_6$D$_6$): δ -53.94 (dq, $^2J_{F-P}$ = 68.4 Hz, $^4J_{F-F'}$ = 8.7 Hz, C*F$_3$*); -50.80 (dq, $^2J_{F-P}$ = 77.0 Hz, $^4J_{F-F'}$ = 8.7 Hz, C*F$_3$'*); ^{31}**P{^{1}H} NMR** (121 MHz, C$_6$D$_6$): δ -2.35 (qq, J_{P-F} = 77.3 Hz, $J_{P-F'}$ = 68.6 Hz, *P*(CF$_3$)$_2$); 4.77 (d, J_{P-P} = 2.6 Hz, *P*HAd); ^{31}**P NMR** (121 MHz, C$_6$D$_6$): δ 4.77 (d, $^1J_{P-H}$ = 204 Hz, *P*HAd); **EA** calcd for C$_{24}$H$_{28}$F$_6$FeP$_2$ (548.09): C 52.58, H 5.15, F 20.79, P 11.30 found C 52.79, H 5.42, F 20.58,

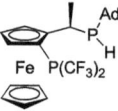

P 11.23; **HRMS** (EI) calcd (m/z) for $C_{24}H_{28}F_6FeP_2$: 548.0914; found: 548.0918 ([M]); 479.0781 ([M]-CF_3); 380.9771 ([M]-PHAd).

Minor Isomer: $^{31}P\{^1H\}$ **NMR** (121 MHz, C_6D_6): δ -5.00 – -4.20 (m, $P(CF_3)_2$); 12.47 (dq, J_{P-P} = 39.2 Hz, J = 15.8 Hz, PHAd); ^{31}P **NMR** (121 MHz, C_6D_6): δ 12.48 (d, $^1J_{P-H}$ = 195 Hz, PHAd); ^{19}F **NMR** (282 MHz, C_6D_6): δ -53.71 (dq, $^2J_{F-P}$ = 67.7 Hz, $^4J_{F-F'}$ = 8.5 Hz, CF_3); -50.73 (dq, $^2J_{F-P}$ = 76.5 Hz, $^4J_{F-F'}$ = 8.4 Hz, CF_3').

(R_C,S_{Fc})-1-[Bis(trifluoromethyl)phosphino]-2-[1-(cyclohexylphosphino)-ethyl]ferrocene (131). To a solution of (R_C,S_{Fc})-1-[bis(trifluoromethyl)- phosphino]-2-[1-(dimethylamino)ethyl]ferrocene (2.00 g, 4.70 mmol, 1 eq.) in 5 mL of degassed acetic acid was added trifluoroacetic acid (350 µL,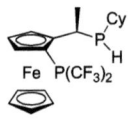
4.70 mmol, 1 eq.). Cyclohexylphosphine (1.29 mL, 9.41 mmol, 2 eq.) was then added, the mixture heated to 90 °C and the red solution stirred at this temperature for 18 h. The mixture was then cooled to 60 °C and the acetic acid removed under reduced pressure. The residue was then diluted with diethyl ether and saturated aqueous $NaHCO_3$ and the organic phase separated and dried over $MgSO_4$. After removal of the solvent, the crude product was purified by FC (silica, pentane:Et_2O 1:0-20:1) to yield the product as a mixture of two diastereoisomers (10:7). Yield: 1.99 g (85%), dark red oil. Major isomer: ^1H **NMR** (300 MHz, C_6D_6): δ 1.03-1.24 (m, 5 H, Cy);1.28-1.75 (m, 5 H, Cy); 1.34 (dd, $^3J_{H-P}$ = 13.3 Hz, ^3J= 7.0 Hz, 3 H, CHMe); 2.99-3.09 (m, 1 H, CHMe); 3.19 (dd, $^1J_{H-P}$ = 199.1 Hz, J = 5.4 Hz, 1 H, PH); 4.00 (s, 5 H, Cp'); 4.06 (s, 1 H, Cp); 4.14 (s, 1 H, Cp); 4.36 (s, 1 H, Cp); ^{19}F **NMR** (282 MHz, C_6D_6): δ -54.05 (dqd, $^2J_{F-P}$ = 63.7 Hz, $^4J_{F-F'}$ = 8.5 Hz, $J_{F-P'}$ = 4.0 Hz, CF_3); -50.78 (dq, $^2J_{F-P}$ = 76.6 Hz, $^4J_{F-F'}$ = 8.6 Hz, CF_3'); $^{31}P\{^1H\}$ **NMR** (121 MHz, C_6D_6): δ -15.79 (d, J_{P-P} = 12.1 Hz, PHCy); -4.62 – -0.97 (m, $P(CF_3)_2$); ^{31}P **NMR** (121 MHz, C_6D_6): δ -15.78 (d, $^1J_{P-H}$ = 201 Hz, PHAd); **HRMS** (EI) calcd (m/z) for $C_{20}H_{24}F_6FeP_2$: 496.0602 ([M]$^+$); found: 496.0607 ([M]$^+$); 427.0652 ([M-CF_3]$^+$); 380.9930 ([M-PHCy]$^+$). Minor Isomer: ^{19}F **NMR** (282 MHz, C_6D_6): δ-53.83 (ddq, $^2J_{F-P}$ = 66.6 Hz, $J_{F-P'}$ = 17.0 Hz, $^4J_{F-F'}$ = 8.5 Hz, CF_3); -50.93 (dq, $^2J_{F-P}$ = 76.4 Hz, $^4J_{F-F'}$ = 8.5 Hz, CF_3'); $^{31}P\{^1H\}$ **NMR** (121 MHz, C_6D_6): δ -10.48 (dq, J_{P-P} = 33.5 Hz, $^2J_{P-F}$ = 16.5 Hz, PHCy); -4.62 – -0.97 (m, $P(CF_3)_2$); ^{31}P **NMR** (121 MHz, C_6D_6): δ-10.47 (d, $^1J_{P-H}$ = 200 Hz, PHAd).

(*R*$_P$,*R*$_C$,*S*$_{Fc}$)-1-[(1-Adamantyl)methylphosphino)]-2-[1-(dicyclohexyl-phosphino)ethyl]–ferrocene. (*R*$_P$,*R*$_C$,*S*$_{Fc}$)-1-[(1-Adamantyl)methyl-phosphino]-2-[(1-dimethylamino)ethyl]ferrocene (**192**) (50 mg, 114 μmol, 1 eq.) and dicyclohexylphosphine (25 μL, 1047 μmol, 1.1 eq.) were 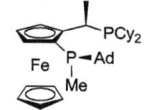 dissolved in 0.5 mL of degassed acetic acid, heated up to 90 °C and stirred at this temperature for 12 h. The solvent was then removed under reduced pressure and the residue diluted with diethyl ether and saturated aqueous NaHCO$_3$ solution. The organic layer was separated and the aqueous phase extracted with diethyl ether. The combined organic phases were dried over MgSO$_4$ and the solvent evaporated in vacuo. The residue was purified by FC (silica, pentane:EtOAc 100:1) and crystallization from ethanol to yield the oxidation sensitive product. Yield: 42 mg (62%), orange crystalline solid. ^{1}H NMR (300 MHz, C$_6$D$_6$): δ 1.09-1.54 (m, 14 H, *Cy*, P*Me*); 1.57-2.00 (m, 29 H, *Ad*, *Cy* CH*Me*); 3.25-3.35 (m, 1 H, C*H*Me); 4.06 (s, 1 H, *Cp*); 4.10 (s, 6 H, *Cp*, *Cp'*); 4.27 (bs, 1 H, *Cp*); 4.44 (bs, 1 H, *Cp*); ^{13}C{^{1}H} NMR (62.9 MHz, C$_6$D$_6$): δ 6.12 (d, $^1J_{C-P}$ = 16.5 Hz, P*Me*); 19.95 (dd, J_{C-P} = 4.4, 1.0 Hz); 25.58 (dd, J_{C-P} = 23.1, 8.2 Hz); 19.95 (dd, J_{C-P} = 4.4, 1.0 Hz); 26.72 (s); 27.24 (d, J_{C-P} = 11.4 Hz); 27.52 (d, J_{C-P} = 7.2 Hz); 27.99 (d, J_{C-P} = 4.7 Hz); 28.15 (d, J_{C-P} = 14.0 Hz); 28.95 (d, J_{C-P} = 8.3 Hz); 30.08 (d, J_{C-P} = 4.3 Hz); 30.44 (d, J_{C-P} = 9.7 Hz); 31.97 (d, J_{C-P} = 23.3 Hz); 32.33 (d, J_{C-P} = 18.8 Hz); 33.08 (d, J_{C-P} = 11.8 Hz); 33.36 (d, J_{C-P} = 21.2 Hz); 33.75 (d, J_{C-P} = 23.8 Hz); 37.23 (s); 40.05 (dd, J_{C-P} = 11.0, 4.8 Hz); 67.69 (s, *Cp*); 69.14 (d, J_{C-P} = 5.7 Hz, *Cp*); 69.32 (s, *Cp'*); 69.63 (dd, J_{C-P} = 10.9, 4.1 Hz, *Cp*); 75.66 (dd, J_{C-P} = 23.4, 3.4 Hz, *Cp*); 103.21 (dd, J_{C-P} = 26.5, 18.2 Hz, *Cp*); ^{31}P{^{1}H} NMR (121 MHz, C$_6$D$_6$): δ -31.19 (d, $J_{P-P'}$ = 7.1 Hz, *P*AdMe); 12.32 (d, $J_{P-P'}$ = 6.7 Hz, *P*AdMe); HRMS (MALDI) calcd (*m/z*) for C$_{35}$H$_{53}$FeP$_2$: 591.2967 ([MH]$^+$); found 591.2977 ([MH]$^+$); 393.1434 ([M-PCy$_2$]$^+$).

(*R*$_C$,*S*$_{Fc}$)-1-[Diphenylphosphino]-2-{1-[(1-adamantyl)phosphinyl]ethyl}-ferrocene (**97**). To a suspension of (*S*$_C$,*R*$_{Fc}$)-1-(diphenylphosphino)-2-[1-(dimethylamino)ethyl]ferrocene P,N-bis-borane complex (**107**) (1.00 g, 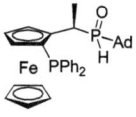 2.13 mmol, 1 eq.) in 2 mL degassed acetic acid was added trifluoroacetic acid (0.16 mL, 2.13 mmol, 1 eq.). 1-adamantylphosphine (897 mg, 5.33 mmol, 2.5 eq.) was then added, the mixture heated to 90 °C and the red solution stirred at this temperature for 14 h. The mixture was then cooled to room temperature, diluted with 20 mL of water and 20 mL of dichloromethane and the phases were separated. The organic phase was then washed with brine, dried over MgSO$_4$ and the solvent removed. FC (silica, pentane:EtOAc

6 Experimental Part

1:0-20:1, then DCM:MeOH 20:1) gave in the last fraction 205 mg (17%) of (R_C,S_{Fc})-1-[diphenylphosphino]-2-{1-[(1-adamantyl)phosphinyl]ethyl}ferrocene as a 3:1 mixture of diastereoisomers. Single crystals of the major isomer suitable for X-ray were obtained by slow diffusion of hexane into a concentrated dichloromethane solution. Major isomer: **^{1}H NMR** (300 MHz, CDCl$_3$): δ 1.42-1.78 (m, 15 H, P(O)H*Ad*, CH*Me*); 1.90 (s, 3 H, P(O)H*Ad*); 3.31(bs, 1 H, C*H*Me); 3.97 (s, 1 H, *Cp*); 4.05 (s, 5 H, *Cp*'); 4.43 (s, 1 H, *Cp*); 4.62 (s, 1 H, *Cp*); 5.43 (d, *J* = 454 Hz, 1 H, P(O)*H*Ad); 7.27 (s, 5 H, P*Ph*$_2$); 7.41 (s, 3 H, P*Ph*$_2$); 7.61 (bs, 2 H, P*Ph*$_2$); **^{13}C{^{1}H} NMR** (62.9 MHz, CDCl$_3$): δ 14.72 (d, $^2J_{C-P}$ = 3.4 Hz, PCH*Me*); 26.55 (d, $^1J_{C-P}$ = 12.7 Hz, P*C*HMe); 27.27 (d, J_{C-P} = 10.3 Hz, *Ad*); 33.12 (s, *Ad*); 35.53 (d, $^1J_{C-P}$ = 61.4 Hz, *Ad*); 36.39 (s, *Ad*); 69.47 (s, *Cp*'); 70.14 (s, *Cp*); 70.84 (d, J_{C-P} = 4.4 Hz, *Cp*); 73.57 (d, J_{C-P} = 4.0 Hz, *Cp*); 96.55 (d, J_{C-P} = 28.1 Hz, *Cp*); 128.11 (s, P*Ph*$_2$); 128.21 (s, P*Ph*$_2$); 128.39 (s, P*Ph*$_2$); 129.33 (s, P*Ph*$_2$); 133.31 (d, J_{C-P} = 19.1 Hz, P*Ph*$_2$); 135.22 (d, J_{C-P} = 21.6 Hz, P*Ph*$_2$); 136.55 (d, J_{C-P} = 6.8 Hz, P*Ph*$_2$); 139.03 (d, J_{C-P} = 8.0 Hz, P*Ph*$_2$); **^{31}P{^{1}H} NMR** (121 MHz, CDCl$_3$): δ 55.79 (s, *P*(O)HAd); -26.84 (s, *P*Ph$_2$); **^{31}P NMR** (121 MHz, CDCl$_3$): δ 55.86 (d, $^1J_{P-H}$ = 454 Hz, *P*(O)HAd); -26.81 (s, *P*Ph$_2$); **EA** calcd for C$_{34}$H$_{38}$FeOP$_2$ (580.46): C 70.35, H 6.60, O 2.76, P 10.67; found: C 69.80, H 6.68, 3.13 P 10.60; **HRMS** (EI) calcd (m/z) for C$_{34}$H$_{38}$FeOP$_2$: 580.1742 ([M]$^+$); found: 563.1741 ([M]$^+$); 445.0561 ([M-Ad]$^+$); 397.0 ([M-PHAd]$^+$).

Minor isomer: **^{31}P{^{1}H} NMR** (121 MHz, CDCl$_3$): δ 52.38 (d, J_{P-P} = 7.0 Hz, *P*(O)HAd); -26.43 (s, J_{P-P} = 7.0 Hz, *P*Ph$_2$); **^{31}P NMR** (121 MHz, CDCl$_3$): δ +52.38 (d, $^1J_{P-H}$ = 427 Hz, *P*(O)HAd); -26.42 (s, *P*Ph$_2$).

(R_C,S_{Fc})-1-Bromo-2-{1-[(1-adamantyl)phosphinyl]ethyl}ferrocene

(101). To a solution of (R_C,S_{Fc})-1-bromo-2-{1-[(1-adamantyl)phosphino]-ethyl}ferrocene (**92**) (100 mg, 218 μmol, 1 eq.) in 1 mL of methanol was added a 30% solution of hydrogen peroxide (33 μL, 327 μmol, 1 eq.) at 0 °C and the mixture allowed to warm up over a period of 60 min. The orange solution was then diluted with dichloromethane (10 mL) and water (5 mL), the organic phase washed with brine, dried over MgSO$_4$ and the solvent removed under reduced pressure to give the product as a mixture of two diastereoisomers (5:1). Yield: 104 mg (100%), orange crystalline solid. Major isomer: **^{1}H NMR** (300 MHz, CDCl$_3$): δ 1.68 (dd, J_{H-P} = 13.5 Hz, *J* = 6.9 Hz, 3 H, CH*Me*); 1.72-1.91 (m, 12 H, *Ad*); 2.06 (b s, 3 H, Ad); 3.09 (b s, 1 H, C*H*Me); 4.16 (s, 1 H,

Cp); 4.22 (s, 6 H, Cp', Cp); 4.47 (s, 1 H, Cp); 4.72 (s, 1 H, Cp); 5.92 (d, $^1J_{P-H}$ = 452 Hz, 1 H, P*H*); ^{31}P{^{1}H} NMR (121 MHz, CDCl$_3$): δ 55.21 (s, *P*(O)HAd); ^{31}P NMR (121 MHz, CDCl$_3$): δ 55.20 (d, $^1J_{P-H}$ = 453 Hz, *P*(O)HAd).

Minor isomer: ^{31}P{^{1}H} NMR (121 MHz, CDCl$_3$): δ 51.99 (s, *P*(O)HAd); ^{31}P NMR (121 MHz, CDCl$_3$): δ 51.91 (d, $^1J_{P-H}$ = 422 Hz, *P*(O)HAd).

(R_C,S_{Fc})-1-[Bis(trifluoromethyl)phosphino]-2-{1-[(1-adamantyl)-phosphinyl]ethyl}ferrocene (111). To a solution of (R_C,S_{Fc})-1-[bis(trifluoromethyl)phosphino]-2-{1-[(1-adamantyl)phosphino]ethyl}-ferrocene (110) (200 mg, 365 µmol, 1 eq.) in 1 mL of acetone was added a

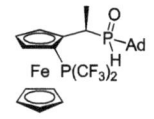

0.1 M solution of hydrogen peroxide (3.65 mL, 365 µmol, 1 eq.) and the mixture stirred for 90 min. The orange solution was then diluted with dichloromethane and water, the organic phase washed with brine, dried over MgSO$_4$ and the solvent removed under reduced pressure to give the product as a mixture of two diastereoisomers (20:1). Yield: 201 mg (98%), orange crystalline solid. Major isomer: ^{1}**H NMR** (300 MHz, C$_6$D$_6$): δ 1.55 (bs, 6 H, *Ad*); 1.59 (dd, J_{H-P} = 13 Hz, J = 7.2 Hz, 3 H, CH*Me*); 1.76-1.84 (m, 9 H, *Ad*); 2.88 (bs, 1 H, C*H*Me); 3.99 (s, 5 H, Cp'); 4.11 (s, 1 H, Cp); 4.33 (s, 1 H, Cp); 4.72 (s, 1 H, Cp); 5.97 (d, J_{P-H} = 444 Hz, 1 H, P*H*); ^{13}C{^{1}H} **NMR** (75.5 MHz, C$_6$D$_6$): δ 15.81 (d, J_{C-P} = 3.3 Hz, PCH*Me*); 19.94 (s, PCHMe); 26.75 (d, J_{C-P} = 10.6 Hz, *Ad*); 27.40 (d, J_{C-P} = 10.2 Hz, *Ad*); 35.80 (s, *Ad*); 36.39 (d, J_{C-P} = 0.9 Hz, *Ad*); 57.40-57.80 (m, Cp); 70.26 (s, Cp'); 71.48 (s, Cp); 72.10 (dd, J_{C-P} = 6.1, 2.6 Hz, Cp); 73.16 (d, J_{C-P} = 2.1 Hz, Cp); 100.04 (d, J_{C-P} = 35.0 Hz, Cp); ^{19}**F NMR** (282 MHz, C$_6$D$_6$): δ -50.57 (dq, J_{F-P} = 77 Hz, $J_{F-F'}$ = 8.5 Hz, C*F*$_3$); -54.32 (dq, J_{F-P} = 70 Hz, $J_{F-F'}$ = 8.5 Hz, C*F*$_3$'); ^{31}P{^{1}H} **NMR** (121 MHz, C$_6$D$_6$): δ 52.61 (d, J_{P-P} = 2.4 Hz, *P*(O)HAd); -4.80 – -1.17 (qq, J_{P-F} = 77, 70 Hz, *P*(CF$_3$)$_2$); ^{31}**P NMR** (121 MHz, C$_6$D$_6$): δ 52.61 (d, J_{P-H} = 446 Hz, *P*(O)HAd); **EA** calcd for C$_{24}$H$_{28}$F$_6$FeOP$_2$ (564.26): C 51.09, H 5.00, F 20.20, P 10.98; found: C 51.36, H 5.21, F 20.47, P 11.02; **HRMS** (ESI) calcd (*m/z*) for C$_{24}$H$_{28}$F$_6$FeOP$_2$: 564.0869; found: 587.0763 ([MNa]$^+$); 565.0937 ([MH]$^+$); 380.9933 ([M-PHAd]$^+$).

6 Experimental Part

6.3 Trifluoromethylphosphines

($1S,2R,3R,S_{Fc}$)-2-(1-Adamantyl)-3-methyl-1-(trifluoromethyl)-2,3-dihydro-1H-ferroceno[c] [1,2]diphosphole (129). To a solution of (R_C,S_{Fc})-1-[bis(trifluoromethyl)phosphino]-2-{1-[(1-adamantyl)phosphino]ethyl}-ferrocene (110) (400 mg, 0.730 mmol, 1 eq.) in 2 mL of tetrahydrofuran was added a solution of t-BuOK (81.9 mg, 0.730 mmol, 1 eq.) and 18-crown-6 (212 mg, 0.803 mmol, 1.2 eq.) in 6 mL of tetrahydrofuran. The orange solution immediately turned dark. The mixture was allowed to stir for ten minutes, quenched with 1 mL of saturated aqueous NH$_4$Cl and diluted with 2 mL water and 20 mL of diethyl ether. The organic phase was separated, dried over MgSO$_4$ and the solvent removed under reduced pressure. FC (Silica, pentane:EtOAc = 100:1-20:1) gave the product as an orange crystalline solid. Yield: 274 mg (79%); ^{1}H NMR (300 MHz, C$_6$D$_6$): δ 1.33 (dd, $^3J_{H-P}$ = 21.0 Hz, ^3J = 7.2 Hz, 3 H, CHMe); 1.61 (bs, 6 H, Ad); 1.88 (bs, 9 H, Ad); 3.43 (pent., J = 6.9 Hz, 1 H, CHMe); 3.95 (s, 5 H, Cp'); 3.98-4.00 (m, 2 H, Cp); 4.20 (s, 1 H, Cp); ^{13}C{^{1}H} NMR (62.9 MHz, C$_6$D$_6$): δ 23.87 (d, J_{C-P} = 34.9 Hz, PCHMe); 28.54 (d, J_{C-P} = 29.3 Hz, Ad); 29.05 (dd, J_{C-P} = 26.0, 2.9 Hz, PCHMe); 32.89 (dd, J_{C-P} = 20.1, 10.4 Hz, Ad); 36.53 (s, Ad); 42.13 (dd, J_{C-P} = 11.0, 7.7 Hz, Ad); 66.70 (s, Cp); 68.11 (d, J_{C-P} = 11.2 Hz, Cp); 70.09 (s, Cp'); 73.15 (d, J = 2.6 Hz, Cp); 76.12 (dd, J_{C-P} = 10.3, 4.5 Hz, Cp); 103.08 (s, Cp); ^{19}F NMR (282 MHz, C$_6$D$_6$): δ -55.30 (dd, $^2J_{F-P}$ = 57.0 Hz, $^3J_{F-P}$ = 19.3 Hz, CF_3); ^{31}P{^{1}H} NMR (121 MHz, C$_6$D$_6$): δ -2.35 (dq, $^1J_{P-P}$ = 207.8 Hz, $^2J_{P-F}$ = 57.2 Hz, PCF$_3$); 68.33 (dq, $^1J_{P-P}$ = 207.4 Hz, $^3J_{P-F}$ = 19.3 Hz, PAd); EA calcd for C$_{23}$H$_{27}$F$_3$FeP$_2$ (478.26): C 57.76, H 5.69, F 11.92, P 12.95; found: C 57.58, H 5.56, F 12.17, P 12.87; HRMS (EI) calcd (m/z) for C$_{23}$H$_{27}$F$_3$FeP$_2$: 478.0884 ([M]$^+$); found: 478.0887 ([M]$^+$).

($1S,2R,3R,S_{Fc}$)-2-Cyclohexyl-3-methyl-1-(trifluoromethyl)-2,3-dihydro-1H-ferroceno[c] [1,2]diphosphole (132). To a solution of (R_C,S_{Fc})-1-[bis(trifluoromethyl)phosphino]-2-[1-(cyclohexylphosphino)ethyl]ferrocene (131) (1060 mg, 2.136 mmol, 1 eq.) in 4 mL of tetrahydrofuran was added a solution of t-BuOK (240 mg, 2.136 mmol, 1 eq.) and 18-crown-6 (621 mg, 2.350 mmol, 1.2 eq.) in 4 mL of tetrahydrofuran. The orange solution immediately turned dark. The mixture was allowed to stir for ten minutes, quenched with 1 mL of saturated aqueous NH$_4$Cl

6 Experimental Part

and diluted with 2 mL water and 20 mL of diethyl ether. The organic phase was separated, dried over MgSO$_4$ and the solvent removed under reduced pressure. FC (Silica, pentane:EtOAc = 100:1-20:1) gave the impure product in the first coloured fraction. Recrystallization from hot ethanol gave the product as an orange crystalline solid. Yield: 364 mg (40%); **^{1}H NMR** (300 MHz, C$_6$D$_6$): δ 1.09-1.39 (m, 5 H, *Cy*); 1.33 (dd, $^3J_{H-P}$ = 20.8 Hz, ^3J = 7.5 Hz, 3 H, CH*Me*); 1.55 (bs, 1 H, *Cy*); 1.66 (bs, 3 H, *Cy*); 2.08-2.23 (m, 2 H, *Cy*); 2.88 (q, ^3J = 6.9 Hz, 1 H, C*H*Me); 3.87 (s, 5 H, *Cp'*); 3.93 (bs, 1 H, *Cp*); 4.07 (t, J = 62.2 Hz, 1 H, *Cp*); 4.16 (bs, 1 H, *Cp*); **^{19}F NMR** (282 MHz, C$_6$D$_6$): δ -54.87 (dd, $^2J_{F-P}$ = 52.7 Hz, $^3J_{F-P}$ = 16.7 Hz, C*F$_3$*); **^{31}P{^{1}H} NMR** (121 MHz, C$_6$D$_6$): δ -16.13 (dq, $^1J_{P-P}$ = 229.5 Hz, $^2J_{P-F}$ = 52.7 Hz, P*CF$_3$*); 47.87 (dq, $^1J_{P-P}$ = 229.5 Hz, $^3J_{P-F}$ = 16.7 Hz, P*Cy*); **EA** calcd for C$_{19}$H$_{23}$F$_3$FeP$_2$ (426.18): C 53.55, H 5.44, F 13.37, P 14.54; found: C 53.46, H 5.45, F 13.43, P 14.57; **HRMS** (EI) calcd (*m/z*) for C$_{19}$H$_{23}$F$_3$FeP$_{22}$: 426.0571 ([M]$^+$); found: 426.0574 ([M]$^+$); 357.0614 ([M-CF$_3$]$^+$); 342.9710 ([M-C$_2$H$_3$F$_3$]$^+$); 273.9756 ([M-C$_7$H$_{11}$F$_3$]$^+$).

(S_{P1},S_{P2},R_C,S_{Fc})-1-[Methyl(trifluoromethyl)phosphino]-2-{1-[(1-adamantyl)methylphosphino]ethyl}ferrocene (133). To a solution of (1*R*,2*S*,3*R*,S_{Fc})-2-(1-adamantyl)-3-methyl-1-(trifluoromethyl)-2,3-dihydro-1*H*-ferroceno-[*c*][1,2]diphosphole (**129**) (300 mg, 0.627 mmol, 1 eq.) in 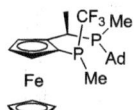 5 mL of dichloromethane was added methyl triflate (71 µL, 0.627 mmol, 1 eq.) and the solution was stirred for 1 h. The solvent was then removed under reduced pressure and the residue washed with 5 mL of diethyl ether. The residue was dissolved in 2 mL of tetrahydrofuran and the solution cooled down to -78 °C. Then, a 1.6 M solution of methyl lithium in diethyl ether (470 µL, 0.753 mmol, 1.2 eq.) was added and the mixture was allowed to warm up to room temperature. The reaction was then quenched with 1 mL of saturated aqueous NH$_4$Cl and diluted with 2 mL of distilled water and 20 mL of diethyl ether. The organic phase was separated, dried over MgSO$_4$ and the solvent removed under reduced pressure. FC (Silica, pentane:EtOAc:Et$_3$N = 50:1:1) gave the product as an orange crystalline solid. Yield: 173 mg (54%); **^{1}H NMR** (300 MHz, C$_6$D$_6$): δ 0.79 (d, $^2J_{H-P}$ = 5.4 Hz, 3 H, PAd*Me*); 1.35 (d, $^2J_{H-P}$ = 4.8 Hz, 3 H, PCF$_3$*Me*); 1.41 (dd, ^3J = 6.9 Hz, $^3J_{H-P}$ = 4.5 Hz, 3 H, CH*Me*); 1.75 (s, 6 H, *Ad*); 1.87 (s, 6 H, *Ad*); 1.98 (s, 3 H, *Ad*); 3.12 (qd, ^3J = 7.2 Hz, J_{H-P} = 2.7 Hz, 1 H, C*H*Me); 3.90 (s, 5 H, *Cp'*); 4.05 (d, *J* = 2.4 Hz, 1 H, *Cp*); 4.11 (s, 1 H, *Cp*); 4.14 (s, 1 H, *Cp*); **^{13}C{^{1}H} NMR** (62.9 MHz, C$_6$D$_6$): δ 0.05 (d, J_{C-P} = 27.6 Hz, PAd*Me*); 7.07-7.29 (m, PCF$_3$*Me*); 18.02 (s, CH*Me*); 24.20 (dd, J_{C-P} = 24.7, 7.5 Hz, C*H*Me); 28.71 (d,

6 Experimental Part

J_{C-P} = 8.0 Hz, *Ad*); 32.78 (d, J_{C-P} = 20.6 Hz, *Ad*); 37.11 (s, *Ad*); 39.54 (d, J_{C-P} = 11.6 Hz, *Ad*); 66.27-66.54 (m, *Cp*); 69.28 (s, *Cp'*); 69.63-69.87 (m, *Cp*); 104.31 (dd, J_{C-P} = 28.9, 23.0 Hz, *Cp*); 131.55 (qd, $^1J_{C-F}$ = 323.5 Hz, $^1J_{C-P}$ = 33.3, *CF₃*); **¹⁹F NMR** (282 MHz, C₆D₆): δ -59.10 (dd, $^2J_{F-P}$ = 57.9, J_{F-P} = 25.4 Hz, *CF₃*); **³¹P{¹H} NMR** (121 MHz, C₆D₆): δ -31.16 (qd, $^2J_{P-F}$ = 57.8 Hz, J_{P-P} = 20.4 Hz, *PCF₃Me*); 4.10 (qd, J_{P-F} = 25.4 Hz, J_{P-P} = 20.5 Hz, *PAdMe*); **EA** calcd for C₂₅H₃₃F₃FeP₂ (508.33): C 59.07, H 6.54, F 11.21, P 12.19; found: C 59.16, H 6.72, F 11.35, P 12.21; **HRMS** (EI) calcd (*m/z*) for C₂₃H₂₇F₃FeP₂: 508.1354 ([M]⁺); found: 508.1350 ([M]⁺); 469.1302 ([M-Me]⁺); 439.1402 ([M-CF₃]⁺).

(S_{P1},S_{P2},R_C,S_{Fc})-1-[Phenyl(trifluoromethyl)phosphino]-2-{1-[(1-adamantyl)methylphosphino]ethyl}ferrocene (134). To a solution of (1*R*,2*S*,3*R*,S_{Fc})-2-(1-adamantyl)-3-methyl-1-(trifluoromethyl)-2,3-dihydro-1*H*-ferroceno-[*c*][1,2]diphosphole **(129)** (200 mg, 0.418 mmol, 1 eq.) in 3 mL of dichloromethane was added methyl triflate (57 µL, 0.502 mmol, 1.2 eq.) and the solution was stirred for 1 h. After removal of the solvent under reduced pressure, the residue was dissolved in 4 mL of tetrahydrofuran and the solution cooled down to -78 °C. Then, a 2 M solution of phenylmagnesium chloride in tetrahydrofuran (251 µL, 0.502 mmol, 1.2 eq.) was added and the mixture was stirred for 1 h at -78 °C and for another 1 h at room temperature. The solvent was removed under reduced pressure and the residue diluted with 5 mL of saturated aqueous NH₄Cl, 10 mL water and 20 mL of diethyl ether. The organic phase was separated, dried over MgSO₄ and the solvent removed under reduced pressure. FC (Silica, pentane:EtOAc:Et₃N = 50:1:1) gave the product as an orange crystalline solid. Yield: 143 mg (60%); **¹H NMR** (300 MHz, CDCl₃): δ 0.89 (d, $^2J_{H-P}$ = 4.8 Hz, 3 H, PAd*Me*); 1.53 (dd, ^3J = 6.8 Hz, $^3J_{P-H}$ = 4.9 Hz, 3 H, CH*Me*); 1.74 (s, 12 H, *Ad*); 1.98 (s, 3 H, *Ad*); 3.10-3.18 (m 1 H, C*H*Me); 3.76 (s, 5 H, *Cp'*); 4.33 (s, 1 H, *Cp*); 4.42 (s, 1 H, *Cp*); 4.51 (s, 1 H, *Cp*); 7.50-7.53 (m, 3 H, *Ph*); 8.01 (t, $^3J_{H-P}$ = 8.2 Hz, *Ph*); **¹³C{¹H} NMR** (62.9 MHz, CDCl₃): δ 0.14 (d, J_{C-P} = 26.4 Hz, PAd*Me*); 18.46 (s, CH*Me*); 23.56 (dd, J_{C-P} = 22.6, 10.5 Hz, *C*HMe); 28.26 (d, J_{C-P} = 8.0 Hz, *Ad*); 32.75 (d, J_{C-P} = 18.8 Hz, *Ad*); 37.09 (s, *Ad*); 39.40 (d, J_{C-P} = 11.0 Hz, *Ad*); 64.36-64.68 (m, *Cp*); 68.92 (dd, J_{C-P} = 7.3, 6.4 Hz, *Cp*); 69.37 (s, *Cp'*); 70.22 (s, *Cp*); 71.58 (d, J_{C-P} = 4.6 Hz, *Cp*); 105.05 (dd, J_{C-P} = 32.4, 23.3 Hz, *Cp*); 128.11 (d, J_{C-P} = 10.0 Hz, *Ph*); 130.23 (qd, $^1J_{C-F}$ = 323.2 Hz, $^1J_{C-P}$ = 26.3 Hz, *CF₃*); 130.99 (d, J_{C-P} = 1.3 Hz, *Ph*); 131.46-131.68 (m, *Ph*); 135.35 (d, J_{C-P} = 25.7 Hz, *Ph*); **¹⁹F NMR** (282 MHz, CDCl₃): δ -54.72 (dd, $^2J_{F-P}$ = 67.5 Hz, *J* = 20.4 Hz, *CF₃*); **³¹P{¹H} NMR**

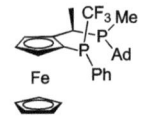

(121 MHz, CDCl$_3$): δ -16.16 (qd, $^2J_{P\text{-}F}$ = 67.3 Hz, $J_{P\text{-}P}$ = 14.5 Hz, PCF$_3$Ph); 4.83 (qd, $^2J_{P\text{-}F}$ = 20.4 Hz, $J_{P\text{-}P}$ = 14.8 Hz, PAdMe); **EA** calcd for C$_{25}$H$_{33}$F$_3$FeP$_2$ (570.40): C 63.17, H 6.18, F 9.99, P 10.86; found: C 62.88, H 6.28, F 10.20, P 10.64; **HRMS** (EI) calcd (m/z) for C$_{30}$H$_{35}$F$_3$FeP$_2$: 570.1510 ([M]$^+$); found: 570.1497 ([M]$^+$); 501.1554 ([M-CF$_3$]$^+$).

(S_{P1},S_{P2},R_C,S_{Fc})-1-[Isopropyl(trifluoromethyl)phosphino]-2-{1-[(1-adamantyl)methylphosphino]ethyl}ferrocene (135). To a solution of (1R,2S,3R,S_{Fc})-2-(1-adamantyl)-3-methyl-1-(trifluoromethyl)-2,3-dihydro-1H-ferroceno-[c][1,2]diphosphole (129) (120 mg, 251 µmol, 1 eq.) in 2 mL

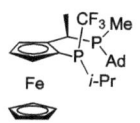

of dichloromethane was added methyl triflate (34 µL, 301 µmol, 1.2 eq.) and the solution was stirred for 1 h. After removal of the solvent under reduced pressure, the residue was dissolved in 2 mL of tetrahydrofuran and the solution cooled down to -78 °C. Then, a 2 M solution of isopropylmagnesium chloride in tetrahydrofuran (251 µL, 0.502 mmol, 1.2 eq.) was added and the mixture was stirred for 1 h at -78 °C and for another 1 h at room temperature. The solvent was removed under reduced pressure and the residue diluted with 3 mL of degassed water and 10 mL of diethyl ether. The organic phase was concentrated and the residue crystallized from hot ethanol to give the oxidation sensitive product as orange crystals. Yield: 86 mg (60%); **^{1}H NMR** (300 MHz, CDCl$_3$): δ 0.92 (d, $^2J_{H\text{-}P}$ = 4.7 Hz, 3 H, PAdMe); 1.39-1.60 (m, 9 H, CHMe_2, CHMe); 1.66 (s, 6 H, Ad); 1.70 (s, 6 H, Ad); 1.94 (s, 3 H, Ad) 2.41-2.56 (m 1 H, CHMe$_2$); 2.89 (q, ^3J = 7.7 Hz, CHMe); 4.23 (s, 5 H, Cp'); 4.39 (bs, 1 H, Cp); 4.41 (t, J = 2.5 Hz; 1 H, Cp); 4.49 (s, 1 H, Cp); **^{13}C{^{1}H} NMR** (75 MHz, CDCl$_3$): δ 0.99 (d, $J_{C\text{-}P}$ = 25.1 Hz, PAdMe); 20.01 (dd, $^2J_{C\text{-}P}$ = 4.6 Hz, J = 1.3 Hz, CHMe_2); 20.20 (s, CHMe); 21.87 (d, $^2J_{C\text{-}P}$ = 25.99 Hz, CHMe_2); 23.22 (dd, $J_{C\text{-}P}$ = 21.6, 8.4 Hz, CHMe); 26.05 (d, $^2J_{C\text{-}P}$ = 11.2 Hz, CHMe$_2$); 28.51 (d, $^3J_{C\text{-}P}$ = 8.0 Hz, Ad); 32.56 (d, $^1J_{C\text{-}P}$ = 17.7 Hz, Ad); 37.02 (s, Ad); 39.23 (d, $^2J_{C\text{-}P}$ = 10.6 Hz, Ad); 66.92-67.93 (m, Cp); 69.21 (d, $^3J_{C\text{-}P}$ = 2.3 Hz, Cp); 69.40 (s, Cp'); 69.55 (d, $^3J_{C\text{-}P}$ = 5.1 Hz, Cp); 71.56 (s, Cp); 104.10 (dd, $^2J_{C\text{-}P}$ = 25.2, 22.8 Hz, Cp); 131.43 (qd, $^1J_{C\text{-}F}$ = 323.5 Hz, $^1J_{C\text{-}P}$ = 39.0, CF_3); **^{19}F NMR** (282 MHz, CDCl$_3$): δ -50.84 (dd, $^2J_{F\text{-}P}$ = 58.5 Hz, $J_{F\text{-}P'}$ = 14.5 Hz, CF_3); **^{31}P{^{1}H} NMR** (121 MHz, CDCl$_3$): δ -1.32 (qd, $^2J_{P\text{-}F}$ = 58.4 Hz, $J_{P\text{-}P'}$ = 10.0 Hz, PCF$_3$^{i}Pr); 2.10 (qd, $J_{P\text{-}F}$ = 14.5 Hz, $J_{P\text{-}P'}$ = 10.0 Hz, PAdMe); **HRMS** (EI) calcd (m/z) for C$_{24}$H$_{30}$F$_3$FeP$_2$: 483.1119 ([M-^{i}Pr]$^+$); found: 493.1115 ([M-^{i}Pr]$^+$); 467.1725 ([M-CF$_3$]$^+$).

(S_{P1},S_{P2},R_C,S_{Fc})-1-[Phenyl(trifluoromethyl)phosphino]-2-[1-(cyclohexyl-methylphosphino)ethyl]ferrocene (136). To a solution of (1R,2S,3R,S_{Fc})-2-cyclohexyl-3-methyl-1-(trifluoromethyl)-2,3-dihydro-1H-ferroceno-[c][1,2]-diphosphole **(132)** (300 mg, 704 µmol, 1 eq.) in 5 mL of dichloromethane 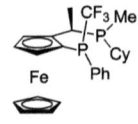 was added methyl triflate (96 µL, 845 µmol, 1.2 eq.) and the solution was stirred for 1 h. After removal of the solvent under reduced pressure, the residue was dissolved in 5 mL of tetrahydrofuran and cooled down to -78 °C. Then, a 2 M solution of phenylmagnesium-chloride in tetrahydrofuran (422 µL, 845 µmol, 1.2 eq.) was added and the mixture was stirred for 1 h at -78 °C and for another 2 h at room temperature The solvent was removed under reduced pressure and the residue diluted with 3 mL of degassed water and 10 mL of diethyl ether. The organic phase was concentrated, transferred into a glove box dissolved it again in diethyl ether and filtered over a short silica plug. After removal of the solvent the product was obtained as an orange crystalline solid. Yield: 280 mg (79%); **^{1}H NMR** (300 MHz, CDCl$_3$): δ 0.82 (d, $^2J_{H-P}$ = 3.1 Hz, 3 H, PCyMe); 1.13-1.53 (m, 9 H, CHMe, Cy); 1.58-1.94 (m, 5 H, Cy); 3.09-3.20 (m, 1 H, CHMe); 3.79 (s, 5 H, Cp'); 4.32 (s, 1 H, Cp); 4.45 (s, 1 H, Cp); 4.54 (s, 1 H, Cp); 7.48-7.58 (m, 3 H, Ph); 8.02 (t, $^3J_{H-P}$ = 7.8 Hz, Ph); **^{13}C{^{1}H} NMR** (75 MHz, CDCl$_3$): δ 3.53 (d, $^1J_{C-P}$ = 23.9 Hz, PCyMe); 14.64 (s, CHMe); 25.76 (dd, J_{C-P} = 19.1, 10.3 Hz, CHMe); 26.43 (s, Cy); 26.72 (d, J_{C-P} = 1.4 Hz, Cy); 26.85 (s, Cy); 28.98 (d, J_{C-P} = 13.8 Hz, Cy); 29.67 (d, J_{C-P} = 15.9 Hz, Cy); 36.48 (d, J_{C-P} = 15.5 Hz, Cy); 65.10-65.34 (m, Cp); 68.49-68.68 (m, Cp); 69.45 (s, Cp'); 70.28 (s, Cp); 71.87 (d, J_{C-P} = 4.2 Hz, Cp); 102.31 (dd, J_{C-P} = 34.0, 16.9 Hz, Cp); 128.36 (d, J_{C-P} = 10.0 Hz, Ph); 130.16 (qd, $^1J_{C-F}$ = 323.1 Hz, $^1J_{C-P}$ = 26.6, CF_3); 130.93 (s, Ph); 131.49-131.71 (m, Ph); 135.77 (d, J_{C-P} = 25.1 Hz, Ph); **^{19}F NMR** (282 MHz, CDCl$_3$): δ -55.19 (dd, $^2J_{F-P}$ = 66.2 Hz, $J_{F-P'}$ = 25.3 Hz, CF_3); **^{31}P{^{1}H} NMR** (121 MHz, CDCl$_3$): δ -16.69 (qd, $^2J_{P-F}$ = 66.2 Hz, $J_{P-P'}$ = 26.8 Hz, PCF$_3$Ph); -8.85 (p, J_{P-F} = $J_{P-P'}$ = 25.5 Hz, PCyMe); **HRMS** (EI) calcd (m/z) for C$_{26}$H$_{31}$F$_3$FeP$_2$: 518.1197 ([M]$^+$); found: 518.1194 ([M]$^+$); 499.1224 ([M-F]$^+$); 449.1250 ([M-CF$_3$]$^+$).

(S_{P1},S_{P2},R_C,S_{Fc})-1-[Isopropyl(trifluoromethyl)phosphino]-2-[1-(cyclohexylmethylphosphino)ethyl]ferrocene (137). To a solution of (1R,2S,3R,S_{Fc})-2-cyclohexyl-3-methyl-1-(trifluoromethyl)-2,3-dihydro-1H-ferroceno-[c][1,2]diphosphole **(132)** (200 mg, 469 µmol, 1 eq.) in 3 mL of dichloromethane was added methyl triflate (63 µL, 563 µmol, 1.2 eq.) and the solution was stirred for 1 h. After removal of the solvent under reduced pressure, the residue was dissolved in 3 mL of tetrahydrofuran and the solution cooled down to -78 °C. Then, a 2 M solution of isopropylmagnesiumchloride in tetrahydrofuran (282 µL, 502 µmol, 1.2 eq.) was added and the mixture was stirred for 1 h at -78 °C and for another 2 h at room temperature. The solvent was removed under reduced pressure and the residue diluted with 3 mL of degassed water and 10 mL of diethyl ether. The organic phase was concentrated and the residue crystallized from hot ethanol to give the oxidation sensitive product as orange crystals. Yield: 140 mg (62%); **^{1}H NMR** (300 MHz, CDCl$_3$): δ 0.81 (d, $^2J_{H-P}$ = 3.5 Hz, 3 H, PCy*Me*); 1.08-1.50 (m, 15 H, CH*Me*, CH*Me$_2$*, *Cy*); 1.55-1.83 (m, 5 H, *Cy*); 2.46-2.61 (m, 1 H, C*H*Me$_2$); 2.87 (q, ^3J = 6.8 Hz, 1 H, C*H*Me); 4.21 (s, 5 H, *Cp'*); 4.35 (bs, 2 H, *Cp*); 4.40 (s, 1 H, *Cp*); **^{13}C{^{1}H} NMR** (75 MHz, CDCl$_3$): δ 3.83 (d, $^1J_{C-P}$ = 23.3 Hz, PCy*Me*); 15.69 (s, CH*Me*); 20.02 (d, $^2J_{C-P}$ = 5.0 Hz, CH*Me$_2$*); 21.59 (d, $^2J_{C-P}$ = 25.5 Hz, CH*Me$_2$*); 24.87-26.01 (m, CHMe, CHMe$_2$); 26.41 (s, *Cy*); 26.77 (s, *Cy*); 26.90 (d, J_{C-P} = 3.6 Hz, *Cy*); 28.92 (d, J_{C-P} = 12.8 Hz, *Cy*); 29.24 (d, J_{C-P} = 16.0 Hz, *Cy*); 36.25 (d, J_{C-P} = 14.9 Hz, *Cy*); 67.57-68.34 (m, *Cp*); 68.78-69.18 (m, *Cp*); 69.43 (s, *Cp*); 69.51 (s, *Cp'*); 72.47 (d, J_{C-P} = 6.7 Hz, *Cp*); 100.56 (dd, J_{C-P} = 21.4, 16.4 Hz, *Cp*); 128.36 (d, J_{C-P} = 10.0 Hz, *Ph*); 130.16 (qd, $^1J_{C-F}$ = 322.9 Hz, $^1J_{C-P}$ = 39.9, *CF$_3$*); **^{19}F NMR** (282 MHz, CDCl$_3$): δ -50.31 (dd, $^2J_{F-P}$ = 58.0 Hz, $J_{F-P'}$ = 19.3 Hz, *CF$_3$*); **^{31}P{^{1}H} NMR** (121 MHz, CDCl$_3$): δ -10.52 (qd, J_{P-F} = 19.1 Hz, $J_{P-P'}$ = 18.9 Hz, *P*CyMe); 0.84 (qd, $^2J_{P-F}$ = 57.9 Hz, $J_{P-P'}$ = 17.5 Hz, *P*CF$_3^i$Pr); **EA** calcd for C$_{23}$H$_{33}$F$_3$FeP$_2$ (586.44): C 57.04, H 6.87, F 11.77, P 12.79; found: C 57.85, H 6.92, F 11.37, P 12.23; **HRMS** (EI) calcd (*m/z*) for C$_{23}$H$_{33}$F$_3$FeP$_2$: 484.1354 ([M]$^+$); found: 484.1364 ([M]$^+$); 441.0808 ([M-^{i}Pr]$^+$); 415.1402 ([M-CF$_3$]$^+$).

Bis(4-cyanophenyl)phenylphosphonite. To a solution of dichlorophenylphosphine (5 mL, 36.85 mmol, 1 eq.) and triethylamine (10.8 mL, 77.38 mmol, 2.1 eq.) in 100 mL of tetrahydrofuran was added 4-cyanophenol (8.78 g, 73.69 mmol, 2 eq.) at -78 °C. The resulting white suspension was allowed to warm up to room temperature and was stirred for 2 h. The precipitate was filtered off and the filtrate concentrated under reduced pressure to yield the crude product, which was used without further purification. Yield: 12.6 g (100%), white solid. **^{1}H NMR** (250 MHz, CDCl$_3$): δ 7.13 (d, J = 8.3 Hz, 4 H, *Ar*); 7.46-7.53 (m, 3 H, *Ph*); 7.58 (d, J = 8.4 Hz, 4 H, *Ar*); 7.76 (t, J = 7.4 Hz, 2 H, *Ph*); **^{13}C{^{1}H} NMR** (62.9 MHz, CDCl$_3$): δ 107.62 (d, J_{C-P} = 1.3 Hz, *Ar*); 118.45 (s, *CN*); 120.66 (d,

J_{C-P} = 8.9 Hz, *Ar*); 128.84 (d, J_{C-P} = 7.3 Hz, *Ph*); 129.96 (d, J_{C-P} = 25.1 Hz, *Ph*); 132.13 (s, *Ph*); 134.17 (s, *Ar*); 137.51 (d, J_{C-P} = 15.5 Hz, *Ph*); 158.15 (d, J_{C-P} = 4.2 Hz, *Ar*); ^{31}P{^{1}H} NMR (101 MHz, CDCl$_3$): δ 159.60 (s); **HRMS** (EI) calcd (*m/z*) for C$_{20}$H$_{13}$N$_2$O$_2$P: 344.0710 ([M]$^+$); found: 344.0702 ([M]$^+$); 243.0436 ([M-PhCN]$^+$); 226.0419 ([M-OPhCN]$^+$).

Bis(4-cyanophenyl)-(1-adamantyl)phosphonite. To a solution of adamantyldichlorophosphine (2.82 g, 11.89 mmol, 1 eq.) and 4-cyanophenol (2.83 g, 23.79 mmol, 2 eq.) in 40 mL of tetrahydrofuran was added triethylamine (3.5 mL, 24.98 mmol, 2.1 eq.) at 0 °C. The resulting white suspension was allowed to warm up to room temperature and was stirred for 10 h. The precipitate was filtered off and the filtrate concentrated under reduced pressure to yield the crude product, which was used without further purification. Yield: 4.8 g (100%), white solid. ^{1}**H NMR** (300 MHz, CDCl$_3$): δ 1.82 (bs, 6 H, *Ad*); 1.91 (bs, 6 H, *Ad*); 2.11 (bs, 3 H, *Ad*); 7.07 (d, *J* = 7.4 Hz, 4 H, *Ar*); 7.59 (d, *J* = 8.7 Hz, 4 H, *Ar*); ^{13}**C{^{1}H} NMR** (75 MHz, CDCl$_3$): δ 27.46 (d, J_{C-P} = 9.7 Hz, *Ad*); 33.38 (d, J_{C-P} = 13.6 Hz, *Ad*); 36.95 (s, *Ad*); 38.30 (d, J_{C-P} = 10.2 Hz, *Ad*); 107.14 (s, *Ar*); 118.48 (s, *CN*); 120.14 (d, J_{C-P} = 9.3 Hz, *Ar*); 134.15 (s, *Ar*); ^{31}**P{^{1}H} NMR** (121 MHz, CDCl$_3$): δ 178.29 (s); **HRMS** (ESI) calcd (*m/z*) for C$_{24}$H$_{27}$N$_3$O$_2$P: 420.1830 ([MNH$_4$]$^+$); found: 420.1830 ([MNH$_4$]$^+$).

2-Naphthyltrifluoromethylphosphine.[222] A suspension of 1-trifluoromethyl-1,2-benziodoxol-3-(1*H*)-one (**122**) (592 mg, 1.873 mmol, 1 eq.) in 3 mL of dichloromethane was added to a solution of 2-naphthylphosphine (300 mg, 1.873 mmol, 1 eq.) in 2 mL dichloromethane at room temperature and the mixture stirred for 3 h. The solvent was then removed under reduced pressure and the residue was suspended in pentane and filtered over a short plug (5 cm) of alumina. After removal of the solvent under reduced pressure the product was obtained as a white solid. Yield: 200 mg (46%). ^{1}**H NMR** (300 MHz, CD$_2$Cl$_2$): δ 5.04 (dq, $^1J_{H-P}$ = 222.3 Hz, $^3J_{H-F}$ = 11.2 Hz, 1 H, P*H*); 7.57 (bs, 2 H, *Np*); 7.65 (bs, 1 H, *Np*); 7.88 (bs, 3 H, *Np*); 8.21 (d, $^3J_{H-P}$ = 9.8 Hz, 1 H, *Np*); ^{13}**C{^{1}H} NMR** (75 MHz, CD$_2$Cl$_2$): δ 121.54-121.44 (m, *Np*); 127.74 (s, *Np*); 127.84 (s, *Np*); 128.11 (s, *Np*); 128.54 (d, J_{C-P} = 5.6 Hz, *Np*); 130.97 (qd, $^1J_{C-F}$ = 319.4 Hz, $^1J_{C-P}$ = 28.6, *CF*$_3$); 131.49 (d, J_{C-P} = 11.2 Hz, *Np*); 133.09 (d, J_{C-P} = 9.9 Hz,

Np); 134.03 (s, Np); 137.65 (d, J_{C-P} = 25.5 Hz, Np); ^{19}F NMR (282 MHz, CD$_2$Cl$_2$): δ -51.80 (dd, $^2J_{F-P}$ = 57.5 Hz, $^3J_{F-H}$ = 11.3 Hz, CF_3); ^{31}P{^{1}H} NMR (121 MHz, CD$_2$Cl$_2$): δ -40.52 (q, $^2J_{P-F}$ = 57.5 Hz);); ^{31}P NMR (121 MHz, CD$_2$Cl$_2$): δ -40.52 (dqdd, $^1J_{P-H}$ = 222.5 Hz, $^2J_{P-F}$ = 57.4 Hz, $^3J_{P-H}$ = 10.0, 6.0 Hz).

2-Naphthylbis(trifluoromethyl)phosphine (143).[157,222] DBU (1.1 mL, 7.49 mmol, 4 eq.) was added dropwise to a suspension of 2-naphthylphosphine (300 mg, 1.87 mmol, 1 eq.) and 1-trifluoromethyl-1,2-benziodoxol-3-(1H)-one (**122**) (1.18 g, 3.75 mmol, 2 eq.) in 5 mL dichloromethane at -78 °C. The yellow solution was stirred for 2 h at -78 °C and was then allowed to warm up to room temperature overnight. The solvent was then removed under reduced pressure and the residue was suspended in pentane and filtered over a silica plug (20 cm). After removal of the solvent under reduced pressure the highly volatile product was obtained as slightly yellow oil. Yield: 210 mg (33%). ^{1}H NMR (300 MHz, CD$_2$Cl$_2$): δ 7.75-7.62 (m, 2 H, Np); 7.81-7.91 (m, 1 H, Np); 7.93-8.07 (m, 3 H, Np); 8.44 (d, J = 14.7 Hz, 1 H, Np); ^{19}F NMR (282 MHz, CD$_2$Cl$_2$): δ -53.48 (d, $^2J_{F-P}$ = 79.0 Hz, CF_3); ^{31}P{^{1}H} NMR (121 MHz, CDCl$_3$): δ 0.68 (hept., $^2J_{P-F}$ = 78.8 Hz).

Bis(trifluoromethyl)phenylphosphine (145).[170] **Method A:** DBU (3.1 mL, 20.60 mmol, 4 eq.) was added dropwise to a suspension of phenylphosphine (567 mg, 5.150 mmol, 1 eq.) and 1-trifluoromethyl-1,2-benziodoxol-3-(1H)-one (**122**) (3.583 g, 11.33 mmol, 2.2 eq.) in 20 mL dichloromethane at -78 °C. The yellow solution was stirred for 2 h at -78 °C and was then allowed to warm up to room temperature overnight. The solvent was then removed under reduced pressure and the residue was suspended in pentane and filtered over a silica plug (20 cm). After removal of the solvent under reduced pressure the highly volatile product was obtained as colourless oil.
Method B: To a solution of bis(4-cyanophenyl)phenylphosphonite (5 g, 14.52 mmol, 1 eq.) in 50 mL diethyl ether and 5 mL acetonitrile was added trimethylsilyltrifluoromethane (4.55 mL, 30.50 mmol, 2.1 eq.) and cesium fluoride (463 mg, 3.05 mmol, 0.21 eq.) at 0 °C. The mixture was allowed to warm up to room temperature and stirred for 5 h. The progress of the reaction was monitored by ^{19}F NMR spectroscopy using a small aliquot of the reaction mixture. If no reaction was observed more trimethylsilyltrifluoromethane and cesium fluoride

6 Experimental Part

were added until the reaction started. After completion was detected the solvent was removed under reduced pressure and the residue was suspended in pentane and filtered through a silica plug (20 cm). After removal of the solvent under reduced pressure the highly volatile product was obtained as colourless oil. Yield: 2.19 g (61%). ^{1}**H NMR** (300 MHz, CDCl$_3$): δ 7.54 (td, J = 7.3, 1.3 Hz, 2 H, *Ph*); 7.67 (td, J = 7.7, 1.3 Hz, 2 H, *Ph*); 7.78 (t, J = 8.7Hz, 2 H, *Ph*); ^{13}**C{^{1}H} NMR** (75 MHz, CDCl$_3$): δ 121.11-121.36 (m, *Ph*); 129.35 (d, J_{C-P} = 10.1 Hz, *Ph*); 133.26 (d, J_{C-P} = 1.5 Hz, *Ph*); 136.03 (d, J_{C-P} = 25.0 Hz, *Ph*); 128.09 (qdq, $^1J_{C-F}$ = 301.0 Hz, $^1J_{C-P}$ = 18.8 Hz, $^3J_{C-F}$ = 5.8 Hz, *CF$_3$*); ^{19}**F NMR** (282 MHz, CDCl$_3$): δ -53.31 (d, $^2J_{F-P}$ = 79.6 Hz, *CF$_3$*); ^{31}**P{^{1}H} NMR** (121 MHz, CDCl$_3$): δ 0.27 (hept., $^2J_{P-F}$ = 79.6 Hz); **HRMS** (EI) calcd (*m/z*) for C$_8$H$_5$F$_6$P: 246.0028 ([M]$^+$); found: 246.0028 ([M]$^+$); 177.0073 ([M-CF$_3$]$^+$); 127.0140 ([M-C$_2$F$_5$]$^+$).

Bis(trifluoromethyl)adamantylphosphine (148). To a solution of bis(4-cyanophenyl)adamantylphosphonite (2.43 g, 6.038 mmol, 1 eq.) in 50 mL diethyl ether and 5 mL acetonitrile was added trimethylsilyltrifluoromethane (2.25 mL, 15.10 mmol, 2.1 eq.) and cesium fluoride (229 mg, 2.09 mmol, 0.21 eq.) at 0 °C. The mixture was allowed to warm up to room temperature and stirred for 5 h. The progress of the reaction was monitored by ^{19}F NMR spectroscopy using a small aliquot of the reaction mixture. If no reaction was observed more trimethylsilyltrifluoromethane and cesium fluoride were added until the reaction started. After completion was detected the solvent was removed under reduced pressure and the residue was suspended in pentane and filtered through a silica plug (20 cm). After removal of the solvent under reduced pressure the highly volatile product was obtained as colourless oil. Yield: 1.34 g (73%). ^{1}**H NMR** (300 MHz, C$_6$D$_6$): 1.39 (bs, 6 H, *Ad*); 1.62 (bs, 6 H, *Ad*); 1.79 (bs, 3 H, *Ad*); ^{13}**C{^{1}H} NMR** (75 MHz, CDCl$_3$): 27.95 (d, J_{C-P} = 9.6 Hz, *Ad*); 35.3-35.9 (m, *Ad*); 35.74 (d, J_{C-P} = 1.1 Hz, *Ad*); 38.47 (d, J_{C-P} = 11.2 Hz, *Ad*); 128.09 (qdq, $^1J_{C-P}$ = 321.1 Hz, $^1J_{C-P}$ = 33.5 Hz, $^3J_{C-P}$ = 6.0 Hz, *CF$_3$*); ^{19}**F NMR** (282 MHz, CDCl$_3$): δ -47.10 (d, $^2J_{F-P}$ = 65.0 Hz, *CF$_3$*); ^{31}**P{^{1}H} NMR** (121 MHz, CDCl$_3$): δ 25.37 (hept., $^2J_{P-F}$ = 65.0 Hz); **HRMS** (EI) calcd (*m/z*) for C$_{12}$H$_{15}$F$_6$P: 246.0028 ([M]$^+$); found: 246.0028 ([M]$^+$).

(S_P,R_C,S_{Fc})-1-[(2-Naphthyl)trifluoromethylphosphino]-2-[1-(dimethylamino)ethyl]-ferrocene ((S_P)-144) and (R_P,R_C,S_{Fc})-1-[(2-Naphthyl)trifluoromethylphosphino]-2-[1-(dimethylamino)ethyl]ferrocene ((R_P)-144). At -78 °C 1.3 M *s*-BuLi (1.1 mL, 1.486 mmol, 1.1 eq.) was added dropwise to a solution of (*R*)-[1-(dimethylamino)ethyl]ferrocene (347 mg, 1.351 mmol, 1 eq.) in Et$_2$O (6 mL). The orange solution was slowly warmed to room temperature and stirred at ambient temperature for 1 h. The red mixture obtained was cooled to 0 °C and added dropwise to a solution of 2-naphthyltrifluoromethylphosphine (143) (200 mg, 0.675 mmol, 0.5 eq.) in Et$_2$O (3 mL). The resulting dark brown mixture was allowed to warm up to room temperature overnight. The mixture was then diluted with diethyl ether and water, the organic layer separated and the aqueous phase extracted with diethyl ether. The combined organic phases were dried over MgSO$_4$ and the solvent evaporated in vacuo. The crude dark oil was purified by FC (silica, pentane:EtOAc:Et$_3$N 100:2:3) to yield the two isomers in the first two coloured fractions.

1) (S_P)-Isomer: Yield: 72 mg (22%), brown crystalline solid. ^{1}H NMR (300 MHz, CDCl$_3$): δ 1.13 (d, ^3J = 6.7 Hz, 3 H, CH*Me*); 1.46 (s, 6 H, N*Me*$_2$); 4.19 (q, ^3J = 6.9 Hz, 1 H, C*H*Me); 4.25 (s, 5 H, *Cp*'); 4.42 (s, 1 H, *Cp*); 4.46 (t, *J* = 2.5 Hz, 1 H, *Cp*); 4.71 (s, 1 H, *Cp*); 7.44-7.50 (m, 3 H, *Np*); 7.72 (d, *J* = 8.4 Hz, 1 H, *Np*); 7.77-7-7.84 (m, 2 H, *Np*); 7.95 (d, *J* = 10.5 Hz, 1 H, *Np*); ^{13}C{^{1}H} NMR (62.9 MHz, CDCl$_3$): δ 7.31 (s, CH*Me*); 37.81 (s, N*Me*$_2$); 57.81 (d, $^3J_{C-P}$ = 4.4 Hz, *C*HMe); 68.59 (s, *Cp*); 69.29 (d, J_{C-P} = 4.7 Hz, *Cp*); 69.40-69.60 (m, *Cp*); 70.26 (s, *Cp*'); 71.53-71.74 (m, *Cp*); 97.63 (d, $^2J_{C-P}$ = 22.1 Hz, *Cp*); 125.96 (s, *Np*); 126.55 (s, *Np*); 126.86 (d, J_{C-P} = 7.3 Hz, *Np*); 127.60 (s, *Np*); 128.21 (s, *Np*); 129.39 (d, J_{C-P} = 16.6 Hz, *Np*); 130.69-130.91 (m, *Np*); 132.23 (qd, $^1J_{C-F}$ = 323.6 Hz, $^1J_{C-P}$ = 34.4 Hz, *C*F$_3$); 132.69 (d, J_{C-P} = 10.7 Hz, *Np*); 133.58 (d, J_{C-P} = 1 Hz, *Np*); 133.63 (d, J_{C-P} = 26.3 Hz, *Np*); ^{19}F NMR (282 MHz, CDCl$_3$): δ -54.22 (d, $^2J_{F-P}$ = 61.4 Hz, *C*F$_3$); ^{31}P{^{1}H} NMR (121 MHz, CDCl$_3$): δ -16.10 (q, $^2J_{P-F}$ = 61.4 Hz, *P*NpCF$_3$); EA calcd for C$_{25}$H$_{25}$F$_3$FeNP (483.30): C 62.13, H 5.21, F 11.79, N 2.90, P 6.41; found: C 62.03, H 5.33, F 11.98, N 2.87, P 6.38; HRMS (MALDI) calcd (*m/z*) for C$_{25}$H$_{26}$F$_3$FeNP: 484.1099 ([MH]$^+$); found: 484.1095 ([MH]$^+$); 439.0520 ([M-NMe$_2$]$^+$).

2) (R_P)-Isomer: Yield: 126 mg (39%), brown crystalline solid. ^{1}H NMR (300 MHz, CDCl$_3$): δ 1.33 (d, ^3J = 6.7 Hz, 3 H, CH*Me*); 2.15 (s, 6 H, N*Me*$_2$); 3.74 (s, 5 H, *Cp*'); 4.07-4.15 (m, 1 H, C*H*Me); 4.41 (s, 1 H, *Cp*); 4.43 (s, 1 H, *Cp*); 4.55 (s, 1 H, *Cp*); 7.56-7.63 (m, 2 H, *Np*); 7.91-8.01 (m, 3 H, *Np*); 8.02-8.09 (m, 1 H, *Np*); 8.54 (d, *J* = 13.0 Hz, 1 H, *Np*); ^{13}C{^{1}H} NMR (62.9 MHz,

CDCl$_3$): δ 10.05 (s, CH*Me*); 39.78 (s, N*Me*$_2$); 57.03 (d, $^3J_{C-P}$ = 7.5 Hz, *CH*Me); 68.00-68.20 (m, *Cp*); 69.97 (d, J_{C-P} = 4.7 Hz, *Cp*); 70.04 (s, *Cp'*); 70.23 (s, *Cp*); 73.11 (d, J_{C-P} = 5.2 Hz, *Cp*); 99.67 (d, $^2J_{C-P}$ = 28.2 Hz, *Cp*); 127.00 (s, *Np*); 127.95 (s, *Np*); 128.10 (s, *Np*); 128.15 (s, *Np*); 128.91 (s, *Np*); 131.05 (d, J_{C-P} = 11.6 Hz, *Np*); 129.95-130.08 (m, *Np*); 132.23 (qd, $^1J_{C-F}$ = 324.3 Hz, $^1J_{C-P}$ = 33.5 Hz, *CF*$_3$); 133.23 (d, J_{C-P} = 14.5 Hz, *Np*); 134.50 (s, *Np*); 137.87 (d, J_{C-P} = 40.7 Hz, *Np*); **^{19}F NMR** (282 MHz, CDCl$_3$): δ -55.29 (d, $^2J_{F-P}$ = 60.1 Hz, C*F*$_3$); **^{31}P{^{1}H} NMR** (121 MHz, CDCl$_3$): δ -12.66 (q, $^2J_{P-F}$ = 61.0 Hz, *P*NpCF$_3$); **EA** calcd for C$_{25}$H$_{25}$F$_3$FeNP (483.30): C 62.37, H 5.50, F 11.72, N 2.99, P 6.13; found: C 62.03, H 5.33, F 11.98, N 2.87, P 6.38; **HRMS** (MALDI) calcd (*m/z*) for C$_{25}$H$_{26}$F$_3$FeNP: 484.1099 ([MH]$^+$); found: 484.1098 ([MH]$^+$); 439.0521 ([M-NMe$_2$]$^+$).

(S_P,R_C,S_{Fc})-1-(Phenyltrifluoromethylphosphino)-2-[1-(dimethylamino)ethyl]ferrocene ((S_P)-147) and (R_P,R_C,S_{Fc})-1-(Phenyltrifluoromethylphosphino)-2-[1-(dimethylamino)-ethyl]ferrocene ((R_P)-147). At -78 °C 1.3 M *s*-BuLi (3.9 mL, 5.052 mmol, 1.1 eq.) was added dropwise to a solution of (*R*)-[1-(dimethylamino)ethyl]ferrocene (1180 mg, 4.592 mmol, 1 eq.) in Et$_2$O (10 mL). The orange solution was slowly warmed to room temperature and stirred at ambient temperature for 1 h. The red mixture obtained was cooled to 0 °C and added to a solution of phenyltrifluoromethylphosphine (680 mg, 2.296 mmol, 0.5 eq.) in Et$_2$O (3 mL) over a time period of 30 min. The resulting dark brown mixture was allowed to warm up to room temperature overnight. The mixture was then diluted with diethyl ether and water, the organic layer separated and the aqueous phase extracted with diethyl ether. The combined organic phases were dried over MgSO$_4$ and the solvent evaporated in vacuo. The crude dark oil was purified by FC (silica, pentane:EtOAc:Et$_3$N 100:2:3) to yield the two isomers in the first two coloured fractions.

1) (S_P)-Isomer: Yield: 358 mg (36%), brown crystalline solid. **^{1}H NMR** (300 MHz, CDCl$_3$): δ 1.15 (d, ^3J = 6.7 Hz, 3 H, CH*Me*); 1.52 (s, 6 H, N*Me*$_2$); 4.17 (q, ^3J = 6.7 Hz, 1 H, C*H*Me); 4.24 (s, 5 H, *Cp'*); 4.42 (bs, 2 H, *Cp*); 4.65 (s, 1 H, *Cp*); 7.24-7.34 (m, 3 H, *Ph*); 7.44 (t, *J* = 7.8 Hz, 2 H, *Ph*); **^{13}C{^{1}H} NMR** (75 MHz, CDCl$_3$): δ 7.32 (s, CH*Me*); 37.70 (s, N*Me*$_2$); 57.81 (d, $^3J_{C-P}$ = 4.4 Hz, *CH*Me); 68.50 (s, *Cp*); 69.03 (d, J_{C-P} = 4.7 Hz, *Cp*); 69.60 (dq, J_{C-P} = 6.2 Hz, J_{C-F} = 3.1 Hz, *Cp*); 70.24 (s, *Cp'*); 71.54 (dq, J_{C-P} = 7.2 Hz, J_{C-F} = 1.9 Hz, *Cp*); 97.51 (d, $^2J_{C-P}$ = 22.1 Hz, *Cp*); 127.53 (d, J_{C-P} = 8.6 Hz, *Ph*); 129.05 (d, J_{C-P} = 1.2 Hz, *Ph*); 134.97 (s, *Ph*); 133.18-133.38 (m, *Ph*); 133.25 (s, *Ph*); 132.22 (qd, $^1J_{C-F}$ = 323.6 Hz, $^1J_{C-P}$ = 34.5 Hz, *CF*$_3$); **^{19}F NMR**

(282 MHz, CDCl$_3$): δ -54.48 (d, $^2J_{F-P}$ = 61.4 Hz, CF$_3$); ^{31}P{^{1}H} NMR (121 MHz, CDCl$_3$): δ -16.74 (q, $^2J_{P-F}$ = 61.4 Hz, PPhCF$_3$); **EA** calcd for C$_{21}$H$_{23}$F$_3$FeNP (433.24): C 58.22, H 5.35, F 13.16, N 3.23, P 7.15; found: C 58.19, H 5.53, F 13.28, N 3.21, P 7.07; **HRMS** (MALDI) calcd (m/z) for C$_{21}$H$_{23}$F$_3$FeNP: 433.0865 ([MH]$^+$); found: 433.0864 ([MH]$^+$); 388.0285 ([M-NMe$_2$]$^+$); 319.0332 ([M-NMe$_2$-CF$_3$]$^+$).

2) (*R*$_P$)-**Isomer:** Yield: 452 mg (45%), brown crystalline solid. ^{1}**H NMR** (300 MHz, CDCl$_3$): δ 1.31 (d, ^3J = 6.7 Hz, 3 H, CH*Me*); 2.13 (s, 6 H, N*Me*$_2$); 3.68 (s, 5 H, *Cp*'); 4.17 (qd, ^3J = 6.7 Hz, J = 2.5 Hz, 1 H, C*H*Me); 4.38-4.43 (m, 2 H, *Cp*); 4.45 (s, 1 H, *Cp*); 7.48-7.56 (m, 3 H, *Ph*); 7.96-8.04 (m, 2 H, *Ph*); ^{13}**C{^{1}H} NMR** (75 MHz, CDCl$_3$): δ 9.64 (s, CH*Me*); 39.38 (s, N*Me*$_2$); 56.57 (d, $^3J_{C-P}$ = 7.5 Hz, *C*HMe); 67.00-67.82 (m, *Cp*); 69.58 (s, *Cp*); 69.64 (s, *Cp*'); 69.60 (dq, J_{C-P} = 6.2 Hz, J_{C-F} = 3.1 Hz, *Cp*); 69.87 (d, J_{C-P} = 1.1 Hz, *Cp*); 72.61 (d, J_{C-P} = 5.3 Hz, *Cp*); 99.30 (d, $^2J_{C-P}$ = 28.3 Hz, *Cp*); 128.31 (d, J_{C-P} = 9.7Hz, *Ph*); 130.65 (d, J_{C-P} = 1.1 Hz, *Ph*); 132.04-132.25 (m, *Ph*); 135.46 (d, J_{C-P} = 1.0 Hz, *Ph*); 135.77 (d, J_{C-P} = 1.0 Hz, *Ph*); 130.28 (qd, $^1J_{C-F}$ = 324.1 Hz, $^1J_{C-P}$ = 33.5 Hz, CF$_3$); ^{19}**F NMR** (282 MHz, CDCl$_3$): δ -55.17 (d, $^2J_{F-P}$ = 61.0 Hz, CF$_3$); ^{31}**P{^{1}H} NMR** (121 MHz, CDCl$_3$): δ -13.53 (q, $^2J_{P-F}$ = 61.4 Hz, PPhCF$_3$); **EA** calcd for C$_{21}$H$_{23}$F$_3$FeNP (433.24): C 58.22, H 5.35, F 13.16, N 3.23, P 7.15; found: C 58.05, H 5.52, F 13.34, N 3.23, P 7.12; **HRMS** (MALDI) calcd (m/z) for C$_{21}$H$_{23}$F$_3$FeNP: 433.0865 ([MH]$^+$); found: 433.0875 ([MH]$^+$); 48.0633 ([M-Me]$^+$); 389.0363 ([M-NMe$_2$]$^+$); 364.0912([M-CF$_3$]$^+$); 319.0337 ([M-NMe$_2$-CF$_3$]$^+$).

(*S*$_P$,*R*$_C$,*S*$_{Fc}$)-1-[(1-Adamantyl)trifluoromethylphosphino]-2-[(1-dimethylamino)ethyl]ferrocene (150). At -78 °C 1.3 M *s*-BuLi (6.6 mL, 8.678 mmol, 1.1 eq.) was added dropwise to a solution of (*R*)-[1-(dimethylamino)ethyl]ferrocene (2029 mg, 7.889 mmol, 1 eq.) in Et$_2$O (15 mL). The orange solution was slowly warmed to room temperature and stirred at ambient temperature for 1 h. The red mixture thus obtained was cooled to -78 °C and was then added over a time period of 30 min to a solution of 1-adamantyltrifluoromethylphosphine (**148**) (680 mg, 2.296 mmol, 0.5 eq.) in Et$_2$O (3 mL) at -78 °C. The resulting dark brown mixture was allowed to warm up to room temperature over night. The mixture was then diluted with diethyl ether and water, the organic layer separated and the aqueous phase extracted with diethyl ether. The combined organic phases were dried over MgSO$_4$ and the solvent evaporated in vacuo. The crude dark oil was purified by FC (silica, pentane:EtOAc:Et$_3$N 100:2:3) to yield the product in the first coloured fraction. The (S)-configuration was confirmed by X-ray crystallography. Yield: 671 mg (35%), brown crystalline solid. **^{1}H NMR** (300 MHz, C$_6$D$_6$): δ 1.08 (d, ^3J = 6.7 Hz, 3 H, CH*Me*); 1.58 (bs, 6 H, *Ad*); 1.69-1.85 (m, 6 H, *Ad*); 2.01 (s, 6 H, N*Me*$_2$); 2.04-2.08 (m, 3 H, *Ad*); 3.89 (qd, ^3J = 6.6 Hz, J = 2.5 Hz, 1 H, C*H*Me); 4.09 (s, 6 H, *Cp, Cp'*); 4.13 (t, J = 2.4 Hz, 1 H, *Cp*); 4.34 (s, 1 H, *Cp*); **^{13}C{^{1}H} NMR** (62.9 MHz, C$_6$D$_6$): $\delta\delta$ 8.18 (s, CH*Me*); 29.26 (d, $J_{\text{C-P}}$ = 8.9 Hz, *Ad*); 36.15 (dq, $^1J_{\text{C-P}}$ = 17.0 Hz, $^3J_{\text{C-F}}$ = 2.0 Hz, *Ad*); 37.03 (d, $J_{\text{C-P}}$ = 1.0 Hz, *Ad*); 39.19 (s, N*Me*$_2$); 39.99 (dq, $^2J_{\text{C-P}}$ = 11.0 Hz, $^4J_{\text{C-F}}$ = 1.3 Hz, *Ad*); 56.47 (d, $^3J_{\text{C-P}}$ = 8.5 Hz, C*H*Me); 69.10 (dd, $J_{\text{C-P}}$ = 21.7, J = 3.4 Hz, *Cp*); 69.13 (d, $J_{\text{C-P}}$ = 5.5 Hz, *Cp*); 69.61 (s, *Cp*); 70.51 (s, *Cp'*); 72.42 (d, $J_{\text{C-P}}$ = 7.0 Hz, *Cp*); 99.01 (d, $^2J_{\text{C-P}}$ = 28.9 Hz, *Cp*); 134.05 (qd, $^1J_{\text{C-F}}$ = 323.4 Hz, $^1J_{\text{C-P}}$ = 44.8 Hz, *CF*$_3$); **^{19}F NMR** (282 MHz, C$_6$D$_6$): δ -45.60 (d, $^2J_{\text{F-P}}$ = 61.1 Hz, *CF*$_3$); **^{31}P{^{1}H} NMR** (121 MHz, C$_6$D$_6$): δ 8.49 (q, $^2J_{\text{P-F}}$ = 61.1 Hz, *P*AdCF$_3$); **EA** calcd for C$_{25}$H$_{33}$F$_3$FeNP (491.36): C 61.11, H 6.77, F 11.60, N 2.85, P 6.30; found: C 61.15, H 6.83, F 11.54, N 2.84, P 6.38; **HRMS** (EI) calcd (*m/z*) for C$_{25}$H$_{33}$F$_3$FeNP: 491.1647 ([M]$^+$); found: 491.1641 ([M]$^+$); 446.1070 ([M-NMe$_2$]$^+$); 422.1696 ([M- CF$_3$]$^+$).

(S_P,R_C,S_{Fc})-1-[(2-Naphthyl)trifluoromethylphosphino]-2-[1-(dicyclohexylphosphino)ethyl]ferrocene (151). (S_P,R_C,S_{Fc})-1-[(2-Naphthyl)trifluoromethylphosphino]-2-[1-(dimethylamino)ethyl]ferrocene ((S_P)-144) (50 mg, 103 µmol, 1 eq.) and dicyclohexylphosphine (25 µL, 124 µmol, 1.2 eq.) were dissolved in 1 mL degassed acetic acid, heated up to 90 °C and stirred at this temperature for 16 h. The solvent was then removed under reduced pressure and the residue purified by FC (silica, pentane:EtOAc 100:1) to yield the oxidation sensitive product as a single isomer. Yield: 54 mg (82%), orange crystalline solid. **^{1}H NMR** (300 MHz, CD$_2$Cl$_2$): 0.59-1.82 (m, 22 H, PC*y*$_2$); δ 1.58 (dd, J = 7.1, 4.2 Hz, 3 H, CH*Me*); 3.26 (q, ^3J = 6.7 Hz, 1 H, C*H*Me); 4.32 (s, 5 H, *Cp'*); 4.56 (bs, 1 H, *Cp*); 4.57 (t, J = 2.4 Hz, 1 H, *Cp*); 4.70 (bs, 1 H, *Cp*); 7.46-7.63 (m, 3 H, *Np*); 7.75-7.87 (m, 3 H, *Np*); 7.98 (d, J = 9.2 Hz, 1 H, *Np*); **^{19}F NMR** (282 MHz, CD$_2$Cl$_2$): δ -53.00 (dd, $^2J_{\text{F-P}}$ = 65.2 Hz, $J_{\text{F-P}}$ = 4.0 Hz, *CF*$_3$); **^{31}P{^{1}H} NMR** (121 MHz, CD$_2$Cl$_2$): δ -16.48 (dq, $J_{\text{P-P'}}$ = 76.8 Hz, $^2J_{\text{P-F}}$ = 65.5 Hz, *P*NpCF$_3$); 16.55 (d, $J_{\text{P-P'}}$ = 76.3 Hz, *P*Cy$_2$); **HRMS** (MALDI) calcd (*m/z*) for C$_{35}$H$_{42}$F$_3$FeP$_2$: 637.2058 ([MH]$^+$); found: 637.2051 ([MH]$^+$); 439.0520 ([M-PCy$_2$]$^+$).

(S_P,R_C,S_{Fc})-1-(Phenyltrifluoromethylphosphino)-2-[1-(dicyclohexyl-phosphino)ethyl]ferrocene ((S_P)-142). (S_P,R_C,S_{Fc})-1-(Phenyltrifluoro-methylphosphino)-2-(1-dimethylaminoethyl)ferrocene ((S_P)-147) (378 mg,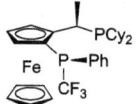
873 µmol, 1 eq.) and dicyclohexylphosphine (211 µL, 1047 µmol, 1.2 eq.) were dissolved in 1 mL degassed acetic acid, heated up to 90 °C and stirred at this temperature for 13 h. The solvent was then removed under reduced pressure and the residue diluted with diethyl ether and saturated aqueous NaHCO$_3$ solution. The organic layer was separated and the aqueous phase extracted with diethyl ether. The combined organic phases were then dried over MgSO$_4$ and the solvent evaporated in vacuo. The residue was purified by FC (silica, pentane:EtOAc 100:1) to yield the oxidation sensitive product as a single isomer. Yield: 389 mg (76%), orange crystalline solid. ^{1}H NMR (300 MHz, CD$_2$Cl$_2$): δ 0.61-1.78 (m, 22 H, PCy_2); 1.51 (dd, J = 7.1, 4.1 Hz, 3 H, CHMe); 3.17 (q, ^3J = 6.9 Hz, 1 H, CHMe); 4.24 (s, 5 H, Cp'); 4.41-4.48 (m, 2 H, Cp); 4.56 (bs, 1 H, Cp); 7.17-7.34 (m, 3 H, Ph); 7.44 (t, J = 7.3 Hz, 1 H, Ph); ^{19}F NMR (282 MHz, CD$_2$Cl$_2$): δ -53.36 (dd, $^2J_{F-P}$ = 65.3 Hz, $J_{F-P'}$ = 4.3 Hz, CF_3); ^{31}P{^{1}H} NMR (121 MHz, CD$_2$Cl$_2$): δ -17.37 (dq, $J_{P-P'}$ = 78.4 Hz, $^2J_{P-F}$ = 65.5 Hz, PPhCF$_3$); 16.53 (d, $J_{P-P'}$ = 78.2 Hz, PCy$_2$); **EA** calcd for C$_{31}$H$_{39}$F$_3$FeP$_2$ (586.44): C 63.49, H 6.70, F 9.72, P 10.56; found: C 63.03, H 6.68, F 9.87, P 10.49; **HRMS** (MALDI) calcd (m/z) for C$_{31}$H$_{40}$F$_3$FeP$_2$: 587.1901 ([MH]$^+$); found: 587.1903 ([MH]$^+$); 389.0354 ([M-PCy$_2$]$^+$).

(R_P,R_C,S_{Fc})-1-(Phenyltrifluoromethylphosphino)-2-[1-(dicyclohexyl-phosphino)ethyl]ferrocene ((R_P)-142). (R_P,R_C,S_{Fc})-1-(Phenyltrifluoro-methylphosphino)-2-(1-dimethylaminoethyl)ferrocene ((R_P)-147) (304 mg,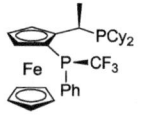
702 µmol, 1 eq.) and dicyclohexylphosphine (170 µL, 842 µmol, 1.2 eq.) were dissolved in 1 mL degassed acetic acid, heated up to 90 °C and stirred at this temperature for 13 h. The solvent was then removed under reduced pressure and the residue dissolved in 3 mL hot ethanol. Cooling to -18 °C gave the oxidation sensitive product as orange crystals. Yield: 340 mg (83%), orange crystalline solid. ^{1}H NMR (300 MHz, CD$_2$Cl$_2$): δ 1.06-1.83 (m, 22 H, PCy_2); 1.55 (dd, J = 6.9, 4.9 Hz, 3 H, CHMe); 3.11-3.21 (m, 1 H, CHMe); 3.71 (s, 5 H, Cp'); 4.39 (bs, 1 H, Cp); 4.44 (bs, 1 H, Cp); 4.57 (bs, 1 H, Cp); 7.46-7.56 (m, 3 H, Ph); 7.99 (t, J = 8.3 Hz, 1 H, Ph); ^{19}F NMR (282 MHz, CD$_2$Cl$_2$): δ -55.28 (dd, $^2J_{F-P}$ = 62.9 Hz, $J_{F-P'}$ = 23.4 Hz, CF_3); ^{31}P{^{1}H} NMR (121 MHz, CD$_2$Cl$_2$): δ -16.67 (qd, $^2J_{P-F}$ = 63.3 Hz, $J_{P-P'}$ = 41.6 Hz, PPhCF$_3$); 18.92 (dq, $J_{P-P'}$ = 40.8 Hz, J_{P-F} = 23.2 Hz, PCy$_2$); **EA** calcd for C$_{31}$H$_{39}$F$_3$FeP$_2$ (586.44): C 63.49, H 6.70, F 9.72, P 10.56; found: C 63.22, H 6.67, F 9.83,

P 10.65; **HRMS** (MALDI) calcd (m/z) for $C_{31}H_{40}F_3FeP_2$: 587.1901 ([MH]$^+$); found: 587.1907 ([MH]$^+$); 389.0360 ([M-PCy$_2$]$^+$).

(S_P,R_C,S_{Fc})-1-[(1-Adamantyl)trifluoromethylphosphino]-2-[1-(dicyclohexylphosphino)-ethyl]ferrocene ((S_P)-152) and (R_P,R_C,S_{Fc})-1-[(1-Adamantyl)trifluoromethylphosphino]-2-[1-(dicyclohexylphosphino)ethyl]ferrocene ((R_P)-152). (S_P,R_C,S_{Fc})-1-[(1-Adamantyl)-trifluoromethylphosphino]-2-(1-dimethylaminoethyl)ferrocene ((S_P)-**150**) (200 mg, 407 μmol, 1 eq.) and dicyclohexylphosphine (100 μL, 488 μmol, 1.2 eq.) were dissolved in 1 mL degassed acetic acid, heated up to 90 °C and stirred at this temperature for 13 h. The solvent was then removed under reduced pressure and the residue purified by FC (silica, pentane:EtOAc: 100:1) to yield the two isomers in the first two coloured fractions.

1) (S_P)-Isomer: Yield: 113 mg (43%), orange crystalline solid. ^{1}**H NMR** (300 MHz, C$_6$D$_6$): δ 1.19-2.04 (m, 34 H, PCy$_2$, Ad); 1.66 (dd, J = 7.2, 4.0 Hz, 3 H, CHMe); 2.19-2.51 (m, 3 H, PCy$_2$); 3.29 (qd, ^{3}J = 7.0 Hz, J = 2.5 Hz, 1 H, CHMe); 4.28 (bs, 1 H, Cp); 4.33 (s, 5 H, Cp'); 4.38 (bs, 1 H, Cp); 4.47 (bs, 1 H, Cp); ^{19}**F NMR** (282 MHz, C$_6$D$_6$): δ -46.28 (d, $^2J_{F-P}$ = 63.9 Hz, CF$_3$); ^{31}**P{^{1}H} NMR** (121 MHz, C$_6$D$_6$): δ 5.46 (qd, $^2J_{P-F}$ = 64.0 Hz, $J_{P-P'}$ = 32.8 Hz, PAdCF$_3$); +12.60 (d, $J_{P-P'}$ = 32.8 Hz, PCy$_2$); **EA** $C_{35}H_{49}F_3FeP_2$ (491.36) calcd C 65.22, H 7.66, F 8.84, P 9.61 found C 65.24, H 7.71, F 8.91, P 9.57; **HRMS** (MALDI) calcd (m/z) for $C_{35}H_{50}F_3FeP_2$: 645.2684 ([MH]$^+$); found: 645.2692 ([MH]$^+$); 447.1150 ([M-PCy$_2$]$^+$).

2) (R_P)-Isomer: Yield: 38 mg (15%), orange crystalline solid. ^{1}**H NMR** (300 MHz, C$_6$D$_6$): δ 1.22-2.05 (m, 31 H, PCy$_2$, Ad); 1.70 (dd, J = 7.0, 5.7 Hz, 3 H, CHMe); 2.41 (bs, 6 H, Ad); 3.26 (qd, ^{3}J = 7.3 Hz, J = 3.0 Hz, 1 H, CHMe); 4.23 (s, 5 H, Cp, Cp'); 4.37 (bs, 1 H, Cp); 4.57 (bs, 1 H, Cp); ^{19}**F NMR** (282 MHz, C$_6$D$_6$): δ -47.59 (dd, $^2J_{F-P}$ = 54.8 Hz, $J_{F-P'}$ = 15.5 Hz, CF$_3$); ^{31}**P{^{1}H} NMR** (121 MHz, C$_6$D$_6$): δ 10.27 (qd, $^2J_{P-F}$ = 55.0 Hz, $J_{P-P'}$ = 40.8 Hz, PAdCF$_3$); 13.71 (dq, $J_{P-P'}$ = 40.2 Hz, J_{P-F} = 15.7 Hz, PCy$_2$); **HRMS** (MALDI) calcd (m/z) for $C_{35}H_{50}F_3FeP_2$: 645.2684 ([MH]$^+$); found: 645.2689 ([MH]$^+$); 661.2620 ([MOH]$^+$).

(S_P,R_C,S_{Fc})-1-(Phenyltrifluoromethylphosphino)-2-[1-(3,5-dimethyl-1H-pyrazole-1-yl)ethyl]ferrocene ((S_P)-153). (S_P,R_C,S_{Fc})-1-(Phenyl-trifluoromethylphosphino)-2-(1-dimethylaminoethyl)ferrocene ((S_P)-147) (200 mg, 464 μmol, 1 eq.) and 3,5-dimethylpyrazole (67 mg, 696 μmol,

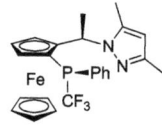

1.5 eq.) were dissolved in 1 mL degassed acetic acid and trifluoroacetic acid (36 μL, 464 μmol, 1 eq.) was added. The mixture was heated up to 90 °C and stirred at this temperature for 20 h. The solvent was then removed under reduced pressure and the residue diluted with diethyl ether and saturated aqueous NaHCO$_3$ solution. The organic layer was separated and the aqueous phase extracted with diethyl ether. The combined organic phases were dried over MgSO$_4$ and the solvent evaporated in vacuo. The residue was purified by FC (silica, pentane:EtOAc:Et$_3$N 100:2:3-100:5:3) to yield the product as a single isomer. Yield: 155 mg (70%), orange powder. ^{1}H NMR (300 MHz, CDCl$_3$): δ 1.75 (d, ^3J = 6.9 Hz, 3 H, CHMe); 1.78 (s, 3 H, MePz); 2.02 (s, 3 H, MePz); 4.32 (s, 5 H, Cp'); 4.57-4.61 (m, 2 H, Cp); 4.83-4.85 (m, 1H, Cp); 4.96 (s, 1 H, Pz); 5.57 (dq, ^3J = 6.9 Hz, J = 2.2 Hz, 1 H, CHMe); 7.01 (t, J = 7.6 Hz, 1 H, Ph); 7.09 (td, J = 7.6, 1.5 Hz, 2 H, Ph); 7.21 (t, J = 7.3 Hz, 1 H, Ph); ^{13}C{^{1}H} NMR (75 MHz, CDCl$_3$): δ 11.30 (d, J_{C-P} = 5.7 Hz, CHMe); 13.33 (s, MePz); 19.87 (s, MePz); 53.10 (d, $^3J_{C-P}$ = 8.5 Hz, CHMe); 67.38-67.56 (m, Cp); 70.10 (s, Cp); 70.44 (s, Cp'); 70.84 (d, J_{C-P} = 5.1 Hz, Cp); 71.60-71.91 (m, Cp); 93.86 (d, $^2J_{C-P}$ = 26.0 Hz, Cp); 104.84 (s, Pz); 127.42 (d, J_{C-P} = 7.7 Hz, Ph); 129.13-129.40 (m, Ph); 129.43 (s, Ph); 131.12 (qd, $^1J_{C-F}$ = 320.5 Hz, $^1J_{C-P}$ = 25.8 Hz, CF_3); 133.16 (d, J_{C-P} = 20.2 Hz, Ph); 136.71 (s, MePz); 146.78 (s, MePz); ^{19}F NMR (282 MHz, CDCl$_3$): δ -54.17 (d, $^2J_{F-P}$ = 69.3 Hz, CF_3); ^{31}P{^{1}H} NMR (121 MHz, CD$_2$Cl$_2$): δ -11.31 (q, $^2J_{P-F}$ = 69.3 Hz, PPhCF$_3$); **EA** calcd for C$_{24}$H$_{24}$F$_3$FeN$_2$P (484.2): C 59.52, H 4.99, F 11.77, N 5.78, P 6.40; found: C 59.69, H 5.10, F 11.88, N 5.76, P 6.43; **HRMS** (MALDI) calcd (m/z) for C$_{24}$H$_{25}$F$_3$FeN$_2$P: 485.1051 ([MH]$^+$); found: 485.1056 ([MH]$^+$); 501.1004([MOH]$^+$); 389.0365 ([M-Pz]$^+$).

(R_P,R_C,S_{Fc})-1-(Phenyltrifluoromethylphosphino)-2-[1-(3,5-dimethyl-1H-pyrazole-1-yl)ethyl]ferrocene ((R_P)-153). (R_P,R_C,S_{Fc})-1-(Phenyl-trifluoromethylphosphino)-2-(1-dimethylaminoethyl)ferrocene

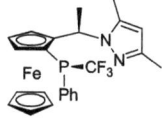

((R_P)-147) (200 mg, 464 μmol, 1 eq.) and 3,5-dimethylpyrazole (67 mg, 696 μmol, 1.5 eq.) were dissolved in 1 mL degassed acetic acid and trifluoroacetic acid (36 μL, 464 μmol, 1 eq.) was added. The mixture was heated up to 90 °C and stirred at this temperature for 20 h. The solvent was then removed under reduced pressure and the residue

6 Experimental Part

diluted with diethyl ether and saturated aqueous $NaHCO_3$ solution. The organic layer was separated and the aqueous phase extracted with diethyl ether. The combined organic phases were dried over $MgSO_4$ and the solvent evaporated in vacuo. The residue was purified by FC (silica, pentane:EtOAc:Et$_3$N 100:2:3-100:5:3) to yield the product as a single isomer. Yield: 82 mg (37%), orange powder. ^{1}H NMR (300 MHz, CDCl$_3$): δ 1.89 (d, ^3J = 6.9 Hz, 3 H, CH*Me*); 2.17 (s, 3 H, *Me*Pz); 2.29 (s, 3 H, *Me*Pz); 3.87 (s, 5 H, *Cp'*); 4.29 (bs, 1 H, *Cp*); 4.53 (t, J = 2.5 Hz, 1H, *Cp*); 4.74 (bs, 1 H, *Cp*); 5.61 (dq, ^3J = 6.8 Hz, J = 2.7 Hz, 1 H, C*H*Me); 5.71 (s, 1 H, *Pz*); 7.46-7.65 (m, 3 H, *Ph*); 8.00 (t, J = 7.9 Hz, 2 H, *Ph*); ^{13}C{^{1}H} NMR (75 MHz, CDCl$_3$): δ 11.51 (d, J_{C-P} = 5.8 Hz, CH*Me*); 13.72 (s, *Me*Pz); 21.32 (s, *Me*Pz); 53.04 (d, $^3J_{C-P}$ = 9.8 Hz, *C*HMe); 65.03-65.79 (m, *Cp*); 69.82 (s, *Cp'*); 71.18 (d, J_{C-P} = 1.1 Hz, *Cp*); 71.42 (d, J_{C-P} = 6.0 Hz, *Cp*); 72.54 (d, J_{C-P} = 4.7 Hz, *Cp*); 71.60-71.91 (m, *Cp*); 96.68 (d, $^2J_{C-P}$ = 31.8 Hz, *Cp*); 104.84 (s, *Pz*); 128.58 (d, J_{C-P} = 10.1 Hz, *Ph*); 129.60 (qd, $^1J_{C-F}$ = 322.6 Hz, $^1J_{C-P}$ = 24.4 Hz, *CF$_3$*); 129.93-130.20 (m, *Ph*); 131.20 (d, J_{C-P} = 1.5 Hz, *Ph*); 135.58 (dq, J_{C-P} = 25.3 Hz, J_{C-F} = 1.3 Hz, *Ph*); 137.64 (s, Me*Pz*); 147.10 (s, Me*Pz*); ^{19}F NMR (282 MHz, CDCl$_3$): δ -57.41 (d, $^2J_{F-P}$ = 71.3 Hz, *CF$_3$*); ^{31}P{^{1}H} NMR (121 MHz, CD$_2$Cl$_2$): δ -17.17 (q, $^2J_{P-F}$ = 71.3 Hz, *P*PhCF$_3$); HRMS (MALDI) calcd (m/z) for C$_{24}$H$_{25}$F$_3$FeN$_2$P: 485.1051 ([MH]$^+$); found: 485.1045 ([MH]$^+$); 501.0997 ([MOH]$^+$); 389.0358 ([M-Pz]$^+$).

(*S*$_P$,*R*$_C$,*S*$_{Fc}$)-1-(Phenyltrifluoromethylphosphino)-2-{1-[(1-adamantyl)-phosphino]ethyl}ferrocene ((*S*$_P$)-167). To a solution of (*S*$_P$,*R*$_C$,*S*$_{Fc}$)-1-(phenyltrifluoromethylphosphino)-2-[1-(dimethylamino)ethyl]ferrocene ((*R*$_P$)-147) (200 mg, 462 µmol, 1 eq.) in 2 mL of degassed acetic acid was added trifluoroacetic acid (34 µL, 462 µmol, 1 eq.). Then 1-adamantylphosphine (155 mg, 923 µmol, 2 eq.) was added, the mixture heated up to 90 °C and the red solution stirred at this temperature for 16 h. The mixture was then cooled to 60 °C and the acetic acid removed under reduced pressure. The residue was diluted with diethyl ether and saturated aqueous NaHCO$_3$, the organic phase separated and dried over MgSO$_4$. After removal of the solvent under reduced pressure, the crude product was purified by FC (silica, pentane:Et$_2$O 1:0-50:1) to yield the product as a mixture of two diastereoisomers (12:1, P2). Yield: 231 mg (90%), orange powder. Major isomer: ^{1}H NMR (300 MHz, C$_6$D$_6$): δ 1.42-1.70 (m, 16 H, *Ad*, CH*Me*); 1.93 (bs, 3 H, *Ad*); 2.28 (d, J_{H-P} = 209 Hz, 1 H, P*H*); 3.38-3.51 (m, 1 H, C*H*Me); 4.18 (s, 6 H, *Cp*, *Cp'*); 4.22 (s, 1 H, *Cp*); 4.54 (bs, 1 H, *Cp*); 6.98 (bs, 3 H, *Ph*); 7.53 (bs, 2 H, *Ph*); ^{19}F NMR (282 MHz, C$_6$D$_6$): δ -53.22 (d, $^2J_{F-P}$ = 69.1 Hz, *CF$_3$*); ^{31}P{^{1}H} NMR (121 MHz,

C$_6$D$_6$): δ -11.72 (q, $^2J_{P\text{-}F}$ = 69.1 Hz, PPhCF$_3$); 3.18 (s, PHAd); ^{31}P NMR (121 MHz, C$_6$D$_6$): δ 3.17 (d, $J_{P\text{-}H}$ = 209 Hz, PHAd); **EA** calcd for C$_{29}$H$_{33}$F$_3$FeP$_2$ (556.37) calcd C 62.61, H 5.98, F 10.24, P 11.13; found: C 62.37, H 5.96, F 10.32, P 11.15; **HRMS** (EI) calcd (m/z) for C$_{29}$H$_{33}$F$_3$FeP$_2$: 556.1354 ([M]$^+$); found: 556.1353 ([M]$^+$); 487.1414 ([M-HCF$_3$]$^+$); 421.0193 ([M-Ad]$^+$).

Minor Isomer: ^{19}**F NMR** (282 MHz, C$_6$D$_6$): δ -53.15 (d, $^2J_{F\text{-}P}$ = 69.3 Hz, CF$_3$); -50.73 (dq, $J_{F\text{-}P}$ = 76.5 Hz, $J_{F\text{-}F'}$ = 8.4 Hz, CF$_3$'); ^{31}**P{^{1}H} NMR** (121 MHz, C$_6$D$_6$): δ -12.45 (qd, $^2J_{P\text{-}F}$ = 70.6 Hz, $J_{P\text{-}P'}$ = 14.9 Hz, PPhCF$_3$); 10.23 (d, $J_{P\text{-}P'}$ = 16.9 Hz, PHAd).

(R_P,R_C,S_{Fc})-1-(Phenyltrifluoromethylphosphino)-2-{1-[(1-adamantyl)-phosphino]ethyl}ferrocene ((R_P)-167). To a solution of (S_P,R_C,S_{Fc})-1-(phenyltrifluoromethylphosphino)-2-[1-(dimethylamino)ethyl]ferrocene ((S_P)-147) (200 mg, 462 µmol, 1 eq.) in 2 mL of degassed acetic acid was added trifluoroacetic acid (34 µL, 462 µmol, 1 eq.). Then 1-adamantylphosphine (155 mg, 923 µmol, 2 eq.) was added, the mixture heated up to 90 °C and the red solution stirred at this temperature for 16 h. The mixture was then cooled to 60 °C and the acetic acid removed under reduced pressure. The residue was diluted with diethyl ether and saturated aqueous NaHCO$_3$, the organic phase separated and dried over MgSO$_4$. After removal of the solvent under reduced pressure, the crude product was purified by FC (silica, pentane:Et$_2$O 1:0-50:1) to yield the product as a mixture of two diastereoisomers (12:1, P2). Yield: 191 mg (75%), orange powder. Major isomer: ^{1}**H NMR** (300 MHz, C$_6$D$_6$): δ 1.60 (dd, $^3J_{H\text{-}P}$ = 15.4 Hz, ^3J = 7.3 Hz, 3 H, CHMe); 1.66 (bs, 6 H, Ad); 1.90 (bs, 3 H, Ad); 1.99 (bs, 6 H, Ad); 3.44 (d, $^1J_{H\text{-}P}$ = 210 Hz, 1 H, PH); 3.65-3.80 (m, 1 H, CHMe); 3.61 (s, 5 H, Cp'); 4.13 (bs, 1 H, Cp); 4.23 (bs, 1 H, Cp); 4.40 (bs, 1 H, Cp); 7.06-7.12 (m, 3 H, Ph); 7.85-7.95 (m, 2 H, Ph); ^{19}**F NMR** (282 MHz, C$_6$D$_6$): δ -55.39 (d, $^2J_{F\text{-}P}$ = 68.5 Hz, CF$_3$); ^{31}**P{^{1}H} NMR** (121 MHz, C$_6$D$_6$): δ -15.59 (q, $^2J_{P\text{-}F}$ = 68.4 Hz, PPhCF$_3$); 2.76 (s, PHAd); ^{31}**P NMR** (121 MHz, C$_6$D$_6$): δ 2.75 (d, $J_{P\text{-}H}$ = 211 Hz, PHAd); **EA** calcd for C$_{29}$H$_{33}$F$_3$FeP$_2$ (556.37) C 62.61, H 5.98, F 10.24, P 11.13 found C 62.32, H 5.97, F 10.26, P 10.87; **HRMS** (EI) calcd (m/z) for C$_{29}$H$_{33}$F$_3$FeP$_2$: 556.1354 ([M]$^+$); found: 556.1353 ([M]$^+$); 487.1414 ([M-HCF$_3$]$^+$); 421.0193 ([M-Ad]$^+$).

Minor Isomer: ^{19}**F NMR** (282 MHz, C$_6$D$_6$): δ -55.32 (d, $^2J_{F\text{-}P}$ = 68.5 Hz, CF$_3$); -50.73 (dq, $J_{F\text{-}P}$ = 76.5 Hz, $J_{F\text{-}F'}$ = 8.4 Hz, CF$_3$'); ^{31}**P{^{1}H} NMR** (121 MHz, C$_6$D$_6$): δ -16.52 (qd,

$^2J_{P-F}$ = 69.4 Hz, $J_{P-P'}$ = 22.3 Hz, PPhCF$_3$); 9.53-10.07 (m, PHAd); ^{31}P NMR (121 MHz, C$_6$D$_6$): δ 9.80 (d, J_{P-H} = 196 Hz, PHAd).

(1*S*,2*R*,3*R*,*S*$_{Fc}$)-2-(1-Adamantyl)-3-methyl-1-phenyl)-2,3-dihydro-1*H*-ferroceno[*c*][1,2]diphosphole and (1*R*,2*S*,3*R*,*S*$_{Fc}$)-2-(1-adamantyl)-3-methyl-1-phenyl)-2,3-dihydro-1*H*-ferroceno[*c*][1,2]diphosphole. To a solution of (*S*$_P$,*R*$_C$,*S*$_{Fc}$)-1-(phenyltrifluoromethylphosphino)-2-{1-[(1-adamantyl)phosphino]-ethyl}ferrocene ((*S*$_P$)-**167**) (251 mg, 451 μmol, 1 eq.) in 1 mL of tetrahydrofuran was added a solution of *t*-BuOK (51 mg, 451 μmol, 1 eq.) and 18-crown-6 (131 mg, 496 μmol, 1.1 eq.) in 1 mL of tetrahydrofuran. The mixture was allowed to stir for 10 min, quenched with 1 mL of saturated aqueous NH$_4$Cl and diluted with 2 mL of water and 20 mL of diethyl ether. The organic phase was separated, dried over MgSO$_4$ and the solvent removed under reduced pressure. FC (Silica, pentane:EtOAc 50:1) gave the product as a mixture of two diastereoisomers (100:6). Yield: 182 mg (83%), orange crystalline solid. Major isomer (**A**): ^{1}H NMR (700 MHz, CDCl$_3$): δ 1.44 (dd, $^3J_{H-P}$ = 19.4 Hz, ^3J = 7.2 Hz, 3 H, CH*Me*); 1.85 (bs, 6 H, *Ad*); 2.02 (bs, 6 H, *Ad*); 2.11 (bs, 3 H, *Ad*); 3.69 (p, *J* = 7.1 Hz, 1 H, C*H*Me); 4.14 (bs, 1 H, *Cp*); 4.21 (s, 5 H, *Cp'*); 4.26 (t, *J* = 2.3 Hz, 1 H, *Cp*); 4.33 (bs, 1 H, *Cp*); 7.21-7.28 (m, 3 H, *Ph*); 7.40 (t, *J* = 7.7 Hz, *Ph*); ^{13}C{^{1}H} NMR (176 MHz, CDCl$_3$): δ 24.13 (d, J_{C-P} = 31.8 Hz, PCH*Me*); 28.67 (d, J_{C-P} = 7.7 Hz, *Ad*); 29.24 (d, J_{C-P} = 25.3 Hz, P*C*HMe); 33.80-34.01 (m, *Ad*); 36.89 (s, *Ad*); 42.24 (t, J_{C-P} = 8.7 Hz, *Ad*); 65.81 (s, *Cp*); 67.77 (d, J_{C-P} = 10.0 Hz, *Cp*); 70.08 (s, *Cp'*); 71.47 (s, *Cp*); 83.32 (s, *Cp*); 102.25 (s, *Cp*); 128.00 (s, *Ph*); 128.27 (d, J_{C-P} = 6.7 Hz, *Ph*); 132.52 (dd, J_{C-P} = 19.9, 6.5, *Ph*); ^{31}P{^{1}H} NMR (121 MHz, CDCl$_3$): δ -40.92 (d, $^1J_{P-P}$ = 198.7 Hz, *P*Ph); 94.86 (dq, $^1J_{P-P}$ = 198.9 Hz, *P*Ad); **EA** calcd for C$_{28}$H$_{32}$FeP$_2$ (486.36): C 68.89, H 6.78, P 12.66; found: C 68.67, H 6.59, P 12.58; **HRMS** (EI) calcd (*m/z*) for C$_{28}$H$_{32}$FeP$_2$: 486.1324 ([M]$^+$); found: 486.1324 ([M]$^+$); 351.0149 ([M-Ad]$^+$).

Minor isomer (**B**): see below.

(1*S*,2*R*,3*R*,*S*$_{Fc}$)-2-(1-Adamantyl)-3-methyl-1-phenyl)-2,3-dihydro-1*H*-ferroceno[*c*][1,2]-diphosphole, (1*R*,2*S*,3*R*,*S*$_{Fc}$)-2-(1-adamantyl)-3-methyl-1-phenyl)-2,3-dihydro-1*H*-ferroceno[*c*][1,2]diphosphole and (1*R*,2*R*,3*R*,*S*$_{Fc}$)-2-(1-adamantyl)-3-methyl-1-phenyl)-2,3-dihydro-1*H*-ferroceno[*c*][1,2]diphosphole. To a solution of (*R*$_P$,*R*$_C$,*S*$_{Fc}$)-1-(phenyl-trifluoromethylphosphino)-2-{1-[(1-adamantyl)phosphino]ethyl} ((*R*$_P$)-**167**) (243 mg, 437 μmol, 1 eq.) in 1 mL of tetrahydrofuran was added a solution of *t*-BuOK (49 mg, 437 μmol,

1 eq.) and 18-crown-6 (127 mg, 480 μmol, 1.1 eq.) in 1 mL of tetrahydrofuran. The mixture was allowed to stir for 10 min, quenched with 1 mL of saturated aqueous NH$_4$Cl and diluted with 2 mL water and 20 mL of diethyl ether. The organic phase was separated, dried over MgSO$_4$ and the solvent removed under reduced pressure. FC (Silica, pentane:EtOAc 50:1) gave the product as three diastereoisomers (6:3:1). Yield: 132 mg (62%), orange crystalline solid. **EA** calcd for C$_{28}$H$_{32}$FeP$_2$ (486.36): C 69.15, H 6.63, P 12.74 found C 68.67, H 6.59, P 12.58; **HRMS** (EI) calcd (m/z) for C$_{28}$H$_{32}$FeP$_2$: 486.1324 ([M]$^+$); found: 486.1324 ([M]$^+$); 351.0149 ([M-Ad]$^+$).

Major Isomer (**B**): ^{1}**H NMR** (700 MHz, CDCl$_3$): δ 1.71 (dd, $^3J_{H-P}$ = 17.4 Hz, ^3J = 6.8 Hz, 3 H, CHMe); 1.79 (bs, 6 H, Ad); 1.90 (d, J = 11.6 Hz, 3 H, Ad); 2.02 (d, J = 10.6 Hz, 3 H, Ad); 2.05 (bs, 3 H, Ad); 2.11 (bs, 3 H, Ad); 3.18 (p, J = 6.9 Hz, 1 H, CHMe); 3.81 (s, 5 H, Cp'); 4.06 (bs, 1 H, Cp); 4.13 (t, J = 2.3 Hz, 1 H, Cp); 4.27 (bs, 1 H, Cp); 7.36-7.50 (m, 3 H, Ph); 7.84 (t, J = 6.7 Hz, Ph); 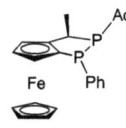 ^{13}**C{^{1}H} NMR** (176 MHz, CDCl$_3$): δ 21.45 (d, J_{C-P} = 25.7 Hz, PCHMe); 25.35 (d, J_{C-P} = 21.5 Hz, PCHMe); 28.76 (d, J_{C-P} = 7.7 Hz, Ad); 33.04 (dd, J_{C-P} = 18.9, 12.3 Hz, Ad); 36.89 (s, Ad); 41.53 (t, J_{C-P} = 9.4 Hz, Ad); 63.03 (s, Cp); 65.80 (d, J_{C-P} = 16.1 Hz, Cp); 69.04 (s, Cp'); 70.99 (d, J_{C-P} = 6.6 Hz, Cp); 84.18 (d, J_{C-P} = 29.6 Hz; Cp); 101.04 (dd, J_{C-P} = 8.7, 2.3 Hz, Cp); 127.51 (s, Ph); 127.71 (d, J_{C-P} = 3.9 Hz, Ph); 131.42 (dd, J_{C-P} = 14.8, 11.8 Hz, Ph); 140.76 (dd, J_{C-P} = 25.4, 11.4 Hz, Ph); ^{31}**P{^{1}H} NMR** (121 MHz, CDCl$_3$): δ -27.92 (d, $^1J_{P-P}$ = 194.5 Hz, PPh); 48.61 (dq, $^1J_{P-P}$ = 194.1 Hz, PAd).

Minor Isomer (**A**): ^{1}**H NMR** (700 MHz, CDCl$_3$): δ 1.44 (dd, $^3J_{H-P}$ = 19.4 Hz, ^3J = 7.2 Hz, 3 H, CHMe); 1.85 (bs, 6 H, Ad); 2.02 (bs, 6 H, Ad); 2.11 (bs, 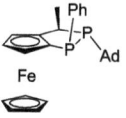 3 H, Ad); 3.69 (p, J = 7.1 Hz, 1 H, CHMe); 4.14 (bs, 1 H, Cp); 4.21 (s, 5 H, Cp'); 4.26 (t, J = 2.3 Hz, 1 H, Cp); 4.33 (bs, 1 H, Cp); 7.21-7.28 (m, 3 H, Ph); 7.40 (t, J = 7.7 Hz, Ph); ^{13}**C{^{1}H} NMR** (176 MHz, CDCl$_3$): δ 24.13 (d, J_{C-P} = 31.8 Hz, PCHMe); 28.67 (d, J_{C-P} = 7.7 Hz, Ad); 29.24 (d, J_{C-P} = 25.3 Hz, PCHMe); 33.80-34.01 (m, Ad); 36.89 (s, Ad); 42.24 (t, J_{C-P} = 8.7 Hz, Ad); 65.81 (s, Cp); 67.77 (d, J_{C-P} = 10.0 Hz, Cp); 70.08 (s, Cp'); 71.47 (s, Cp); 83.32 (s, Cp); 102.25 (s, Cp); 128.00 (s, Ph); 128.27 (d, J_{C-P} = 6.7 Hz, Ph); 132.52 (dd, J_{C-P} = 19.9, 6.5 Hz, Ph); ^{31}**P{^{1}H} NMR** (121 MHz, CDCl$_3$): δ -40.92 (d, $^1J_{P-P}$ = 198.7 Hz, PPh); 94.86 (d, $^1J_{P-P}$ = 198.9 Hz, PAd).

6 Experimental Part

Isomer **C**: ^{31}P{^{1}H} NMR (121 MHz, CDCl$_3$): δ -33.47 (d, $^1J_{P-P}$ = 220.4 Hz, *P*Ph); 56.10 (d, $^1J_{P-P}$ = 220.4 Hz, *P*Ad).

6.4 Transition-Metal Complexes

[PdCl$_2$(89)] (112). A solution of (R_C,S_{Fc})-1-[diphenylphosphino]-2-{1-[(1-adamantyl)phosphino]-ethyl}ferrocene (**89**) (300 mg, 53 mmol, 1 eq.) in 2 mL of dichloromethane was added to a solution of [PdCl$_2$(COD)] (204 mg, 53 mmol, 1 eq.) in 2 mL of dichloromethane and the resulting deep red solution stirred for 30 min. The solvent was then removed and the red residue washed with pentane (5 mL) and diethyl ether (5 mL). The dark red crystalline solid was dried under vacuum. Mixture of two isomers (10:1). Yield: 367 mg (93%); **^{1}H NMR** (300 MHz, CDCl$_3$): δ 1.51-1.65 (m, 6 H, CH*Me*, PH*Ad*); 1.80-1.95 (m, 12 H, P*Ad*); 3.66 (s, 5 H, *Cp*'); 3.73-3.84 (m, 1 H, C*H*Me,); 4.55 (s, 2 H, *Cp*); 4.71 (s, 1 H, *Cp*); 5.35 (d, $^1J_{P-H}$ = 396 Hz, 1 H, P*H*Ad); 7.34-7.45 (m, 3 H, *Ph*); 7.54 (bs, 3 H, *Ph*); 7.76-7.82 (m, 2 H, *Ph*); 8.06-8.13 (m, 2 H, *Ph*); **^{31}P{^{1}H} NMR** (121 MHz, CD$_2$Cl$_2$): δ 13.71 (d, $^3J_{P-P}$ = 8.1 Hz *P*Ph$_2$); 56.49 (d, $^3J_{P-P}$ = 7.9 Hz, PH*Ad*); **^{31}P NMR** (121 MHz, CD$_2$Cl$_2$): δ 13.71 (bs, $^3J_{P-P}$ = 8.1 Hz *P*Ph$_2$); 56.49 (d, $^1J_{P-H}$ = 403 Hz, P*H*Ad); **HRMS** (MALDI) calcd (*m/z*) for C$_{34}$H$_{38}$ClFeP$_2$Pd: 705.0527 ([M-Cl]$^+$); found: 705.0539 ([M-Cl]$^+$).

[PdCl(89)]$_2$ (113). Method A: To a suspension of [Pd(II)Cl$_2$(**89**)] (**112**) (150 mg, 202 μmol, 1 eq.) in 2 mL of tetrahydrofuran was added a solution of KOtBu (204 mg, 202 μmol, 1 eq.) in 1 mL of tetrahydrofuran and the resulting deep red solution stirred for 30 minutes. The solvent was then removed and the red residue recrystallized from DCM/hexane. Yield: 91 mg (64%).

Method B: To a solution of [PdCl$_2$(MeCN)$_2$] (136 mg, 354 μmol, 1 eq.) in tetrahydrofuran was added (R_C,S_{Fc})-1-[diphenylphosphino]-2-{1-[(1-adamantyl)phosphino]-ethyl}ferrocene

(89) (200 mg, 354 μmol, 1 eq.) and the bright red solution stirred for 15 min. Then, a solution of KOtBu (48 mg, 425 μmol, 1.2 eq.) in tetrahydrofuran was added and the resulting deep red solution stirred for 30 min. The solvent was then removed and the red residue recrystallized from DCM/hexane. Yield: 209 mg (84%). ^{1}H NMR (300 MHz, CD$_2$Cl$_2$): δ 1.43 (bs, 6 H, PAd); 1.58 (bs, 3 H, PAd); 2.02 (bs, 6 H, PAd); 2.35 (dd, $^3J_{H-P}$ = 14.3 Hz, ^3J = 7.2 Hz, 3 H, Me); 3.63 (s, 5 H, Cp'); 3.90-4.00 (m, 1 H, CHMe); 4.49 (s, 1 H, Cp); 4.71 (s, 1 H, Cp); 4.78 (s, 1 H, Cp); 7.33-7.40 (m, 3 H, Ph); 7.47-7.55 (m, 3 H, Ph); 7.74-7.80 (m, 2 H, Ph); 8.14-8.21 (m, 2 H, Ph); ^{31}P{^{1}H} NMR (121 MHz, CD$_2$Cl$_2$): AA'XX' spinsystem with two sets of signals around: δ -4.05 (PPh_2) and -75.19 (PAd); **EA** calcd for C$_{68}$H$_{74}$P$_4$Cl$_2$Fe$_2$Pd$_2$ (1410.67): C 57.90, H 5.29, P 8.78, Cl 5.03; found: C 57.18, H 5.36, P 8.77, Cl 5.58; **HRMS** (MALDI) calcd (m/z) for C$_{68}$H$_{74}$ClFe$_2$P$_4$Pd$_2$ ([M-Cl]$^+$): 1375.122; found: 1375.123 ([M-Cl]$^+$); 1339.153 ([M-Cl$_2$]$^+$).

115. A solution of (R_C,S_{Fc})-1-[diphenylphosphino]-2-{1-[(1-adamantyl)-phosphino]ethyl}ferrocene (**89**) (200 mg, 0.354 mmol, 1 eq.) in 2 mL of dichloromethane was added to a solution of [Pd$_2$(μ-Cl)$_2$(η^3-C$_3$H$_5$)$_2$] (64.8 mg, 0.177 mmol, 0.5 eq.) in 2 mL of dichloromethane and the resulting deep red solution stirred for 2 h. The solvent was then removed and the red residue dissolved in 1 mL of dichloromethane. Addition of pentane precipitates an orange powder, which was crystallized from DCM/hexane. Yield: 125 mg (45%); ^{1}H NMR (500 MHz, CD$_2$Cl$_2$): δ 1.63-1.72 (m, 6 H, PAd); 1.89 (d, $^3J_{H-P}$ = 12.0 Hz, ^3J = 7.3 Hz, 3 H, CHMe); 1.91-2.00 (m, 9 H, PAd); 2.69-2.76 (m, 1 H, CH$_2$CHCH$_2$); 3.29 (dq, $^2J_{H-P}$ = 10.2 Hz, ^3J = 7.2 Hz; CHMe); 3.62 (s, 5 H, Cp'); 4.10-4.16 (m, 1 H, CH$_2$CHCH$_2$); 4.49 (s, 2 H, Cp); 4.74 (s, 1 H, Cp); 5.53 (d, ^3J = 10.3 Hz, 1 H, CH$_2$CHCH$_2$); 5.56 (d, ^3J = 17.1 Hz, 1 H, CH$_2$CHCH$_2$); 6.52-6.61 (m, 1 H, CH$_2$CHCH$_2$); 7.41-7.45 (m, 2 H, Ph); 7.52-7.55 (m, 1 H, Ph); 7.63-7.70 (m, 5 H, Ph); 8.21-8.25 (m, 2 H, Ph); ^{13}C{^{1}H} NMR (62.9 MHz, C$_6$D$_6$): δ 14.48 (d, $^2J_{C-P}$ = 7.0 Hz, CHMe); 25.52 (d, $^1J_{C-P}$ = 25.8 Hz, CH$_2$CHCH$_2$); 28.43 (d, J_{C-P} = 8.9 Hz, Ad); 31.68 (dd, J_{C-P} = 17.4, 8.8 Hz, CHMe); 36.03 (s, Ad); 40.21 (s, Ad); 41.85 (d, $^1J_{C-P}$ = 14.7 Hz, Ad); 68.97 (d, J_{C-P} = 7.6 Hz, Cp); 69.18 (d, J_{C-P} = 8.5 Hz, Cp); 70.61 (s, Cp'); 73.85 (dd, J_{C-P} = 50.2, 10.0 Hz, Cp); 75.22 (d, J_{C-P} = 3.5 Hz, Cp); 91.76 (dd, J_{C-P} = 17.5, 4.7 Hz, Cp); 119.76 (d, J_{C-P} = 12.0 Hz, CH$_2$CHCH$_2$); 127.12 (d, J_{C-P} = 11.4 Hz, PPh_2); 127.86 (d, J_{C-P} = 12.0 Hz, PPh_2); 130.53 (s, PPh_2); 130.56 (s, PPh_2); 131.06 (d, J_{C-P} = 3.9 Hz, PPh_2); 131.27 (d, J_{C-P} = 2.7 Hz, CH$_2$CHCH$_2$); 134.50 (d, J_{C-P} = 9.2 Hz, PPh_2); 135.05 (s, PPh_2); 135.84 (d, J_{C-P} = 12.7 Hz,

P*Ph*₂); ³¹P{¹H} NMR (121 MHz, CD₂Cl₂): δ 19.07 (d, ³J_{P-P} = 2.2 Hz, P*Ph*₂); 76.54 (d, ³J_{P-P} = 2.2 Hz, P*Ad*); **EA** calcd for C₃₇H₄₂Cl₂P₂Fe*CH₂Cl₂ (866.79): C 52.66, H 5.12, P 7.15, Cl 16.36; found: C 52.94, H 5.17, P 7.36, Cl 16.20; **HRMS** (MALDI) calcd (*m/z*) for C₃₇H₄₁FeP₂Pd ([M-H]⁺): 709.1077; found: 709.1090 ([M-H]⁺).

[Rh(89)(COD)]PF₆ (116). (*S*C,*R*Fc)-1-[Diphenylphosphino]-2-{1-[(1-adamantyl)phosphino]ethyl}ferrocene (**89**) (30 mg, 53.1 μmol, 1.05 eq.) and [Rh(COD)₂]PF₆ (23.5mg, 50.6 μmol, 1 eq.) were dissolved in 3 mL dichloromethane and the deep red solution stirred for 1 h. The

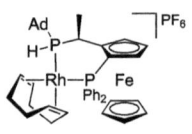

solvent was then partly removed and the solution overlayered with pentane. After 12 h the solvent was decanted and the residue washed with diethyl ether and pentane. Yield: n.d. red crystalline solid. **¹H NMR** (300 MHz, CD₂Cl₂): δ 1.81-2.19 (m, 17 H, *Ad*, *COD*); 1.88 (dd, ³J_{H-P} = 16.1 Hz, ³J = 6.9 Hz, 3 H, CH*Me*); 2.30-2.66 (m, 6 H, *COD*); 3.56-3.67 (m, 1 H, C*H*Me); 3.71 (s, 6 H, *COD*, *Cp'*); 4.23 (bs, 1 H, *Cp*); 4.48 (bs, 1 H, *Cp*); 4.67 (bs, 1 H, *COD*); 4.74 (s, 1 H, *Cp*); 5.21 (dd, ¹J_{H-P} = 337 Hz, *J* = 12.2 Hz, 1 H, P*H*); 5.95 (bs, 1 H, *COD*); 6.03 (bs, 1 H, *COD*); 7.02-7.14 (m, 2 H, *Ph*); 7.33-7.41 (m, 2 H, *Ph*); 7.42-7.51 (m, 1 H, *Ph*); 7.71-7.78 (m, 3 H, *Ph*); 8.18-8.38 (m, 2 H, *Ph*); ³¹P{¹H} NMR (121 MHz, CD₂Cl₂): δ -144.35 (hept., ¹J_{P-F} = 711 Hz, *PF₆*); 19.81 (dd, ¹J_{P-Rh}= 139 Hz, ²J_{P-P} = 43.8 Hz, P*Ph*₂); 29.48 (bd, ¹J_{P-Rh} = 113 Hz, P*HAd*); **HRMS** (MALDI) calcd (*m/z*) for C₄₂H₅₀FeP₂Rh: 775.1787 ([M]⁺); found: 775.1785 ([M]⁺); 667.9864 ([M-COD]⁺).

[PdCl₂(111)] (117). To a solution of (*S*C,*R*Fc)-1-[bis(trifluoromethyl)-phosphino]-2-{1-[(1-adamantyl)phosphinyl]ethyl}ferrocene (**111**) (100 mg, 177 μmol, 1 eq.) in 1 mL of dichloromethane was added a solution of [PdCl₂(COD)] (50.6 mg, 177 μmol, 1 eq.) in 2 mL of dichloromethane and the resulting deep red solution stirred for 30 minutes. The solvent was then removed and the red residue washed with pentane and diethyl ether. The dark red crystalline solid was dried under reduced pressure. Yield: 105 mg (80%); **¹H NMR** (300 MHz, CD₂Cl₂): δ 1.63 (dd, ³J_{H-P} = 12.6 Hz, ³J_{H-H} = 6.6 Hz, 3 H, CH*Me*); 1.82 (bs, 6 H, P*Ad*); 2.15 (bs, 3 H, P*Ad*); 2.25 (bs, 6 H, P*Ad*); 3.09-3.14 (m, 1 H, C*H*Me,); 4.39 (s, 5 H, *Cp'*); 4.83 (s, 2 H, *Cp*); 4.85 (s, 1 H, *Cp*); **¹⁹F NMR** (282 MHz, CD₂Cl₂): δ -54.98 (dq, ²J_{F-P} = 76.1 Hz, ⁴$J_{F-F'}$ = 8.0 Hz, C*F₃*); -51.76 (dq, ²J_{F-P} = 70.5 Hz, ⁴$J_{F-F'}$ = 8.0 Hz, C*F₃'*); ³¹P{¹H} NMR (121 MHz, CD₂Cl₂): δ 51.9-54.8 (m,

$P(CF_3)_2$); 157.3 (d, $^2J_{P-P}$ = 10 Hz, PAd); **HRMS** (MALDI) calcd (m/z) for $C_{24}H_{28}FeOP_2Pd$ ([M-HCl$_2$]): 668,9831; found: 668.9844 ([M-HCl$_2$]).

[Rh(111)(COD)]PF$_6$ (118). To a solution of (R_C,S_{Fc})-1-[bis(trifluoromethyl)phosphino]-2-{1-[(1adamantyl)phosphinyl]-ethyl}ferrocene (**111**) (100 mg, 177 µmol, 1.1 eq.) in 1 mL of dichloromethane was added a solution of [Rh(COD)$_2$]PF$_6$ (75 mg, 168 µmol, 1 eq.) in 2 mL of dichloromethane and the resulting reddish brown solution stirred for 15 h. The mixture was filtered over a short plug of celite and the product precipitated by addition of pentane. The solvent was decanted and the brown residue washed with pentane and diethyl ether. Yield: 109 mg (74%); **^{1}H NMR** (300 MHz, CD$_2$Cl$_2$): δ 1.67-2.87 (m, 23 H, PAd, COD); 1.99 (dd, $^3J_{H-P}$ = 20 Hz, $^3J_{H-H}$ = 7.4 Hz, 3 H, CHMe); 4.33 (bs, 1 H, COD); 4.41-4.43 (m, 1 H, CHMe,); 4.51 (s, 5 H, Cp'); 4.65 (bs, 1 H, COD); 4.81 (s, 1 H, Cp); 4.93 (s, 1 H, Cp); 4.96 (s, 1 H, Cp); 5.42 (bs, 1 H, COD); 5.63 (bs, 1 H, COD); 6.23 (d, $^1J_{H-P}$ = 401 Hz, 1 H, PH); **^{19}F NMR** (282 MHz, CD$_2$Cl$_2$): δ -72.82 (d, $^1J_{F-P}$ = 711 Hz, PF$_6$); -54.85 (dq, $^2J_{F-P}$ = 76.1 Hz, $^4J_{F-F'}$ = 8.0 Hz, CF$_3$); -51.72 (dq, $^2J_{F-P}$ = 70.3 Hz, $^4J_{F-F'}$ = 7.7 Hz, CF$_3$'); **^{31}P{^{1}H} NMR** (121 MHz, CD$_2$Cl$_2$): δ -144.20 (hept., $^1J_{P-F}$ = 711 Hz, PF$_6$); 42.53-46.28 (m, $P(CF_3)_2$); +62.21 (s, PAd);

198. To (R_P,R_C,S_{Fc})-1-[((1-adamantyl)methylphosphino)]-2-[1-(dicyclohexylphosphino)ethyl]ferrocene (20 mg, 34 µmol, 1.1 eq.) and [PtCl$_2$(COD)] (11.5 mg, 31 µmol, 1 eq.) was added 1 mL of dichloromethane and the orange solution stirred for 10 minutes. The solvent was removed under reduced pressure and the red residue washed with pentane and diethyl ether.The yellow solid was dried under vacuum. Yield: 26 mg (99%); **^{1}H NMR** (300 MHz, CD$_2$Cl$_2$): δ 1.08-1.51 (m, 6 H, Cy); 1.61 (dd, J = 11.9, 7.4 Hz, 3 H, CHMe); 1.56-2.31 (m, 29 H, Ad, Cy); 2.18 (d, $^2J_{H-P}$ = 11.3 Hz, 3 H, PMe); 2.48 (bs, 1 H, Cy); 3.01-3.30 (m, 1 H, Cy); 3.53-3.80 (m, 1 H, CHMe); 4.32 (s, 5 H, Cp'); 4.53 (bs, 1 H, Cp); 4.61 (bs, 1 H, Cp); 4.65 (bs, 1 H, Cp); **^{31}P{^{1}H} NMR** (121 MHz, C$_6$D$_6$): δ -0.85 (dd, $^1J_{P-Pt}$ = 3482 Hz, $^2J_{P-P}$ = 18.1 Hz, PAdMe); 30.02 (dd, $^1J_{P-Pt}$ = 3536 Hz, $^2J_{P-P}$ = 18.1 Hz, PCy$_2$); **HRMS** (MALDI) calcd (m/z) for $C_{35}H_{52}ClFeP_2Pt$: 821.2228 ([M-Cl]$^+$); found: 821.2223 ([M-Cl]$^+$).

[PdCl₂(134)] (138). (S_{P1},S_{P2},R_C,S_{Fc})-1-[Phenyl(trifluoromethyl)phosphino]-2-{1-[(1-adamantyl)methylphosphino]ethyl}ferrocene (**134**) (18.0 mg, 32 μmol, 1.1 eq.) and [PdCl₂(COD)] (8.2 mg, 29 μmol, 1 eq.) were dissolved in dichloromethane and the deep red solution stirred for 10 minutes. The solvent was then removed and the red residue washed with pentane and diethyl ether. The dark red crystalline solid was dried under vacuum. Yield: 18 mg (84%); **¹H NMR** (300 MHz, CD₂Cl₂): δ 1.53 (d, $^2J_{H-P}$ = 11.1 Hz, 3 H, PAd*Me*); 1.60-1.91 (m, 18 H, P*Ad*Me, CH*Me*); 2.69 (bs, 1 H, C*H*Me); 4.52 (s, 5 H, *Cp'*); 4.65 (s, 1 H, *Cp*); 4.80 (s, 1 H, *Cp*); 4.82 (s, 1 H, *Cp*); 7.76 (s, 3 H, *Ph*); 8.63-8.69 (m, 2 H, *Ph*); **¹⁹F NMR** (282 MHz, CD₂Cl₂): δ -53.58 (d, $^2J_{F-P}$ = 68.7, C*F₃*); **³¹P{¹H} NMR** (121 MHz, CD₂Cl₂): δ 49.26 (q, $^2J_{P-F}$ = 68.6 Hz, *P*CF₃Ph); 85.01 (s, *P*AdMe); **HRMS** (MALDI) calcd (*m/z*) for C₃₀H₃₅ClF₃FeP₂Pd: 711.0243 ([M-Cl]⁺); found: 711.0250 ([M-Cl]⁺).

[PtCl₂(133)] (139). To (S_{P1},S_{P2},R_C,S_{Fc})-1-[methyl(trifluoromethyl)phosphino]-2-{1-[(1-adamantyl)methylphosphino]ethyl}ferrocene (**133**) (50 mg, 98 mmol, 1 eq.) and [PtCl₂(COD)] (37 mg, 98 mmol, 1 eq.) was added 8 mL of dichloromethane. Immediately, a yellow precipitate was observed. The solvent was removed and the residue dried under vacuum. (No spectroscopic data were measured due to the poor solubility of the product). **EA** C₂₅H₃₃Cl₂F₃FeP₂Pt (774.31) calcd C 38.78, H 4.30, Cl 9.16, F 7.36, P 8.00 found C 38.65, H 4.43, Cl 9.31, F 7.42, P 7.90; **HRMS** (MALDI) calcd (*m/z*) for C₂₅H₃₃ClF₃FeP₂Pt ([M-Cl]⁺): 739.0689; found: 739.0683 ([M-Cl]⁺).

[PtCl(133)PPh₃]PF₆ (140). **139** (30 mg, 36 mmol, 1eq.) and triphenylphosphine (9.4 mg, 36 μmol, 1 eq.) were suspended in 5 mL of tetrahydrofuran. A solution of thallium(I) hexafluorophosphate (12.5 mg, 36 μmol, 1 eq.) in 1 mL of tetrahydrofuran was then added

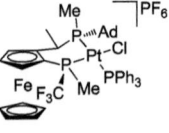

and the mixture stirred for 2 h during which a red solution and a white precipitate was obtained. The mixture was filtered over celite, the solvent removed under reduced pressure and the red residue washed with pentane (5 mL) and diethyl ether (5 mL). The red crystalline solid was dried under vacuum. Yield: 30 mg (70%); **¹H NMR** (300 MHz, CD₂Cl₂): δ 1.51

(dd, $^2J_{H-P}$ = 10.5 Hz, J = 1.8 Hz, 3 H, PAd*Me*); 1.81 (dd, $^3J_{H-P}$ = 12.0 Hz, ^3J = 6.8 Hz, 3 H, CH*Me*); 1.88 (bs, 6 H, P*Ad*Me); 2.19 (bs, 9 H, P*Ad*Me); 2.49 (d, $^2J_{H-P}$ = 10.4 Hz, 3 H, PCF$_3$*Me*); 3.43 (bs, 1 H, C*H*Me,); 4.44 (s, 5 H, *Cp'*); 4.48 (bs, 1 H, *Cp*); 4.67 (bs, 2 H, *Cp*); 7.55-7.60 (m, 9 H, P*Ph$_3$*); 7.69-7.75 (m, 6 H, P*Ph$_3$*); ^{19}**F NMR** (282 MHz, CD$_2$Cl$_2$): δ -72.80 (d, $^1J_{F-P}$ = 711 Hz, P*F$_6$*); -58.02 (dd, $^2J_{F-P}$ = 74.7 Hz, $^3J_{F-Pt}$ = 39.6 Hz, PC*F$_3$*Me); ^{31}**P{^{1}H} NMR** (121 MHz, C$_6$D$_6$): δ -144.22 (hept., $^1J_{P-F}$ = 711.1 Hz, P*F$_6$*); 19.64 (dqdd, $^1J_{P-Pt}$ = 3613.9 Hz, $^2J_{P-F}$ = 74.7 Hz, $^2J_{P-P}$ = 26.6, 15.4 Hz, *P*CF$_3$Me); 20.24 (ddd, $^1J_{P-Pt}$ = 2304.5 Hz, $^2J_{P-P}$ = 367.8, 14.9 Hz, *P*Ph$_3$); 60.08 (ddd, $^1J_{P-t}$ = 2333.7 Hz, $^2J_{P-P}$ = 368.4, 26.7 Hz, *P*AdMe); **HRMS** (MALDI) calcd (*m/z*) for C$_{43}$H$_{48}$ClF$_3$FeP$_3$Pt: 1001.1607 ([M]$^+$); found: 1001.1588 ([M]$^+$); 966.1964 ([M-Cl]$^+$); 881.1933 ([MH-FeCp]$^+$); 739.0681 ([M-PPh$_3$]$^+$).

[Rh(137)(COD)]PF$_6$ (141). (*S*$_{P1}$,*S*$_{P2}$,*R*$_C$,*S*$_{Fc}$)-1-[Isopropyl(trifluoro-methyl)phosphino]-2-[1-(cyclohexylmethylphosphino)ethyl]-ferrocene (**137**) (30 mg, 61.9 µmol, 1.1 eq.) and [Rh(COD)$_2$]PF$_6$

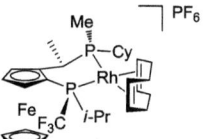

(26.1 mg, 56.3 µmol, 1 eq.) were dissolved in 2 mL dichloromethane and the deep red solution stirred for 1 h. The solvent was then partly removed and the product precipitated by addition of pentane. The orange precipitate was then washed with pentane and diethyl ether. Yield: n.d. orange powder. ^{1}**H NMR** (300 MHz, CD$_2$Cl$_2$): δ 1.23-1.72 (m, 19 H, COD, *Cy*); 1.07-1.63 (m, 9 H, COD, PC*y$_2$*); 1.65-2.60 (m, 20 H, COD, PC*y$_2$*); 1.26 (d, $^2J_{H-P}$ = 8.2 Hz, 3 H, P*Me*); 1.42 (dd, $^3J_{H-P}$ = 14.1 Hz, ^3J = 7.0 Hz, 3 H, P^i*Pr*); 1.66 (dd, $^3J_{H-P}$ = 15.4 Hz, ^3J = 7.1 Hz, 3 H, P^i*Pr*); 1.76 (dd, $^3J_{H-P}$ = 19.6 Hz, ^3J = 7.2 Hz, 3 H, CH*Me*); 2.88-3.11 (m, 1 H, C*H*Me); 4.38 (s, 5 H, *Cp'*); 4.66 (bs, 1 H, *Cp*); 4.69 (bs, 1 H, *Cp*); 4.70 (s, 1 H, *Cp*); 5.00 (bs, 1 H, *COD*); 5.23 (bs, 1 H, *COD*); 5.43 (bs, 1 H, *COD*); 5.68 (bs, 1 H, *COD*); ^{19}**F NMR** (282 MHz, CD$_2$Cl$_2$): δ -73.37 (d, $^1J_{F-P}$ = 711 Hz, P*F$_6$*); -51.15 (d, $^2J_{F-P}$ = 55.0 Hz, C*F$_3$*); ^{31}**P{^{1}H} NMR** (121 MHz, CD$_2$Cl$_2$): δ -144.35 (hept., $^1J_{P-F}$ = 711 Hz, P*F$_6$*); 35.59 (dd, $^1J_{P-Rh}$= 137 Hz, $^2J_{P-P}$ = 42.6 Hz, *P*CyMe); 44.13-49.29 (m, *P*CF$_3^i$Pr); **EA** calcd for C$_{31}$H$_{45}$F$_9$FeP$_3$Rh (840.36): C 44.31, H 5.40, F 20.35, P 11.06; found: C 44.42, H 5.54, F 20.09, P 10.83; **HRMS** (MALDI) calcd (*m/z*) for C$_{31}$H$_{45}$F$_3$FeP$_2$Rh: 695.1348 ([M]$^+$); found: 695.1352 ([M]$^+$).

[PtCl₂(151)] (154). To (S_P,R_C,S_Fc)-1-[(2-naphthyl)trifluoromethyl-phosphino]-2-[1-(dicyclohexyl-phosphino)ethyl]ferrocene (151) (18 mg, 28 μmol, 1.1 eq.) and [PtCl₂(COD)] (9.6 mg, 26 μmol, 1 eq.) was added 1 mL of dichloromethane. The solvent was partly removed and the remaining red solution was filtered over celite. The solvent was removed under reduced pressure and the red residue washed with pentane. The pink orange solid was dried under vacuum. Yield: 24 mg (100%); **¹H NMR** (300 MHz, CD₂Cl₂): δ 1.41-1.58 (m, 9 H, Cy); 1.59-2.14 (m, 13 H, Cy, CH*Me*); 2.20-2.37 (m, 1 H, Cy); 2.44 (bs, 1 H, Cy); 3.26-3.57 (m, 2 H, Cy, C*H*Me,); 3.70 (bs, 1 H, Cp); 4.35 (s, 5 H, Cp'); 4.41 (bs, 1 H, Cp); 4.75 (bs, 1 H, Cp); 7.59-7.80 (m, 3 H, Np); 7.89-8.09 (m, 3 H, Np); 8.42 (d, J = 13.9 Hz, 1 H, Np); **¹⁹F NMR** (282 MHz, CD₂Cl₂): δ -56.11 (dd, $^2J_{F\text{-}P}$ = 75.9 Hz, $^3J_{F\text{-}Pt}$ = 54.7 Hz, PC*F*₃Ph); **³¹P{¹H} NMR** (121 MHz, CD₂Cl₂): δ 15.06 (dqd, $^1J_{P\text{-}Pt}$ = 3794 Hz, $^2J_{P\text{-}F}$ = 76.0 Hz, $^2J_{P\text{-}P}$ = 15.5 Hz, PCF₃Ph); 55.90 (dd, $^1J_{P\text{-}Pt}$ = 3387 Hz, $^2J_{P\text{-}P}$ = 15.5 Hz, PCy₂); **HRMS** (MALDI) calcd (*m/z*) for C₃₅H₄₁Cl₂F₃FeNaP₂Pt: 925.0896 ([MNa]⁺); found: 925.0895 ([MNa]⁺); 867.1402 ([M-Cl]⁺).

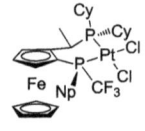

PtCl₂(((S_P)-152))] (155). To (S_P,R_C,S_Fc)-1-[(1-adamantyl)trifluoromethyl-phosphino]-2-[1-(dicyclohexyl-phosphino)ethyl]ferrocene ((S_P)-152) (30 mg, 46.5 μmol, 1.1 eq.) and [PtCl₂(COD)] (15.8 mg, 42.3 μmol, 1 eq.) was added 1 mL of dichloromethane and the red solution stirred for 10 min. The solvent was removed under reduced pressure and the red residue washed with diethyl ether and pentane. Yield: 38 mg (97%); orange crystalline solid. **¹H NMR** (300 MHz, CD₂Cl₂): δ 0.89-1.04 (m, 2 H, Cy); 1.30-1.46 (m, 4 H, Cy); 1.53-2.03 (m, 12 H, Cy); 1.74 (bs, 6 H, Ad); 1.77 (dd, J = 11.8, 7.1 Hz, CH*Me*); 2.08 (bs, 3 H, Ad); 2.18 (bs, 6 H, Ad); 2.21-2.56 (m, 3 H, Cy); 2.95-3.24 (m, 1 H, Cy); 3.79-4.02 (m, 1 H, C*H*Me,); 4.34 (s, 5 H, Cp'); 4.70 (bs, 1 H, Cp); 4.79 (bs, 1 H, Cp); 4.96 (bs, 1 H, Cp); **¹⁹F NMR** (282 MHz, CD₂Cl₂): δ -47.34 (dd, $^2J_{F\text{-}P}$ = 56.5Hz, $^3J_{F\text{-}Pt}$ = 51.6 Hz, PC*F*₃Ph); **³¹P{¹H} NMR** (121 MHz, C₆D₆): δ 30.88 (dqd, $^1J_{P\text{-}Pt}$ = 3747 Hz, $^2J_{P\text{-}F}$ = 57.6 Hz, $^2J_{P\text{-}P}$ = 16.5 Hz, PCF₃Ph); 33.48 (dd, $^1J_{P\text{-}Pt}$ = 3206 Hz, $^2J_{P\text{-}P}$ = 16.4 Hz, PCy₂); **EA** calcd for C₃₅H₄₉Cl₂F₃FeP₂Pt (910.55): C 46.17, H 5.42, Cl 7.79, F 6.26, P 6.80; found: C 46.45, H 5.52, Cl 7.62, F 6.45, P 6.81; **HRMS** (MALDI) calcd (*m/z*) for C₃₅H₄₉Cl₂F₃FeNaP₂Pt: 933.1522 ([MNa]⁺); found: 933.1504 ([MNa]⁺); 875.1889 ([M-Cl]⁺).

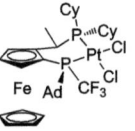

[PdCl₂((S_P)-142)] **(156).** (S_P,R_C,S_Fc)-1-[Phenyl(trifluoromethyl)-phosphino]-2-[1-(dicyclohexylphosphino)ethyl]ferrocene ((S_P)-**142**) (16.0 mg, 27 μmol, 1.1 eq.) and [PdCl₂(COD)] (7.1 mg, 25 μmol, 1 eq.) were dissolved in dichloromethane and the deep red solution stirred for 10 min. 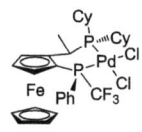 The solvent was then removed and the red residue washed with pentane and diethyl ether. The dark red crystalline solid was dried under vacuum. Yield: 18 mg (80%); **¹H NMR** (300 MHz, CD₂Cl₂): δ 1.05-2.08 (m, 19 H, P*Cy₂*); 1.88 (dd, $^3J_{H-P}$ = 11.6 Hz, ^3J = 7.1 Hz, 3 H, CH*Me*); 2.20-2.35 (m, 2 H, P*Cy₂*); 3.10-3.32 (m, 2 H, P*Cy₂*, C*H*Me); 3.70 (s, 1 H, *Cp*); 4.30 (s, 5 H, *Cp'*); 4.45 (s, 1 H, *Cp*); 4.71 (s, 1 H, *Cp*); 7.45-7.53 (m, 2 H, *Ph*); 7.54-7.60 (m, 1 H, *Ph*); 8.74-8.84 (m, 2 H, *Ph*); **¹⁹F NMR** (282 MHz, CD₂Cl₂): δ -54.26 (d, $^2J_{F-P}$ = 76.0 Hz, C*F₃*); **³¹P{¹H} NMR** (121 MHz, CD₂Cl₂): δ 33.01 (qd, $^2J_{P-F}$ = 76.06 Hz, $^2J_{P-P}$ = 4.3 Hz, *P*CF₃Ph); 85.01 (d, $^2J_{P-P}$ = 4.2 Hz, *P*Cy₂).

[PdCl₂((R_P)-142)] **(157).** (R_P,R_C,S_Fc)-1-(Phenyltrifluoromethylphosphino)-2-[1-(dicyclohexylphosphino)ethyl]ferrocene ((S_P)-**142**) (50 mg, 85 μmol, 1.1 eq.) and [PdCl₂(COD)] (22.1 mg, 78 μmol, 1 eq.) were dissolved in 1 mL of dichloromethane and the deep red solution stirred for 10 min. The 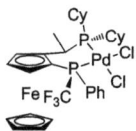 red solution was overlayered with pentane to yield the product as deep red crystals. Yield: 45 mg (70%); **¹H NMR** (300 MHz, CDCl₃): δ 1.01-1.21 (m, 4 H, P*Cy₂*); 1.25-2.16 (m, 15 H, P*Cy₂*); 1.79 (dd, $^3J_{H-P}$ = 11.3 Hz, ^3J = 7.2 Hz, 3 H, CH*Me*); 2.19-2.37 (m, 2 H, P*Cy₂*); 2.82-3.18 (m, 2 H, P*Cy₂*, C*H*Me); 3.99 (s, 5 H, *Cp'*); 4.64 (s, 1 H, *Cp*); 4.68 (s, 1 H, *Cp*); 5.00 (s, 1 H, *Cp*); 7.58-7.68 (m, 3 H, *Ph*); 8.42 (dd, *J* = 13.3, 7.0 Hz, 2 H, *Ph*); **¹⁹F NMR** (282 MHz, CDCl₃): δ -50.59 (d, $^2J_{F-P}$ = 71.8 Hz, C*F₃*); **³¹P{¹H} NMR** (121 MHz, CDCl₃): δ 40.13 (qd, $^2J_{P-F}$ = 71.8 Hz, $^2J_{P-P}$ = 2.6 Hz, *P*CF₃Ph); +84.39 (d, $^2J_{P-P}$ = 2.7Hz, *P*Cy₂); **HRMS** (MALDI) calcd (*m/z*) for C₃₁H₃₉ClF₃FeP₂Pd: 727.0557 ([M-Cl]⁺); found: 727.0563 ([M-Cl]⁺).

[Rh((S_P)-142)COD)]PF₆ **(158).** (S_P,R_C,S_Fc)-1-[(Phenyl)trifluoromethylphosphino]-2-[1-(dicyclohexylphosphino)ethyl]ferrocene ((S_P)-**142**) (40 mg, 68 μmol, 1.1 eq.) and [Rh(COD)₂]PF₆ (28.7 mg, 62 μmol, 1 eq.) were dissolved in 3 mL dichloromethane and the deep red 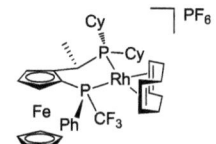 solution stirred for 1 h. The solvent was then partly removed and the product precipitated by addition of pentane. The orange precipitate was then washed with pentane and diethyl ether.

6 Experimental Part

Yield: n.d. orange powder. **^{1}H NMR** (300 MHz, CD$_2$Cl$_2$): δ 0.61-0.98 (m, 1 H, PCy$_2$); 1.07-1.63 (m, 9 H, COD, PCy$_2$); 1.65-2.60 (m, 20 H, COD, PCy$_2$); 1.91 (dd, $^3J_{H-P}$ = 10.8 Hz, ^3J = 7.2 Hz, 3 H, CHMe); 2.20-2.35 (m, 2 H, PCy$_2$); 3.08-3.27 (m, 1 H, CHMe); 3.96 (s, 1 H, Cp); 4.21 (bs, 1 H, COD); 4.30 (s, 5 H, Cp'); 4.58 (s, 1 H, Cp); 4.62 (bs, 1 H, COD); 4.84 (s, 1 H, Cp); 5.37 (bs, 1 H, COD); 5.64 (bs, 1 H, COD); 7.46-7.70 (m, 5 H, Ph); **^{19}F NMR** (282 MHz, CD$_2$Cl$_2$): δ -73.47 (d, $^1J_{F-P}$ = 711 Hz, PF$_6$); -54.26 (d, $^2J_{F-P}$ = 70.6 Hz, CF$_3$); **^{31}P{^{1}H} NMR** (121 MHz, CD$_2$Cl$_2$): δ -144.37 (hept., $^1J_{P-F}$ = 711 Hz, PF$_6$); 36.43 (dqd, $^1J_{P-Rh}$ = 159 Hz, $^2J_{P-F}$ = 70.6 Hz, $^2J_{P-P}$ = 32.5 Hz, PCF$_3$Ph); 56.28 (dd, $^1J_{P-Rh}$ = 139 Hz, $^2J_{P-P}$ = 32.5 Hz, PCy$_2$); **HRMS** (MALDI) calcd (m/z) for C$_{39}$H$_{51}$F$_3$FeP$_2$Rh: 797.1817 ([M]$^+$); found: 797.1812 ([M]$^+$).

[Rh(Cl)((S_P)-142)(CO)] (160). (S_P,R_C,S_{Fc})-1-(Phenyltrifluoromethylphosphino)-2-[1-(dicyclohexylphosphino)ethyl]ferrocene ((S_P)-142) (30 mg, 51.2 µmol, 2.1 eq.) and [Rh$_2$(µ-Cl)$_2$(CO)$_4$] (9.5 mg, 24.4 µmol, 1 eq.) were dissolved in 2 mL of dichloromethane and the deep red solution

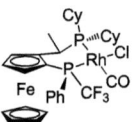

stirred for 10 minutes. The mixture was then filtered over celite, the solvent removed and the product washed with diethyl ether. Yield: 24 mg (65%) red crystalline solid. **^{1}H NMR** (300 MHz, CD$_2$Cl$_2$): δ 1.00-2.14 (m, 20 H, PCy$_2$); 1.77 (dd, J = 9.8, 7.2 Hz, 3 H, CHMe); 2.23-2.36 (m, 1 H, PCy$_2$); 2.88-3.07 (m, 1 H, PCy$_2$); 3.15-3.31 (m, 1 H, CHMe); 3.80 (s, 1 H, Cp); 4.35 (s, 5 H, Cp'); 4.38 (s, 1 H, Cp); 4.66 (s, 1 H, Cp); 7.47-7.65 (m, 3 H, Ph); 7.94-8.14 (m, 2 H, Ph); **^{19}F NMR** (282 MHz, CD$_2$Cl$_2$): δ -58.96 (d, $^2J_{F-P}$ = 75.3 Hz, $^3J_{F-Rh}$ = 3.0 Hz, CF$_3$); **^{31}P{^{1}H} NMR** (121 MHz, CD$_2$Cl$_2$): δ 50.41 (dqd, $^1J_{P-Rh}$ = 175 Hz, $^2J_{P-F}$ = 75.2 Hz, $^2J_{P-P}$ = 43.8 Hz, PCF$_3$Ph); 58.23 (dd, $^1J_{P-Rh}$ = 123 Hz, $^2J_{P-P}$ = 43.8 Hz, PCy$_2$); **IR** (neat): υ 2015.1 cm^{-1} (s, CO). **EA** calcd for C$_{32}$H$_{40}$ClF$_3$FeOP$_2$Rh (753.82): C 50.99, H 5.35, Cl 4.70, F 7.56 P 8.22; found: C 50.82, H 5.32, Cl 4.74, F 7.67 P 8.50.

[Rh(Cl)((R_P)-142)(CO)] (161). (S_P,R_C,S_{Fc})-1-(Phenyltrifluoromethylphosphino)-2-[1-(dicyclohexylphosphino)ethyl]ferrocene ((S_P)-142) (30 mg, 51.2 µmol, 2.1 eq.) and [Rh$_2$(µ-Cl)$_2$(CO)$_4$] (9.5 mg, 24.4 µmol, 1 eq.) were dissolved in 2 mL of dichloromethane and the deep red solution stirred for

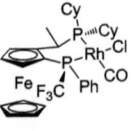

10 min. The mixture was then filtered over celite, the solvent removed and the product washed with diethyl ether. Yield: 28 mg (76%) red crystalline solid. **^{1}H NMR** (300 MHz,

CD$_2$Cl$_2$): δ 1.04-2.09 (m, 20 H, PCy_2); 2.24-2.34 (m, 1 H, PCy_2); 1.71 (dd, J = 9.8, 7.2 Hz, 3 H, CHMe); 2.86-3.12 (m, 2 H, CHMe, PCy_2); 3.81 (s, 5 H, Cp'); 4.52 (s, 1 H, Cp); 4.61 (s, 1 H, Cp); 5.00 (s, 1 H, Cp); 7.52-7.69 (m, 3 H, Ph); 8.30-8.55 (m, 2 H, Ph); ^{19}F NMR (282 MHz, CD$_2$Cl$_2$): δ -58.41 (d, $^2J_{\text{F-P}}$ = 70.7 Hz, CF_3); ^{31}P{^{1}H} NMR (121 MHz, CD$_2$Cl$_2$): δ 52.68 (dd, $^1J_{\text{P-Rh}}$ = 122 Hz, $^2J_{\text{P-P}}$ = 43.5 Hz, PCy_2); 54.62 (dqd, $^1J_{\text{P-Rh}}$ = 175 Hz, $^2J_{\text{P-F}}$ = 71.6 Hz, $^2J_{\text{P-P}}$ = 43.5 Hz, PCF$_3$Ph); IR (neat): υ 2018.2 cm^{-1} (s, CO). EA calcd for C$_{32}$H$_{40}$ClF$_3$FeOP$_2$Rh (753.82): C 50.99, H 5.35, Cl 4.70, F 7.56 P 8.22; found: C 51.04, H 5.34, Cl 4.78, F 7.69 P 8.44.

[Rh(Cl)((120)(CO)] (162). (R_C,S_{Fc})-1-[Bis(trifluoromethyl)phosphino]-2-[1-(dicyclohexylphosphino)ethyl]ferrocene (**120**) (30 mg, 51.2 μmol, 2.1 eq.) and [Rh$_2$(μ-Cl)$_2$(CO)$_4$] (9.6 mg, 24.7 μmol, 1 eq.) were dissolved in 2 mL of dichloromethane and the deep red solution stirred for 10 min. The mixture was then filtered over celite, the solvent removed and the product washed with diethyl ether. Yield: 26 mg (70%) red crystalline solid. ^{1}H NMR (300 MHz, CD$_2$Cl$_2$): δ 1.06-2.93 (m, 19 H, PCy_2); 1.79 (dd, J = 10.1, 7.2 Hz, 3 H, CHMe); 1.98-2.23 (m, 2 H, PCy_2); 2.87-3.07 (m, 1 H, PCy_2); 3.11-3.30 (m, 1 H, CHMe); 4.36 (s, 5 H, Cp'); 4.66 (bs, 1 H, Cp); 4.81 (bs, 2 H, Cp); ^{19}F NMR (282 MHz, CD$_2$Cl$_2$): δ -60.59 (dqd, $^2J_{\text{F-P}}$ = 80.2 Hz, $^4J_{\text{F-F'}}$ = 8.7 Hz, $^3J_{\text{F-Rh}}$ = 3.4 Hz, CF_3); -58.42 (dqd, $^2J_{\text{F-P}}$ = 76.7 Hz, $^4J_{\text{F-F'}}$ = 8.7 Hz, $^3J_{\text{F-Rh}}$ = 3.0 Hz, CF_3); ^{31}P{^{1}H} NMR (121 MHz, CD$_2$Cl$_2$): δ 56.92 (dd, $^1J_{\text{P-Rh}}$ = 118 Hz, $^2J_{\text{P-P}}$ = 46.5 Hz, PCy_2); 60.74-65.76 (m, PCF$_3$Ph); IR (neat): υ 2029.3 cm^{-1} (s, CO).

6.5 Tridentate Phosphine Ligands of the Pigiphos-Type

(S_P)-{(R_C)-1-[(S_{Fc})-2-(Bis(trifluoromethyl)phosphino)-ferrocenyl]ethyl}-{(R_C)-1-[(S_{Fc})-2-(diphenylphosphino)-ferrocenyl]ethyl}adamantylphosphine (179). Under argon atmosphere, (R_C,S_{Fc})-1-[bis(trifluoromethyl)phosphino]-2-{1-[(1-adamantyl)phosphino]ethyl}ferrocene (**110**) (690 mg, 1.278 mmol, 1 eq.) and (R_C,S_{Fc})-1-(diphenylphosphino)-2-(1-acetoxyethyl)ferrocene (**100**) (690 mg, 1.534 mmol, 1.2 eq.) were dissolved in 6 mL of degassed acetic acid and the dark red mixture

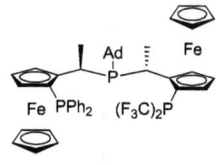

6 Experimental Part

was stirred for 20 h at room temperature. The solvent was then removed under reduced pressure, the residue diluted with saturated aqueous NaHCO$_3$ and diethyl ether and the phases were separated. The aqueous phase was extracted with diethyl ether, the organic phases were combined and washed with saturated aqueous NaHCO$_3$. After drying over magnesium sulfate, the solvent was removed under reduced pressure and the residue purified by FC (silica, pentane:EtOAc 100:1-50:1) to yield 1085 mg (91%) of the product as a 9:2 mixture of diastereoisomers. The major isomer in pure form was obtained by recrystallization from EtOAc. Major isomer: **^{1}H NMR** (300 MHz, C$_6$D$_6$): δ 1.60 (s, 6 H, P*Ad*), 1.64-1.71 (m, 6 H, C*HMe*); 1.82 (s, 9 H, P*Ad*); 3.04-3.15 (m, 1 H, C*H*Me); 3.55-3.66 (m, 1 H, C*H*Me); 3.81 (s, 5 H, *Cp'*); 3.93 (s, 1 H, *Cp*); 4.03 (s, 1 H, *Cp*); 4.14 (s, 6 H, *Cp'*, *Cp*); 4.17 (s, 1 H, *Cp*); 4.23 (s, 1 H, *Cp*); 4.36 (s, 1 H, *Cp*); 7.06-7.14 (m, 6 H, P*Ph$_2$*); 7.54 (d, J = 7.0 Hz, 2 H, P*Ph$_2$*); 7.73-7.81 (m, 2 H, P*Ph$_2$*); **^{19}F NMR** (282 MHz, C$_6$D$_6$): δ -53.94 (ddq, $^2J_{F-P}$ = 66.5 Hz, $J_{F-P'}$ = 11.0 Hz, $^4J_{F-F'}$ = 8.5 Hz, C*F$_3$*); -51.00 (dq, $^2J_{F-P}$ = 77.0 Hz, $^4J_{F-F'}$ = 8.7 Hz, C*F$_3$'*); **^{31}P{^{1}H} NMR** (121 MHz, C$_6$D$_6$): δ -25.68 (d, J_{P-P} = 44.0 Hz, P*Ph$_2$*); -5.69 – -1.89 (m, *P*(CF$_3$)$_2$); 35.65 - 36.59 (m, R$_2$*P*Ad); **EA** calcd for C$_{48}$H$_{49}$F$_6$Fe$_2$P$_3$ (944.52): C 61.04, H 5.23, F 12.07, P 9.84; found: C 61.18, H 5.43, F 12.19, P 9.88; **HRMS** (MALDI) calcd (*m/z*) for C$_{48}$H$_{49}$F$_6$Fe$_2$P$_3$: 945.1724 ([MH]$^+$); found: 954.1742 ([MH]$^+$).

Minor isomer: **^{19}F NMR** (282 MHz, C$_6$D$_6$): δ -51.92 (ddq, $^2J_{F-P}$ = 84.0 Hz, $J_{F-P'}$ = 29.6 Hz, $^4J_{F-F'}$ = 8.3 Hz, C*F$_3$*); -51.17 (dm, $^2J_{F-P}$ = 75.8 Hz, C*F$_3$'*); **^{31}P{^{1}H} NMR** (121 MHz, C$_6$D$_6$): δ -25.65 (s, P*Ph$_2$*); -8.50 – -5.50 (m, *P*(CF$_3$)$_2$); 42.40 – 44.50 (m, R$_2$*P*Ad).

(S_P)-{(R_C)-1-[(S_{Fc})-2-(Bis(trifluoromethyl)phosphino)-ferrocenyl]ethyl}-{(R_C)-1-[(S_{Fc})-2-(di-*tert*-butylphosphino)-ferrocenyl]ethyl}adamantylphosphine (180). Under argon atmosphere, (R_C,S_{Fc})-1-[bis(trifluoromethyl)phosphino]-2-{1-[(1-adamantyl)phosphino]ethyl}ferrocene (**110**) (229 mg, 418 μmol, 1 eq.) and (R_C,S_{Fc})-1-[di-(*tert*-butyl)phosphino]-2-[1-acetoxyethyl]ferrocene (209 mg, 501 μmol, 1.2 eq.) were dissolved in 1 mL of degassed acetic acid and the dark red mixture was stirred for 15 h at 60 °C. The solvent was then removed under reduced pressure, the residue diluted with saturated aqueous NaHCO$_3$ and diethyl ether and the phases were separated. The aqueous phase was extracted with ether, the organic phases combined and washed with saturated aqueous NaHCO$_3$. After drying over MgSO$_4$, the solvent was removed

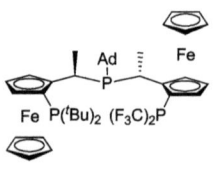

under reduced pressure and the residue purified by FC (silica, pentane:EtOAc 100:1-20:1) to yield the product as a single isomer in the second coloured fraction. Yield: 127 mg (34%), orange crystalline solid. **¹H NMR** (300 MHz, C₆D₆): δ 0.95 (d, $^3J_{H-P}$ = 11.1 Hz, 9 H, CMe_3); 1.53 (d, $^3J_{H-P}$ = 11.9 Hz, 9 H, CMe_3); 1.65-1.71 (m, 9 H, CHMe, PHAd); 1.92-1.96 (m, 6 H, CHMe, PHAd); 2.16 (bs, 6 H, PHAd); 3.05-3.15 (m, 1 H, CHMe); 3.27 (q, ^3J = 6.7 Hz 1 H, CHMe); 4.01 (s, 1 H, Cp); 4.05 (s, 6 H, Cp', Cp); 4.09 (s, 6 H, Cp', Cp); 4.40 (s, 2 H, Cp); 4.45 (s, 1 H, Cp); **¹⁹F NMR** (282 MHz, C₆D₆): δ -53.94 (ddq, $^2J_{F-P}$ = 60.4 Hz, $J_{F-P'}$ = 25.5 Hz, $^4J_{F-F'}$ =8.4 Hz, CF_3); -51.00 (dqd, $^2J_{F-P}$ = 76.0 Hz, $^4J_{F-F'}$ = 8.4 Hz, $J_{F-P'}$ = 3.9 Hz, CF_3'); **³¹P{¹H} NMR** (121 MHz, C₆D₆): δ 13.64 (dd, J_{P-P} = 12.3, 4.9 Hz, P^tBu); -6.87 – -3.85 (m, P(CF_3)₂); 40.08 – 41.76 (m, R₂PAd); **HRMS** (MALDI) calcd (m/z) for C₄₄H₅₇F₆Fe₂P₃: 905.2350 ([MH]⁺); found: 905.2367 ([MH]⁺).

(S_P)-{(R_C)-1-[(S_{Fc})-2-(Bis(trifluoromethyl)phosphino)-ferrocenyl]ethyl}-{(S_C)-1-[(R_{Fc})-2-(diphenylphosphino)-ferrocenyl]ethyl}adamantylphosphine (182). Under argon 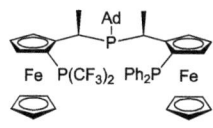 atmosphere, (R_C,S_{Fc})-1-[bis(trifluoromethyl)phosphino]-2-{1-[(1-adamantyl)phosphino]ethyl}ferrocene (**110**) (800 mg, 1.459 mmol, 1 eq.) and (S_C,R_{Fc})-1-(diphenylphosphino)-2-(1-acetoxyethyl)ferrocene (800 mg, 1.751 mmol, 1.2 eq.) were dissolved in 6 mL of degassed acetic acid and the dark red mixture was stirred for 20 h at room temperature during which a precipitate was formed. The precipitate was filtered off, washed with methanol and dried under reduced pressure to yield 507 mg (37%) of the product as a single diastereoisomer. The filtrate was concentrated under reduced pressure, the residue diluted with saturated aqueous NaHCO₃ and diethyl ether and the phases were separated. The aqueous phase was extracted with ether, the organic phases combined and washed with saturated aqueous NaHCO₃. After drying over MgSO₄, the solvent was removed under reduced pressure and the residue purified by FC (silica, pentane:EtOAc 100:1-10:1) to yield in the second orange fraction 492 mg (36%) of the product as a single diastereoisomer. Yield: 999 mg (73%), orange crystalline solid, crystals suitable for X-ray measurement were obtained by slow diffusion of MeOH into a concentrated dichloromethane solution. Major isomer: **¹H NMR** (300 MHz, C₆D₆): δ 1.52 (bs, 9 H, PHAd), 170-1.83 (m, 12 H, PHAd, CHMe); 3.27-3.34 (m, 1 H 2.81, CHMe); 3.54-3.62 (m, 1 H, CHMe); 3.73 (s, 5 H, Cp'); 4.02 (s, 5 H, Cp'); 4.07 (t, J = 2.40 Hz, 6 H, Cp); 4.09-4.13 (m, 2 H, Cp); 4.28 (s, 1 H, Cp); 4.52 (s, 1 H, Cp); 4.67 (s, 1 H, Cp); 7.02-7.13 (m, 6 H, PPh_2); 7.55 (t, J = 7.0 Hz, 2 H, PPh_2); 7.73 (t,

J = 6.7 Hz, 2 H, PPh_2); ^{19}**F NMR** (282 MHz, C_6D_6): δ -51.66 (ddq, $^2J_{F-P}$ = 55.3 Hz, $J_{F-P'}$ = 26.4 Hz, $^4J_{F-F'}$ = 8.2 Hz, CF_3); -51.05 (dm, $^2J_{F-P}$ = 76.2 Hz, CF_3'); ^{31}**P{^{1}H} NMR** (121 MHz, C_6D_6): δ -27.03 (d, J_{P-P} = 16.2 Hz, PPh_2); -7.89 – -4.40 (m, P(CF$_3$)$_2$); 46.21 (dm, $J_{P-P'}$ = 105 Hz, R$_2$PAd); **EA** calcd for $C_{48}H_{49}F_6FeP_3$ (944.52): C 61.04, H 5.23, F 12.07, P 9.84; found: C 60.85, H 5.28, F 12.22, P 9.77; **HRMS** (MALDI) calcd (m/z) for $C_{48}H_{49}F_6FeP_3$: 945.1724 ([MH]$^+$); found: 954.1742 ([MH]$^+$).

(R_P)-{(S_C)-1-[(R_{Fc})-2-(Diphenylphosphino)ferrocenyl]ethyl}-{(S_C)-1-[(R_{Fc})-2-(dicyclohexylphosphino)ferrocenyl]ethyl}-adamantylphosphine (181). Under argon atmosphere, (S_C,R_{Fc})-1-[dicyclohexylphosphino]-2-{1-[(1-adamantyl)phosphino]ethyl}-ferrocene (91) (337 mg, 584 µmol, 1 eq.) and (S_C,R_{Fc})-2-[1-acetoxyethyl]-1-[diphenylphosphino]ferrocene (320 mg, 701 µmol, 1.2 eq.) were dissolved in 1 mL of degassed acetic acid and the dark red mixture was stirred for 20 h at room temperature. The solvent was then removed under reduced pressure, the residue diluted with saturated aqueous NaHCO$_3$ and diethyl ether and the phases were separated. The aqueous phase was extracted with diethyl ether, the organic phases combined and washed with saturated aqueous NaHCO$_3$. After drying over MgSO$_4$, the solvent was removed under reduced pressure and the residue purified by FC (silica, pentane:EtOAc 40:1-10:1) to yield the impure product in the first fraction. Recrystallization from EtOAc yielded the pure product as a single isomer. Yield: 267 mg (47%), orange crystalline solid. ^{1}**H NMR** (500 MHz, C_6D_6): δ 1.15-1.50 (m, 9 H), 1.55-1.90 (m, 28 H); 1.99-2.01 (m, 1 H, Cy); 2.13-2.22 (m, 4 H, Cy); 2.27-2.29 (m, 1 H, Cy); 2.43 (bs, 1 H, CHMe); 3.36 (bs, 1 H, CHMe); 3.78-3.88 (m, 6 H, Cp', Cp); 3.97 (s, 1 H, Cp); 4.11 (s, 1 H, Cp); 4.18 (s, 1 H, Cp); 4.25 (s, 1 H, Cp); 4.37 (s, 6 H, Cp', Cp); 7.11-7.14 (m, 4 H, PPh_2); 7.22 (t, J = 7.0 Hz, 2 H, PPh_2); 7.62 (t, J = 6.8 Hz, 2 H, PPh_2); 7.80 (bs, 2 H, PPh_2); ^{31}**P{^{1}H} NMR** (202 MHz, C_6D_6): δ -23.76 (d, J_{P-P} = 71.0 Hz, PPh_2); -15.76 (s, PCy_2); 30.37 (d, J_{P-P} = 71.5 Hz, R$_2$PAd); **HRMS** (MALDI) calcd (m/z) for $C_{58}H_{71}Fe_2P_3$: 973.3542 ([MH]$^+$); found: 973.3533 ([MH]$^+$).

**Bis{(R_C)-1-[(R_P,S_{Fc})-2-((1-napthyl)phenylphosphino)-
ferrocenyl]ethyl}adamantylphosphine (195).** (S_P,R_C,S_{Fc})-1-[(1-
Naphthyl)phenylphosphino]-2-[(1-acetoxy)ethyl]ferrocene (193)
(337 mg, 66 µmol, 1 eq.) and 1-adamantylphosphine (56 mg, 333
µmol, 0.5 eq.) were suspended in 10 mL of degassed acetic acid

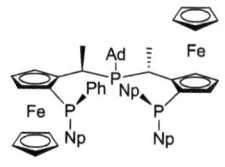

and the mixture stirred at room temperature for 16 h. Then, diethyl ether and saturated aqueous NaHCO$_3$ solution were added, the organic layer was separated and the aqueous phase extracted with diethyl ether. The combined organic phases were then dried over MgSO$_4$ and the solvent evaporated in vacuo. The residue was purified by FC (silica, pentane:Et$_2$O 20:1) to afford the product as a mixture of two diastereoisomers. Recrystallization from EtOAc yielded the product as a single diastereoisomer. Yield: 289 mg (82%), orange crystalline solid. ^{1}H NMR (300 MHz, CD$_2$Cl$_2$): δ 1.35-1.55 (m, 12 H, *Ad*); 1.61 (bs, 3 H, *Ad*); 1.77 (dd, *J* = 7.3, 4.2 Hz, 3 H, CH*Me*); 2.07 (t, *J* = 6.7 Hz, 3 H, CH*Me*); 3.04-3.23 (m, 1 H, C*H*Me); 3.46 (s, 5 H, *Cp'*); 3.51 (s, 5 H, *Cp'*); 3.57-3.69 (m, 1 H, C*H*Me); 3.90-3.99 (m, 2 H, *Cp*); 4.15 (bs, 1 H, *Cp*); 4.24 (bs, 1 H, *Cp*); 4.25-4.28 (m, 1 H, *Cp*); 4.38 (bs, 1 H, *Cp*); 7.09-7.31 (m, 8 H, P*NpPh*); 7.35-7.76 (m, 10 H, P*NpPh*); 7.81-7.99 (m, 4 H, P*NpPh*); 9.36-9.41 (m, 1 H, P*NpPh*); 9.45-9.50 (m, 1 H, P*NpPh*); ^{31}P{^{1}H} NMR (121 MHz, CDCl$_3$): δ -44.64 (d, $J_{P-P'}$ = 69.3 Hz, *P*PhNp); -42.97 (d, $J_{P-P'}$ = 4.6 Hz, *P*PhNp'); +33.61 (dd, $J_{P-P'}$ = 69.2, 4.7 Hz, R$_2$*P*Ad).

**Bis{(R_C)-1-[(R_P,S_{Fc})-2-((2-methoxyphenyl)phenylphosphino)-
ferrocenyl]ethyl}cyclohexylphosphine (196).** (S_P,R_C,S_{Fc})-1-[(2-
Methoxyphenyl)phenylphosphino]-2-[(1-acetoxy)ethyl]ferrocene
(194) (1.152 g, 2.37 mmol, 1 eq.) and cyclohexylphosphine (162
µL, 1.18 mmol, 0.5 eq.) were suspended in 4 mL of degassed acetic
acid and the mixture stirred at room temperature for 19 h. Then, diethyl ether and saturated aqueous NaHCO$_3$ solution were added, the organic layer was separated and the aqueous phase extracted with diethyl ether. The combined organic phases were then dried over MgSO$_4$ and the solvent evaporated in vacuo. The residue was purified by FC (silica, cyclohexane:EtOAc 20:1) to afford the product as a single diastereoisomer. Yield: 667 mg (58%), orange crystalline solid. ^{1}H NMR (300 MHz, CDCl$_3$): δ 0.63-0.84 (m, 1 H, *Cy*); 0.90-1.13 (m, 4 H, *Cy*); 1.15-1.69 (m, 6 H, *Cy*); 1.56 (dd, *J* = 6.9, 4.5 Hz, 3 H, CH*Me*); 1.64 (t, *J* = 7.6 Hz, 3 H, CH*Me*); 3.13-3.24 (m, 1 H, C*H*Me); 2.69-2.92 (m, 1 H, C*H*Me); 3.85 (s, 5 H, *Cp'*); 3.92 (s,

1 H, *Cp*); 3.93 (s, 6 H, O*Me*); 3.96 (s, 6 H, *Cp, Cp'*); 4.09 (bs, 1 H, *Cp*); 4.12 (bs, 1 H, *Cp*); 4.24 (bs, 1 H, *Cp*); 4.30 (bs, 1 H, *Cp*); 6.86 (m, 4 H, P*AnPh*); 7.12-7.26 (m, 8 H, P*AnPh*); 7.25-7.45 (m, 6 H, P*AnPh*); ^{31}P{^{1}H} NMR (121 MHz, CDCl$_3$): δ -41.92 (d, $J_{\text{P-P'}}$ = 8.9 Hz, *P*PhNp); -41.73 (d, $J_{\text{P-P'}}$ = 27.2 Hz, *P*PhNp'); +5.31 (dd, $J_{\text{P-P'}}$ = 26.6, 8.7 Hz, R$_2$*P*Cy); **HRMS** (MALDI) calcd (*m/z*) for C$_{56}$H$_{60}$Fe$_2$O$_2$P$_3$: 969.2501 ([MH]$^+$); found: 969.2501 ([MH]$^+$).

Bis{(R_C)-1-[(R_P,S_Fc)-2-((1-adamantyl)methylphosphino)-ferrocenyl]ethyl}cyclohexylphosphine (197). (R_P,R_C,S_Fc)-1-[(1-Adamantyl)methylphosphino]-2-[(1-dimethylamino)ethyl]-ferrocene (**192**) (400 mg, 915 µmol, 1 eq.) and trifluoroacetic acid (136 µL, 1829 µmol, 2 eq.) were dissolved in 1 mL of degassed acetic acid. To the dark red solution was added cyclohexylphosphine (62 µL, 458 µmol, 0.5 eq.) and the mixture heated up to 80 °C and stirred at this temperature for 20 h. The solvent was then removed under reduced pressure and the residue diluted with diethyl ether and saturated aqueous NaHCO$_3$ solution. The organic layer was separated and the aqueous phase extracted with diethyl ether. The combined organic phases were then dried over MgSO$_4$ and the solvent evaporated in vacuo. The residue was purified by FC (silica, pentane:EtOAc 100:1) to yield the impure product in the second coloured fraction. Recrystallization from EtOAc afforded the oxidation sensitive product as a single diastereoisomer. Yield: 180 mg (44%), yellow powder. ^{1}**H NMR** (250 MHz, C$_6$D$_6$): δ 1.18-2.24 (m, 53 H, *Cy, Ad*, CH*Me*, P*Me*); 3.13-3.39 (m, 1 H, C*H*Me); 3.97 (s, 1 H, *Cp*); 4.03 (s, 1 H, *Cp*); 4.09 (s, 7 H, *Cp, Cp'*); 4.15 (s, 5 H, *Cp'*); 4.43 (s, 1 H, *Cp*); 4.66 (s, 1 H, *Cp*); ^{31}P{^{1}H} **NMR** (101 MHz, C$_6$D$_6$): δ -31.52 (d, $J_{\text{P-P'}}$ = 11.7 Hz, *P*AdMe); -31.08 (d, $J_{\text{P-P'}}$ = 7.2 Hz, *P*AdMe); 10.00 (dd, $J_{\text{P-P'}}$ = 11.4, 7.3 Hz, R$_2$*P*Cy); **EA** calcd for C$_{52}$H$_{71}$Fe$_2$P$_3$ (900.75): C 69.34, H 7.94, P 10.32; found: C 69.23, H 7.99, P 10.13; **HRMS** (MALDI) calcd (*m/z*) for C$_{52}$H$_{72}$Fe$_2$P$_3$: 901.3542 ([MH]$^+$); found: 901.3545 ([MH]$^+$).

Bis{(R_C)-1-[(S_P,S_{Fc},)-2-(phenyltrifluoromethylphosphino)-ferrocenyl]ethyl}cyclohexylphosphine (199). (S_P,R_C,S_{Fc})-1-(Phenyltrifluoromethylphosphino)-2-[1-(dimethylamino)ethyl]-ferrocene ((S_P)-**142**) (400 mg, 923 µmol, 1 eq.) and trifluoroacetic acid (69 µL, 923 µmol, 1 eq.) were dissolved in 1 mL of degassed

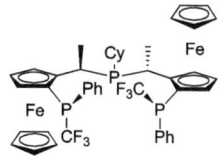

acetic acid. To the dark red solution was added cyclohexylphosphine (63 µL, 462 µmol, 0.5 eq.) and the mixture heated up to 70 °C and stirred at this temperature for 15 h. The solvent was then removed under reduced pressure and the residue diluted with diethyl ether and saturated aqueous NaHCO$_3$ solution. The organic layer was separated and the aqueous phase extracted with diethyl ether. The combined organic phases were then dried over MgSO$_4$ and the solvent evaporated in vacuo. The residue was purified by FC (silica, pentane:EtOAc 100:1-50:1) to yield the oxidation sensitive product as a single diasteroisomer. Yield: 315 mg (76%), orange powder. **^{1}H NMR** (300 MHz, CDCl$_3$): δ 0.01-0.17 (m, 1 H, Cy); 0.45-1.10 (m, 8 H, Cy); 1.15-1.30 (m, 2 H, Cy); 1.35 (dd, J = 6.8, 5.5 Hz, 3 H, CH*Me*); 1.53 (dd, J = 9.7, 7.2 Hz, 3 H, CH*Me*); 2.49-2.60 (m, 1 H, C*H*Me); 2.98 (q, J = 7.0 Hz, 1 H, C*H*Me); 4.05 (bs, 1 H, Cp); 4.16 (bs, 1 H, Cp); 4.23 (s, 6 H, Cp, Cp'); 4.30 (s, 5 H, Cp'); 4.33 (s, 1 H, Cp); 4.57 (s, 1 H, Cp); 7.25-7.41 (m, 6 H, Ph); 7.45-7.56 (m, 4 H, Ph); **^{19}F NMR** (282 MHz, CDCl$_3$): δ -53.63 (d, $^2J_{F-P}$ = 70.7 Hz, CF_3); -53.26 (dd, $^2J_{F-P}$ = 67.4 Hz, $J_{F-P'}$ = 2.8 Hz, CF_3'); **^{31}P{^{1}H} NMR** (121 MHz, CDCl$_3$): δ -16.08 (qd, $^2J_{P-F}$ = 67.2 Hz, $J_{P-P'}$ = 59.0 Hz, PPhCF$_3$); -14.05 (qd, $^2J_{P-F}$ = 70.4 Hz, $J_{P-P'}$ = 17.0 Hz, PPhCF$_3$'); 20.19 (dd, $J_{P-P'}$ = 61.2, 17.0 Hz, R$_2$$P$Cy); **EA** calcd for C$_{44}H_{45}F_6Fe_2P_3$ (892.45): C 59.22, H 5.08, F 12.77, P 10.41; found: C 59.23, H 5.27, F 12.66, P 10.36; **HRMS** (MALDI) calcd (*m/z*) for C$_{44}$H$_{46}$F$_6$Fe$_2$P$_3$: 893.1411 ([MH]$^+$); found: 893.1397 ([MH]$^+$).

Bis{(R_C)-1-[(R_P,S_{Fc})-2-(phenyltrifluoromethylphosphino)-ferrocenyl]ethyl}cyclohexylphosphine (200). (S_P,R_C,S_{Fc})-1-(Phenyltrifluoromethylphosphino)-2-[1-(dimethylamino)ethyl]-ferrocene ((R_P)-**142**) (400 mg, 923 µmol, 1 eq.) and trifluoroacetic acid (69 µL, 923 µmol, 1 eq.) were dissolved in 1 mL of degassed acetic acid. To the dark red solution was added cyclohexylphosphine (63 µL, 462 µmol, 0.5 eq.) and the mixture heated up to 70 °C and stirred at this temperature for 15 h. The solvent was then removed under reduced pressure and the residue diluted with diethyl ether and saturated aqueous NaHCO$_3$ solution. The organic layer was separated and the aqueous

6 Experimental Part

phase extracted with diethyl ether. The combined organic phases were then dried over MgSO$_4$ and the solvent evaporated in vacuo. The residue was purified by FC (silica, pentane:EtOAc 100:1-50:1) to yield the oxidation sensitive product as a single diastereoisomer. Yield: 289 mg (70%), orange powder. ^{1}**H NMR** (300 MHz, CDCl$_3$): δ 0.85-1.05 (m, 1 H, Cy); 1.10-1.32 (m, 5 H, Cy); 1.46-1.75 (m, 11 H, Cy, CHMe); 3.03-3.14 (m, 1 H, CHMe); 3.17-3.27 (m, 1 H, CHMe); 3.75 (s, 5 H, Cp'); 3.79 (s, 5 H, Cp'); 4.31 (bs, 1 H, Cp); 4.34 (bs, 2 H, Cp); 4.40 (s, 1 H, Cp); 4.45 (s, 1 H, Cp); 4.51 (s, 1 H, Cp); 7.50-7.60 (m, 6 H, Ph); 7.98-8.09 (m, 4 H, Ph); ^{19}**F NMR** (282 MHz, CDCl$_3$): δ -55.16 (dd, $^2J_{F-P}$ = 67.1 Hz, $J_{F-P'}$ = 11.5 Hz, CF$_3$); -54.76 (dd, $^2J_{F-P}$ = 64.5 Hz, $J_{F-P'}$ = 18.5 Hz, CF$_3$'); ^{31}**P{^{1}H} NMR** (121 MHz, CDCl$_3$): δ -14.31 (qd, $^2J_{P-F}$ = 64.9 Hz, $J_{P-P'}$ = 16.6 Hz, PPhCF$_3$); -16.01 (bq, $^2J_{P-F}$ = 66.5 Hz, PPhCF$_3$'); 23.19-24.19 (m, R$_2$PCy).

6.6 Transition-Metal Complexes of Pigiphos Ligands

[PdCl(179)]PF$_6$ (183). (S$_P$)-{(R$_C$)-1-[(S$_{Fc}$)-2-(Bis(trifluoro-methyl)phosphino)ferrocenyl]ethyl}-{(R$_C$)-1-[(S$_{Fc}$)-2-(diphenylphosphino)ferrocenyl]ethyl}adamantylphosphine (**179**) (80 mg, 85 µmol, 1 eq.) and [PdCl$_2$(COD)] (22 mg, 85 µmol, 1 eq.) were dissolved in 5 mL of dichloromethane. The deep red

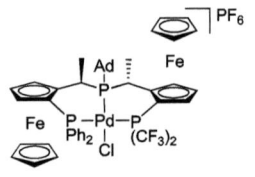

solution was stirred for 10 min before TlPF$_6$ (33 mg, 93 µmol, 1.1 eq.) was added. The dark mixture was filtered over celite, the solvent was partly removed and the solution overlayered with pentane. After 12 h the liquid was decanted and the residue washed with ether and dried under vacuum. Yield: 77 mg (74%), dark purple crystalline solid. ^{1}**H NMR** (300 MHz, CD$_2$Cl$_2$): δ 1.59-1.71 (m, 6 H, PAd); 1.97-2.25 m, 15 H, PAd, CHMe); 3.30-3.50 (m, 1H, CHMe); 3.63 (bs, 6 H, Cp', CHMe); 4.49 (bs, 5 H, Cp'); 4.62 (bs, 1 H, Cp); 4.85 (bs, 1 H, Cp); 4.98 (bs, 1 H, Cp); 5.04 (bs, 1 H, Cp); 5.10 (bs, 1 H, Cp); 7.39-7.79 (m, 10 H, PPh$_3$); ^{19}**F NMR** (282 MHz, CD$_2$Cl$_2$): δ -73.06 (d, $^1J_{F-P}$ = 710.8 Hz, PF$_6$); -53.60 – -52.10 (m, CF$_3$); -48.00 – -46.00(m, CF$_3$'); ^{31}**P{^{1}H} NMR** (121 MHz, CD$_2$Cl$_2$): δ -144.25 (hept., $^1J_{P-F}$ = 710.6 Hz, PF$_6$); 15.15 (bd, J_{P-P} = 450 Hz, PPh$_2$); 49.19 (bd, J_{P-P} = 447 Hz, P(CF$_3$)$_2$); 117.35 – 121.15 (m, PAd); **HRMS** (MALDI) calcd (m/z) for C$_{48}$H$_{49}$Cl$_2$F$_6$FeP$_3$Pd: 1085.0386 ([M]$^+$); found: 1085.0420 ([M]$^+$).

6 Experimental Part

[NiCl(179)]PF$_6$ (184). A green solution of nickel(II) chloride hexahydrate (50.3 mg, 212 µmol, 1 eq.) in ethanol (1 mL) was added to an orange solution of (S$_P$)-{(R$_C$)-1-[(S$_{Fc}$)-2-(bis-(trifluoromethyl)phosphino)ferrocenyl]ethyl}-{(R$_C$)-1-[(S$_{Fc}$)-2-(diphenylphosphino)ferrocenyl]ethyl}adamantylphosphine **(179)**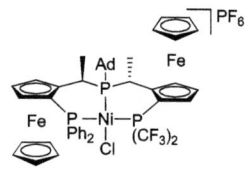
(200 mg, 212 µmol, 1 eq.) in dichloromethane (1 mL). The dark solution was stirred for 10 min before TlPF$_6$ (81.4 mg, 233 µmol, 1.1 eq.) was added. The dark blue mixture was filtered over celite, the solvent was partly removed and the solution overlayered with pentane. After 12 h the liquid was decanted and the residue washed with diethyl ether and dried under vacuum. Yield: 200 mg (80%), dark blue crystalline solid. ^{1}**H NMR** (300 MHz, CD$_2$Cl$_2$): δ 1.69-1.84 (m, 6 H, P*Ad*); 1.98 (dd, $^3J_{H-P}$ = 12.8 Hz, ^3J = 7.3 Hz, 3 H, CH*Me*); 2.05 (bs, 6 H, P*Ad*); 2.10 (dd, $^3J_{H-P}$ = 11.3 Hz, ^3J = 7.0 Hz, 3 H, CH*Me*); 2.45 (bs, 3 H, P*Ad*); 3.10-3.20 (m, 2 H, C*H*Me, C*H*Me); 3.90 (s, 5 H, *Cp'*); 4.03 (s, 1 H, *Cp*); 4.51 (s, 5 H, *Cp'*); 4.59 (s, 1 H, *Cp*); 4.86 (s, 1 H, *Cp*); 4.91 (bs, 1 H, *Cp*); 5.08 (s, 1 H, *Cp*); 5.12 (s, 1 H, *Cp*); 7.50-7.68 (m, 8 H, P*Ph$_2$*); 7.79-8.00 (m, 2 H, P*Ph$_2$*); ^{19}**F NMR** (282 MHz, CD$_2$Cl$_2$): δ -73.01 (d, $^1J_{F-P}$ = 710.8 Hz, P*F$_6$*); -65.94 (bd, $^2J_{P-F}$ = 65.9 Hz, PC*F$_3$*); -48.65 (bd, $^2J_{P-F}$ = 67.7 Hz, PC*F$_3$*'); ^{31}**P{^{1}H} NMR** (121 MHz, CD$_2$Cl$_2$): δ -144.28 (hept., $^1J_{P-F}$ = 710.9 Hz, P*F$_6$*); 8.76 (dd, J_{P-P} = 369.9, 61.8 Hz, P*Ph$_2$*); 45.28 (dm, J_{P-P} = 358.5 Hz, *P*(CF$_3$)$_2$); 110.71 (bs, *P*Ad); **HRMS** (MALDI) calcd (*m/z*) for C$_{48}$H$_{49}$ClF$_6$Fe$_2$NiP$_3$: 1037.0689 ([M]$^+$); found: 1037.0674 ([M]$^+$).

[NiCl(181)]PF$_6$ (185). A green solution of nickel(II) chloride hexahydrate (4.8 mg, 21 µmol, 1 eq.) in ethanol (1 mL) was added to an orange solution of (R$_P$)-{(S$_C$)-1-[(R$_{Fc}$)-2-(diphenylphosphino)ferrocenyl]ethyl}-{(S$_C$)-1-[(R$_{Fc}$)-2-(dicyclohexylphosphino)ferrocenyl]ethyl}adamantylphosphine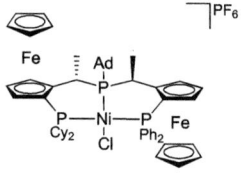
(181) (20 mg, 21 µmol, 1 eq.) in dichloromethane (1 mL). The dark solution was stirred for 10 min before TlPF$_6$ (7.9 mg, 23 µmol, 1.1 eq.) was added. The dark blue mixture was filtered over celite, the solvent was removed and the residue washed with diethyl ether and dried under vacuum. Yield: 23 mg (92%), dark blue crystalline solid. ^{1}**H NMR** (500 MHz, CD$_2$Cl$_2$): δ 0.94-1.05 (m, 1 H, *Cy*, *Ad*); 1.07-1.18 (m, 1 H, *Cy*); 1.16-2.18 (m, 35 H, *Cy*); 2.24-2.34 (m, 1 H, *Cy*); 2.49 (d, *J* = 11.7 Hz, 1 H, *Cy*); 2.55 (d, *J* = 11.1 Hz, 3 H); 2.63 (d, *J* = 12.8 Hz, 1 H, *Cy*); 3.00 (q, ^3J = 6.2 Hz, 1 H, C*H*Me); 3.13 (q, ^3J = 6.0 Hz, 1 H, C*H*Me); 3.84 (bs, 1 H, *Cp*); 3.98 (s, 5 H, *Cp'*); 4.40 (s, 5 H, *Cp'*); 4.42 (bs, 1 H, *Cp*); 4.58 (bs, 1 H, *Cp*);

4.71 (bs, 1 H, *Cp*); 4.80 (bs, 1 H, *Cp*); 4.82 (bs, 1 H, *Cp*); 7.50-7.60 (m, 5 H, P*Ph₂*); 7.63 (t, *J* = 7.4 Hz, 1 H, P*Ph₂*); 7.72-8.79 (m, 4 H, P*Ph₂*); ^{19}**F NMR** (282 MHz, CD$_2$Cl$_2$): δ -73.01 (d, $^1J_{F\text{-}P}$ = 711 Hz, PF_6).

In ^{31}P NMR no signals are visible at 298 K, but two isomers are visible at 203 K: Isomer A: ^{31}**P{^{1}H} NMR** (121 MHz, CD$_2$Cl$_2$): δ -143.55 (hept., $^1J_{P\text{-}F}$ = 711 Hz, PF_6); 5.37 (dd, $J_{P\text{-}P'}$ = 299, 63.4 Hz, P*Ph₂*); 24.91 (dd, $J_{P\text{-}P'}$ = 303, 51.0 Hz, P*Cy₂*); 111.42 (dd, $J_{P\text{-}P'}$ = 58.8, 51.0 Hz, P*Ad*); Isomer B: ^{31}**P{^{1}H} NMR** (121 MHz, CD$_2$Cl$_2$): δ -143.55 (hept., $^1J_{P\text{-}F}$ = 711 Hz, PF_6); 5.01 (dd, $J_{P\text{-}P'}$ = 303, 55.8 Hz, P*Ph₂*); 17.40 (dd, $J_{P\text{-}P'}$ = 303, 44.3 Hz, P*Cy₂*); 111.42 (dd, $J_{P\text{-}P'}$ = 58.8, 51.0 Hz, P*Ad*); **EA** calcd for C$_{58}$H$_{71}$ClF$_6$Fe$_2$NiP$_4$ (1211.92): C 57.48, H 5.90, Cl 2.93 F 9.41, P 10.22; found: C 57.81, H 6.40; **HRMS** (MALDI) calcd (*m/z*) for C$_{58}$H$_{71}$ClFe$_2$P$_3$Ni: 1065.2507 ([M]$^+$); found: 1065.2489 ([M]$^+$); 1030.2810 ([M-Cl]$^+$).

[NiCl(195)]PF$_6$ (201). A green solution of nickel(II) chloride hexahydrate (12.1 mg, 50.8 μmol, 1 eq.) in ethanol (1 mL) was added to an orange solution of bis{(*R*$_C$)-1-[(*S*$_{Fc}$,*S*$_P$)-2-((1-napthyl)phenylphosphino)ferrocenyl]ethyl}cyclohexylphosphine (**195**) (50 mg, 50.8 μmol, 1 eq.) in toluene (1 mL).

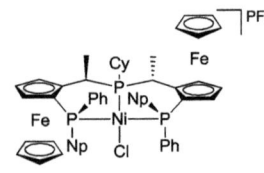

The dark solution was stirred for 10 min before TlPF$_6$ (19.5 mg, 55.9 μmol, 1.1 eq.) was added. The solvent was removed, the dark purple residue suspended in dichloromethane and filtered over celite. The solvent was partly removed and the solution overlayered with pentane. After 12 h the liquid was decanted and the residue washed with ether and dried under vacuum. Yield: 200 mg (80%), deep purple crystalline solid. ^{1}**H NMR** (500 MHz, CD$_2$Cl$_2$): δ 0.75-2.1 (m, 21 H, P*Ad*, C*HMe*); 3.18 (bs, 6 H, C*HMe*, *Cp'*); 3.36 (bs, 1 H, C*HMe*); 4.56 (s, 2 H, *Cp*); 4.59 (s, 5 H, *Cp'*); 4.70 (s, 2 H, *Cp*); 4.98 (s, 1 H, *Cp*); 5.21 (bs, 1 H, *Cp*); 7.23-7.85 (m, 18 H, P*NpPh*); 7.91 (bs, 1 H, P*NpPh*); 8.04 (d, *J* = 8.0 Hz, 1 H, P*NpPh*); 8.14 (d, *J* = 7.3 Hz, 1 H, P*NpPh*); 8.27 (d, *J* = 6.8 Hz, 2 H, P*NpPh*); 9.93 (bs, 1 H, P*NpPh*); ^{19}**F NMR** (188 MHz, CD$_2$Cl$_2$): δ -73.35 (d, $^1J_{F\text{-}P}$ = 710.5 Hz, PF_6); ^{31}**P{^{1}H} NMR** (203 MHz, CD$_2$Cl$_2$): δ -144.28 (hept., $^1J_{P\text{-}F}$ = 710.9 Hz, PF_6); +10.29 (bs, P*NpPh*); 11.79 (bs, P*NpPh'*); 101.12 (bs, P*Ad*); **EA** calcd for C$_{66}$H$_{63}$ClF$_6$Fe$_2$NiP$_4$ (1299.95): C 60.98, H 4.88, Cl 2.73 F 8.77, P 10.32; found: C 60.79, H 4.94, Cl 2.53, F 9.45, P 9.30; **HRMS** (MALDI) calcd (*m/z*) for C$_{66}$H$_{63}$Fe$_2$NiP$_3$ 1118.2192 ([M-Cl]$^+$); found: 1118.2176 ([M-Cl]$^+$).

[NiCl(196)]PF₆ (202). A green solution of nickel (II) chloride hexahydrate (19.7 mg, 83 μmol, 1 eq.) in ethanol (1 mL) was added to an orange solution of bis{(R_C)-1-[(S_P,S_{Fc})-2-((2-methoxyphenyl)phenylphosphino)ferrocenyl]ethyl}-cyclohexylphosphine (**196**) (80.3 mg, 83 μmol, 1 eq.) in

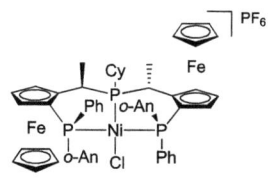

dichloromethane (1 mL). The dark solution was stirred for 20 min before TlPF₆ (31.9 mg, 91 μmol, 1.1 eq.) was added. The solvent was removed, the dark red residue suspended in dichloromethane and filtered over celite. The solvent was partly removed and the solution overlayered with pentane. After 24 h the liquid was decanted and the residue washed with diethyl ether and dried under vacuum. Yield: 92 mg (92%), red crystalline solid. Two isomers are detectable at room temperature. **¹H NMR** (500 MHz, CD₂Cl₂): δ 0.85-2.14 (m, 16 H, CH*Me*, P*Cy*); 2.41 (bs, 1 H); 2.64 (bs, 1 H); 3.54 (bs, 6 H); 3.57-3.69 (m, 5 H); 3.80-3.93 (m, 5 H); 4.21-4.37 (m, 2 H); 4.56-4.94 (m, 5 H, *Cp*); 6.89-7.97 (m, 18 H, P*AnPh*); **¹⁹F NMR** (282 MHz, CD₂Cl₂): δ -73.28 (d, $^1J_{F-P}$ = 711 Hz, P*F₆*); **³¹P{¹H} NMR** (203 MHz, CD₂Cl₂): δ -144.19 (hept., $^1J_{P-F}$ = 711 Hz, P*F₆*); -10.23 (bs, P*AnPh*); 60.88 (bs, P*Cy*); **EA** calcd For C₅₆H₅₉ClF₆Fe₂NiO₂P₄ (1207.80) C 55.69, H 4.92, Cl 2.94 F 9.44, P 10.26; found: C 55.85, H 5.11, Cl 3.10, F 9.61, P 9.98; **HRMS** (MALDI) calcd (*m/z*) for C₅₆H₅₉ClFe₂NiO₂P₃: 1026.178 ([M-Cl]⁺); found: 1026.179 ([M-Cl]⁺).

Minor isomer: δ 72.50 (bs, P*Cy*)

[NiCl(199)]PF₆ (293). A green solution of nickel(II) chloride hexahydrate (7.9 mg, 34 μmol, 1 eq.) in ethanol (1 mL) was added to an orange solution of bis{(R_C)-1-[(S_{Fc},S_P)-2-(phenyl-trifluoromethylphosphino)ferrocenyl]ethyl}cyclohexyl-phosphine (**199**) (30 mg, 34 μmol, 1 eq.) in dichloromethane

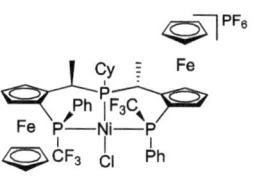

(1 mL). The dark red solution was stirred for 10 min before TlPF₆ (12.9 mg, 37 μmol, 1.1 eq.) was added. The mixture was filtered over celite and the solvent removed under reduced pressure. The dark red residue was washed with diethyl ether and dried under vacuum. Yield: 30 mg (79%), dark red crystalline solid. **¹H NMR** (300 MHz, CD₂Cl₂): δ 0.46-0.78 (m, 1 H, *Cy*); 0.95-1.14 (m, 1 H, *Cy*); 1.23-1.45 (m, 2 H, *Cy*); 1.57-1.76 (m, 4 H, *Cy*, CH*Me*); 1.78-1.97 (m, 3 H, *Cy*); 2.08 (bs, 3 H, CH*Me*); 2.45 (d, *J* = 11.5 Hz, 1 H, *Cy*); 2.94 (d, *J* = 10.5 Hz, 1 H, *Cy*); 3.27-3.40 (m, 2 H, C*H*Me, *Cy*); 3.52-3.65 (m, 1 H, C*H*Me); 4.07 (bs,

6 Experimental Part

1 H, *Cp*); 4.39 (bs, 1 H, *Cp*); 4.42 (s, 5 H, *Cp'*); 4.44 (s, 5 H, *Cp'*); 4.71 (bs, 1 H, *Cp*); 4.81 (bs, 1 H, *Cp*); 4.93 (bs, 1 H, *Cp*); 5.11 (bs, 1 H, *Cp*); 7.45 (t, J = 7.4 Hz, 2 H, P*Ph$_2$*); 7.51-7.66 (m, 5 H, P*Ph$_2$*); 7.68-7.78 (m, 3 H, P*Ph$_2$*); **^{19}F NMR** (282 MHz, CD$_2$Cl$_2$): δ -72.97 (d, $^1J_{\text{F-P}}$ = 711 Hz, P*F$_6$*); -52.19 (bd, $^2J_{\text{P-F}}$ = 67.7 Hz, PC*F$_3$*); -52.11 (bd, $^2J_{\text{P-F}}$ = 66.0 Hz, PC*F$_3$'*); **^{31}P{^{1}H} NMR** (121 MHz, CD$_2$Cl$_2$): δ -144.23 (hept., $^1J_{\text{P-F}}$ = 711 Hz, *PF$_6$*); 20.34 (dt, $^2J_{\text{P-P}}$ = 291 Hz, J = 65 Hz, *P*PhCF$_3$); 29.94 (dt, $^2J_{\text{P-P}}$ = 297 Hz, J = 65 Hz, *P*PhCF$_3$'); +80.25 (bt, $^2J_{\text{P-P}}$ = 67.6 Hz, *P*Cy); **HRMS** (MALDI) calcd (*m/z*) for C$_{48}$H$_{45}$ClF$_6$Fe$_2$NiP$_3$: 985.0375 ([M]$^+$); found: 985.0375 ([M]$^+$); 950.0719 ([M-Cl]$^+$); 893.1421 ([MH-NiCl]$^+$).

7 Literature

[1] Baechler, R. D.; Mislow, K. *J. Am. Chem. Soc.* **1970**, *92*, 3090.
[2] Osborn, J. A.; Wilkinson, G.; Young, J. F. *Chem. Commun.* **1965**, 17.
[3] Knowles, W. S.; Sabacky, M. J. *Chem. Commun.* **1968**, 1445.
[4] Knowles, W. S.; Sabacky, M. J.; Vineyard, B. D.; Weinkauff, D. J. *J. Am. Chem. Soc.* **1975**, *97*, 2567.
[5] Vineyard, B. D.; Knowles, W. S.; Sabacky, M. J.; Bachman, G. L.; Weinkauff, D. J. *J. Am. Chem. Soc.* **1977**, *99*, 5946.
[6] Knowles, W. S. *Angew. Chem., Int. Ed.* **2002**, *41*, 1999.
[7] Dang, T. P.; Kagan, H. B. *J. Chem. Soc., Chem. Commun.* **1971**, 481.
[8] Kagan, H. B.; Dang, T. P. *J. Am. Chem. Soc.* **1972**, *94*, 6429.
[9] Miyashita, A.; Yasuda, A.; Takaya, H.; Toriumi, K.; Ito, T.; Souchi, T.; Noyori, R. *J. Am. Chem. Soc.* **1980**, *102*, 7932.
[10] Burk, M. J.; Feaster, J. E.; Nugent, W. A.; Harlow, R. L. *J. Am. Chem. Soc.* **1993**, *115*, 10125.
[11] Burk, M. J.; Gross, M. F.; Martinez, J. P. *J. Am. Chem. Soc.* **1995**, *117*, 9375.
[12] Togni, A.; Breutel, C.; Schnyder, A.; Spindler, F.; Landert, H.; Tijani, A. *J. Am. Chem. Soc.* **1994**, *116*, 4062.
[13] Harvey, J. S.; Giuffredi, G. T.; Gouverneur, V. *Org. Lett.* **2010**, *12*, 1236.
[14] Cahn, R. S.; Ingold, C.; Prelog, V. *Angew. Chem., Int. Ed.* **1966**, *5*, 385.
[15] Prelog, V.; Helmchen, G. *Angew. Chem., Int. Ed. Engl.* **1982**, *21*, 567.
[16] Levin, C. C. *J. Am. Chem. Soc.* **1975**, *97*, 5649.
[17] Mislow, K. *Trans. N.Y. Acad. Sci.* **1973**, *35*, 227.
[18] Bent, H. A. *Chem. Rev.* **1961**, *61*, 275.
[19] Allred, A. L.; Rochow, E. G. *J. Inorg. Nucl. Chem.* **1958**, *5*, 264.
[20] Grabulosa, A.; Granell, J.; Muller, G. *Coord. Chem. Rev.* **2007**, *251*, 25.
[21] Pietrusiewicz, K. M.; Zablocka, M. *Chem. Rev.* **1994**, *94*, 1375.
[22] Crepy, K. V. L.; Imamoto, T. *New Aspects in Phosphorus Chemistry III* **2003**, *229*, 1.
[23] Horner, L.; Balzer, W. D. *Tetrahedron Lett.* **1965**, 1157.
[24] Naumann, K.; Zon, G.; Mislow, K. *J. Am. Chem. Soc.* **1969**, *91*, 7012.
[25] Imamoto, T.; Kikuchi, S.; Miura, T.; Wada, Y. *Org. Lett.* **2001**, *3*, 87.
[26] Valentine, D.; Blount, J. F.; Toth, K. *J. Org. Chem.* **1980**, *45*, 3691.

7 Literature

[27] Henson, P. D.; Naumann, K.; Mislow, K. *J. Am. Chem. Soc.* **1969**, *91*, 5645.

[28] Ohff, M.; Holz, J.; Quirmbach, M.; Borner, A. *Synthesis* **1998**, 1391.

[29] Carboni, B.; Monnier, L. *Tetrahedron* **1999**, *55*, 1197.

[30] Imamoto, T.; Oshiki, T.; Onozawa, T.; Kusumoto, T.; Sato, K. *J. Am. Chem. Soc.* **1990**, *112*, 5244.

[31] McKinstry, L.; Livinghouse, T. *Tetrahedron Lett.* **1994**, *35*, 9319.

[32] Meisenheimer, J.; Lichtenstadt, L. *Chem. Ber.* **1911**, *44*, 356.

[33] Meisenheimer, J.; Casper, J.; Horing, M.; Lauter, W.; Lichtenstadt, L.; Samuel, W. *Justus Liebigs Ann. Chem.* **1926**, *449*, 213.

[34] Otsuka, S.; Nakamura, A.; Kano, T.; Tani, K. *J. Am. Chem. Soc.* **1971**, *93*, 4301.

[35] Tani, K.; Brown, L. D.; Ahmed, J.; Ibers, J. A.; Yokota, M.; Nakamura, A.; Otsuka, S. *J. Am. Chem. Soc.* **1977**, *99*, 7876.

[36] Wild, S. B. *Coord. Chem. Rev.* **1997**, *166*, 291.

[37] Roberts, N. K.; Wild, S. B. *J. Am. Chem. Soc.* **1979**, *101*, 6254.

[38] Imamoto, T.; Tsuruta, H.; Wada, Y.; Masuda, H.; Yamaguchi, K. *Tetrahedron Lett.* **1995**, *36*, 8271.

[39] Tsuruta, H.; Imamoto, T. *Tetrahedron: Asymmetry* **1999**, *10*, 877.

[40] Ward, T. R.; Venanzi, L. M.; Albinati, A.; Lianza, F.; Gerfin, T.; Gramlich, V.; Tombo, G. M. R. *Helv. Chim. Acta* **1991**, *74*, 983.

[41] Nudelman, A.; Cram, D. J. *J. Am. Chem. Soc.* **1968**, *90*, 3869.

[42] Korpium, O.; Mislow, K. *J. Am. Chem. Soc.* **1967**, *89*, 4784.

[43] Lewis, R. A.; Korpium, O.; Mislow, K. *J. Am. Chem. Soc.* **1967**, *89*, 4786.

[44] Korpium, O.; Lewis, R. A.; Chickos, J.; Mislow, K. *J. Am. Chem. Soc.* **1968**, *90*, 4842.

[45] Imamoto, T.; Kusumoto, T.; Suzuki, N.; Sato, K. *J. Am. Chem. Soc.* **1985**, *107*, 5301.

[46] Death, N. J.; Ellis, K.; Smith, D. J. H.; Trippett, S. *J. Chem. Soc., Chem. Commun.* **1971**, 714.

[47] Mikolajczyk, M.; Omelanczuk, J.; Perlikowska, W. *Tetrahedron* **1979**, *35*, 1531.

[48] Oshiki, T.; Hikosaka, T.; Imamoto, T. *Tetrahedron Lett.* **1991**, *32*, 3371.

[49] Koide, Y.; Sakamoto, A.; Imamoto, T. *Tetrahedron Lett.* **1991**, *32*, 3375.

[50] Juge, S.; Stephan, M.; Laffitte, J. A.; Genet, J. P. *Tetrahedron Lett.* **1990**, *31*, 6357.

[51] Juge, S.; Stephan, M.; Merdes, R.; Genet, J. P.; Halutdesportes, S. *J. Chem. Soc., Chem. Commun.* **1993**, 531.

[52] Li, J. L.; Beak, P. *J. Am. Chem. Soc.* **1992**, *114*, 9206.

[53] Nettekoven, U.; Kamer, P. C. J.; van Leeuwen, P. W. N. M.; Widhalm, M.; Spek, A. L.; Lutz, M. *J. Org. Chem.* **1999**, *64*, 3996.

[54] Maienza, F.; Spindler, F.; Thommen, M.; Pugin, B.; Malan, C.; Mezzetti, A. *J. Org. Chem.* **2002**, *67*, 5239.

[55] Rippert, A. J.; Linden, A.; Hansen, H. J. *Helv. Chim. Acta* **2000**, *83*, 311.

[56] Moulin, D.; Darcel, C.; Juge, S. *Tetrahedron: Asymmetry* **1999**, *10*, 4729.

[57] Moulin, D.; Bago, S.; Bauduin, C.; Darcel, C.; Juge, S. *Tetrahedron: Asymmetry* **2000**, *11*, 3939.

[58] Humbel, S.; Bertrand, C.; Darcel, C.; Bauduin, C.; Juge, S. *Inorg. Chem.* **2003**, *42*, 420.

[59] Bauduin, C.; Moulin, D.; Kaloun, E. B.; Darcel, C.; Juge, S. *J. Org. Chem.* **2003**, *68*, 4293.

[60] Juge, S.; Genet, J. P. *Tetrahedron Lett.* **1989**, *30*, 2783.

[61] Michaelis, A.; Kaehne, R. *Chem. Ber.* **1898**, *31*, 1048.

[62] Arbuzov, B. A. *J. Russ. Phys. Chem. Soc.* **1906**, *38*, 687.

[63] Arbuzov, B. A. *Pure Appl. Chem.* **1964**, *9*, 307.

[64] Corey, E. J.; Chen, Z. L.; Tanoury, G. J. *J. Am. Chem. Soc.* **1993**, *115*, 11000.

[65] Chodkiewicz, W.; Guillerm, D.; Jore, D.; Mathieu, E.; Wodzki, W. *J. Organomet. Chem.* **1984**, *269*, 107.

[66] Segi, M.; Nakamura, Y.; Nakajima, T.; Suga, S. *Chem. Lett.* **1983**, 913.

[67] Muci, A. R.; Campos, K. R.; Evans, D. A. *J. Am. Chem. Soc.* **1995**, *117*, 9075.

[68] Byrne, L. T.; Engelhardt, L. M.; Jacobsen, G. E.; Leung, W. P.; Papasergio, R. I.; Raston, C. L.; Skelton, B. W.; Twiss, P.; White, A. H. *J. Chem. Soc., Dalton Trans.* **1989**, 105.

[69] Strohmann, C.; Strohfeldt, K.; Schildbach, D.; McGrath, M. J.; O'Brien, P. *Organometallics* **2004**, *23*, 5389.

[70] Imamoto, T.; Watanabe, J.; Wada, Y.; Masuda, H.; Yamada, H.; Tsuruta, H.; Matsukawa, S.; Yamaguchi, K. *J. Am. Chem. Soc.* **1998**, *120*, 1635.

[71] Yamanoi, Y.; Imamoto, T. *J. Org. Chem.* **1999**, *64*, 2988.

[72] Nagata, K.; Matsukawa, S.; Imamoto, T. *J. Org. Chem.* **2000**, *65*, 4185.

[73] Ohashi, A.; Imamoto, T. *Tetrahedron Lett.* **2001**, *42*, 1099.

[74] Miura, T.; Yamada, H.; Kikuchi, S.; Imamoto, T. *J. Org. Chem.* **2000**, *65*, 1877.

[75] Imamoto, T.; Sugita, K.; Yoshida, K. *J. Am. Chem. Soc.* **2005**, *127*, 11934.

[76] Wolfe, B.; Livinghouse, T. *J. Am. Chem. Soc.* **1998**, *120*, 5116.

7 Literature

[77] Kovacik, I.; Wicht, D. K.; Grewal, N. S.; Glueck, D. S.; Incarvito, C. D.; Guzei, I. A.; Rheingold, A. L. *Organometallics* **2000**, *19*, 950.

[78] Scriban, C.; Glueck, D. S. *J. Am. Chem. Soc.* **2006**, *128*, 2788.

[79] Blank, N. F.; Moncarz, J. R.; Brunker, T. J.; Scriban, C.; Anderson, B. J.; Amir, O.; Glueck, D. S.; Zakharov, L. N.; Golen, J. A.; Incarvito, C. D.; Rheingold, A. L. *J. Am. Chem. Soc.* **2007**, *129*, 6847.

[80] Scriban, C.; Glueck, D. S.; Golen, J. A.; Rheingold, A. L. *Organometallics* **2007**, *26*, 5124.

[81] Korff, C.; Helmchen, G. *Chem. Commun.* **2004**, 530.

[82] Chan, V. S.; Bergman, R. G.; Toste, F. D. *J. Am. Chem. Soc.* **2007**, *129*, 15122.

[83] Chan, V. S.; Stewart, I. C.; Bergman, R. G.; Toste, F. D. *J. Am. Chem. Soc.* **2006**, *128*, 2786.

[84] Chan, V. S.; Chiu, M.; Bergman, R. G.; Toste, F. D. *J. Am. Chem. Soc.* **2009**, *131*, 6021.

[85] For reviews see: (a) Richards, C. J.; Locke, A. J. *Tetrahedron: Asymmetry* **1998**, *9*, 2377. (b) Colacot, T.J. *Chem. Rev.* **2003**, *103*, 3101. (c) Dai, L. X.; Tu, T.; You, S. L.; Deng, W. P.; Hou, X. L. *Acc. Chem. Res.* **2003**, *36*, 659. (d) Barbaro, P.; Bianchini, C.; Giambastiani, G.; Parisel, S. L. *Coord. Chem. Rev.* **2004**, *248*, 2131.

[86] Togni, A.; Breutel, C.; Soares, M. C.; Zanetti, N.; Gerfin, T.; Gramlich, V.; Spindler, F.; Rihs, G. *Inorg. Chim. Acta* **1994**, *222*, 213.

[87] Lotz, M.; Polborn, K.; Knochel, P. *Angew. Chem., Int. Ed.* **2002**, *41*, 4708.

[88] Perea, J. J. A.; Borner, A.; Knochel, P. *Tetrahedron Lett.* **1998**, *39*, 8073.

[89] Sturm, T.; Weissensteiner, W.; Spindler, F. *Adv. Synth. Catal.* **2003**, *345*, 160.

[90] Sawamura, M.; Hamashima, H.; Sugawara, M.; Kuwano, R.; Ito, Y. *Organometallics* **1995**, *14*, 4549.

[91] Boaz, N. W.; Debenham, S. D.; Mackenzie, E. B.; Large, S. E. *Org. Lett.* **2002**, *4*, 2421.

[92] Nettekoven, U.; Widhalm, M.; Kamer, P. C. J.; vanLeeuwen, P. W. N. M. *Tetrahedron: Asymmetry* **1997**, *8*, 3185.

[93] Maienza, F.; Worle, M.; Steffanut, P.; Mezzetti, A.; Spindler, F. *Organometallics* **1999**, *18*, 1041.

[94] Nettekoven, U.; Widhalm, M.; Kamer, P. C. J.; van Leeuwen, P. W. N. M.; Mereiter, K.; Lutz, M.; Spek, A. L. *Organometallics* **2000**, *19*, 2299.

[95] Nettekoven, U.; Widhalm, M.; Kalchhauser, H.; Kamer, P. C. J.; van Leeuwen, P. W. N. M.; Lutz, M.; Spek, A. L. *J. Org. Chem.* **2001**, *66*, 759.

[96] Oohara, N.; Katagiri, K.; Imamoto, T. *Tetrahedron: Asymmetry* **2003**, *14*, 2171.

[97] Gambs, C.; Consiglio, G.; Togni, A. *Helv. Chim. Acta* **2001**, *84*, 3105.

[98] Chen, W. P.; Mbafor, W.; Roberts, S. M.; Whittall, J. *J. Am. Chem. Soc.* **2006**, *128*, 3922.

[99] Chen, W. P.; Roberts, S. M.; Whittall, J.; Steiner, A. *Chem. Commun.* **2006**, 2916.

[100] Riant, O.; Samuel, O.; Flessner, T.; Taudien, S.; Kagan, H. B. *J. Org. Chem.* **1997**, *62*, 6733.

[101] Chen, W. P.; McCormack, P. J.; Mohammed, K.; Mbafor, W.; Roberts, S. M.; Whittall, J. *Angew. Chem., Int. Ed.* **2007**, *46*, 4141.

[102] Noyori, R.; Ohkuma, T.; Kitamura, M.; Takaya, H.; Sayo, N.; Kumobayashi, H.; Akutagawa, S. *J. Am. Chem. Soc.* **1987**, *109*, 5856.

[103] Tang, W. J.; Zhang, X. M. *Angew. Chem., Int. Ed.* **2002**, *41*, 1612.

[104] Gridnev, I. D.; Yasutake, M.; Higashi, N.; Imamoto, T. *J. Am. Chem. Soc.* **2001**, *123*, 5268.

[105] Yamano, T.; Taya, N.; Kawada, H.; Huang, T.; Imamoto, T. *Tetrahedron Lett.* **1999**, *40*, 2577.

[106] Wang, C. J.; Tao, H. Y.; Zhang, X. M. *Tetrahedron Lett.* **2006**, *47*, 1901.

[107] Imamoto, T.; Nishimura, M.; Koide, A.; Yoshida, K. *J. Org. Chem.* **2007**, *72*, 7413.

[108] Robin, F.; Mercier, F.; Ricard, L.; Mathey, F.; Spagnol, M. *Chem.--Eur. J.* **1997**, *3*, 1365.

[109] Hayashi, T., in *Catalytic Asymmetric Synthesis* (Ed.: I. Ojima), Wiley-VCH, New York, **1993**.

[110] Trost, B. M.; VanVranken, D. L. *Chem. Rev.* **1996**, *96*, 395.

[111] Consiglio, G.; Waymouth, R. M. *Chem. Rev.* **1989**, *89*, 257.

[112] Tsuruta, H.; Imamoto, T. *Synlett* **2001**, 999.

[113] Taylor, A. M.; Altman, R. A.; Buchwald, S. L. *J. Am. Chem. Soc.* **2009**, *131*, 9900.

[114] Imamoto, T.; Saitoh, Y.; Koide, A.; Ogura, T.; Yoshida, K. *Angew. Chem., Int. Ed.* **2007**, *46*, 8636.

[115] *Stereochemistry of Organic Compounds*, Eliel, E. L.; Wilen, S. H.; Mander, L. N. (Ed). New York, **1994**.

[116] Buono, G.; Chiodi, O.; Wills, M. *Synlett* **1999**, 377.

[117] Brunel, J. M.; Buono, G. *New Aspects in Phosphorus Chemistry I* **2002**, *220*, 79.

7 Literature

[118] Vedejs, E.; Daugulis, O.; Diver, S. T. *J. Org. Chem.* **1996**, *61*, 430.

[119] Vedejs, E.; Daugulis, O. *J. Am. Chem. Soc.* **2003**, *125*, 4166.

[120] Shaw, S. A.; Aleman, P.; Vedejs, E. *J. Am. Chem. Soc.* **2003**, *125*, 13368.

[121] Camponovo, F.; Ph. D. Thesis Nr. 18199, ETH Zurich, **2009**.

[122] *Ferrocenes: Homogeneous Catalysis, Organic Synthesis, Materials Science*, Togni, A.; Hayashi, T. (Ed). Wiley-VCH, Weinheim, **1995**.

[123] *Ferrocenes: Ligands, Materials and Biomolecules*, Weinheim, **2008**.

[124] Gokel, G. W.; Ugi, I. K. *J. Chem. Educ.* **1972**, *49*, 294.

[125] Solvias AG, Basel; www.solvias.ch

[126] Hayashi, T.; Yamamoto, K.; Kumada, M. *Tetrahedron Lett.* **1974**, 4405.

[127] Barbaro, P.; Togni, A. *Organometallics* **1995**, *14*, 3570.

[128] Schlögl, K., in *Top. Stereochem.*, John Wiley & Sons, Inc., **1967**, 39-91.

[129] Ackermann, L. *Synthesis* **2006**, 1557.

[130] Ackermann, L., in *Phosphorus Ligands in Asymmetric Catalysis* (Ed.: A. Börner), Wiley-VCH, Weinheim, **2008**, 831.

[131] Haynes, R. K.; Freeman, R. N.; Mitchell, C. R.; Vonwiller, S. C. *J. Org. Chem.* **1994**, *59*, 2919.

[132] Haynes, R. K.; Au-Yeung, T. L.; Chan, W. K.; Lam, W. L.; Li, Z. Y.; Yeung, L. L.; Chan, A. S. C.; Li, P.; Koen, M.; Mitchell, C. R.; Vonwiller, S. C. *Eur. J. Org. Chem.* **2000**, 3205.

[133] Leyris, A.; Nuel, D.; Giordano, L.; Achard, M.; Buono, G. *Tetrahedron Lett.* **2005**, *46*, 8677.

[134] Wolf, C.; Lerebours, R. *J. Org. Chem.* **2003**, *68*, 7077.

[135] Khanapure, S. P.; Garvey, D. S. *Tetrahedron Lett.* **2004**, *45*, 5283.

[136] Dai, W. M.; Yeung, K. K. Y.; Leung, W. H.; Haynes, R. K. *Tetrahedron: Asymmetry* **2003**, *14*, 2821.

[137] Jiang, X. B.; Minnaard, A. J.; Hessen, B.; Feringa, B. L.; Duchateau, A. L. L.; Andrien, J. G. O.; Boogers, J. A. F.; de Vries, J. G. *Org. Lett.* **2003**, *5*, 1503.

[138] Landert, H.; Spindler, F.; Wyss, A.; Blaser, H. U.; Pugin, B.; Ribourduoille, Y.; Gschwend, B.; Ramalingam, B.; Pfaltz, A. *Angew. Chem., Int. Ed.* **2010**, *49*, 6873.

[139] Butti, P.; Ph. D. Thesis Nr. 18839, ETH Zurich, **2009**.

[140] Barbaro, P.; Bianchini, C.; Togni, A. *Organometallics* **1997**, *16*, 3004.

[141] Hayashi, T.; Mise, T.; Fukushima, M.; Kagotani, M.; Nagashima, N.; Hamada, Y.; Matsumoto, A.; Kawakami, S.; Konishi, M.; Yamamoto, K.; Kumada, M. *Bull. Chem. Soc. Jpn.* **1980**, *53*, 1138.

[142] Fadini, L.; Togni, A. *Helv. Chim. Acta* **2007**, *90*, 411.

[143] Hiney, R. M.; Higham, L. J.; Muller-Bunz, H.; Gilheany, D. G. *Angew. Chem., Int. Ed.* **2006**, *45*, 7248.

[144] Henderson, W.; Alley, S. R. *J. Organomet. Chem.* **2002**, *656*, 120.

[145] Brynda, M. *Coord. Chem. Rev.* **2005**, *249*, 2013.

[146] Rauk, A.; Allen, L. C.; Mislow, K. *Angew. Chem., Int. Ed.* **1970**, *9*, 400.

[147] Bader, A.; Pabel, M.; Wild, S. B. *J. Chem. Soc., Chem. Commun.* **1994**, 1405.

[148] Hintermann, L.; Togni, A. *Helv. Chim. Acta* **2000**, *83*, 2425.

[149] Sondenecker, A.; Ph. D. Thesis Nr. 18681, ETH Zurich, **2009**.

[150] Ireland, T.; Perea, J. J. A.; Knochel, P. *Angew. Chem., Int. Ed.* **1999**, *38*, 1457.

[151] Gischig, S.; Ph. D. Thesis Nr. 16270, ETH Zürich, **2005**.

[152] Holt, J.; Maj, A. M.; Schudde, E. P.; Pietrusiewicz, K. M.; Sieron, L.; Wieczorek, W.; Jerphagnon, T.; Arends, I. W. C. E.; Hanefeld, U.; Minnaard, A. J. *Synthesis* **2009**, 2061.

[153] Koller, R.; Togni, A., unpublished results.

[154] Adams, J. J.; Lau, A.; Arulsamy, N.; Roddick, D. M. *Inorg. Chem.* **2007**, *46*, 11328.

[155] Velazco, E. J.; Caffyn, A. J. M.; Le Goff, X. F.; Ricard, L. *Organometallics* **2008**, *27*, 2402.

[156] Adams, J. J.; Arulsamy, N.; Roddick, D. M. *Inorg. Chim. Acta* **2009**, *362*, 2056.

[157] Armanino, N.; Koller, R.; Togni, A. *Organometallics* **2010**, *29*, 1771.

[158] Albinati, A.; Lianza, F.; Pasquali, M.; Sommovigo, M.; Leoni, P.; Pregosin, P. S.; Ruegger, H. *Inorg. Chem.* **1991**, *30*, 4690.

[159] Leoni, P. *Organometallics* **1993**, *12*, 2432.

[160] Buil, M. L.; Esteruelas, M. A.; Onate, E.; Ruiz, N. *Organometallics* **1998**, *17*, 3346.

[161] Pelczar, E. M.; Nytko, E. A.; Zhuravel, M. A.; Smith, J. M.; Glueck, D. S.; Sommer, R.; Incarvito, C. D.; Rheingold, A. L. *Polyhedron* **2002**, *21*, 2409.

[162] Sasamori, T.; Kawai, M.; Takeda, N.; Tokitoh, N. *Chem. Lett.* **2008**, *37*, 1192.

[163] Pankratov, A. N.; Sheenson, V. A.; Mushtakova, S. P.; Gribov, L. A. *J. Struct. Chem.* **1986**, *27*, 199.

[164] Glaser, R.; Kountz, D. J.; Waid, R. D.; Gallucci, J. C.; Meek, D. W. *J. Am. Chem. Soc.* **1984**, *106*, 6324.

7 Literature

[165] Orpen, A. G.; Connelly, N. G. *Organometallics* **1990**, *9*, 1206.

[166] Grobe, J.; Levan, D.; Szameitat, J. *J. Organomet. Chem.* **1985**, *289*, 341.

[167] Grobe, J.; Levan, D.; Meyring, W.; Krebs, B.; Dartmann, M. *J. Organomet. Chem.* **1988**, *346*, 361.

[168] Hoge, B.; Thosen, C.; Pantenburg, I. *Chem.--Eur. J.* **2006**, *12*, 9019.

[169] Panne, P.; Naumann, D.; Hoge, B. *J. Fluorine Chem.* **2001**, *112*, 283.

[170] Murphy-Jolly, M. B.; Lewis, L. C.; Caffyn, A. J. M. *Chem. Commun.* **2005**, 4479.

[171] Koller, R.; Ph. D. Thesis Nr. 19219, ETH Zurich, **2010**.

[172] Eisenberger, P.; Kieltsch, I.; Armanino, N.; Togni, A. *Chem. Commun.* **2008**, 1575.

[173] Sondenecker, A.; Cvengros, J.; Aardoom, R.; Togni, A. *Eur. J. Org. Chem.* **2010**.

[174] For examples, see: a) Grobe, J.; Grosspietsch, T.; Levan, D.; Schulze, J.; Krebs, B.; Dartmann, M., *J. Organomet. Chem.* **1990**, *385*, 255; b) Alder, R. W.; Ganter, C.; Harris, C. J.; Orpen, A. G., *J. Chem. Soc., Chem. Commun.* **1992**, 1170; c) Alder, R. W.; Ellis, D. D.; Hogg, J. K.; Martin, A.; Orpen, A. G.; Taylor, P. N., *Chem. Commun.* **1996**, 537; d) Mizuta, T.; Kunikata, S.; Miyoshi, K., *J. Organomet. Chem.* **2004**, *689*, 2624; e) Miluykov, V.; Bezkishko, I.; Zagidullin, A.; Sinyashin, O.; Lonnecke, P.; Hey-Hawkins, E., *Eur. J. Org. Chem.* **2009**, 1269.

[175] Kauffmann, T.; Antfang, E.; Olbrich, J. *Chem. Ber.-Recl.* **1985**, *118*, 1022.

[176] Alder, R. W.; Ganter, C.; Harris, C. J.; Orpen, A. G. *J. Chem. Soc., Chem. Commun.* **1992**, 1170.

[177] Alder, R. W.; Ellis, D. D.; Hogg, J. K.; Martin, A.; Orpen, A. G.; Taylor, P. N. *Chem. Commun.* **1996**, 537.

[178] Alder, R. W.; Canter, C.; Gil, M.; Gleiter, R.; Harris, C. J.; Harris, S. E.; Lange, H.; Orpen, A. G.; Taylor, P. N. *J. Chem. Soc., Perkin Trans. 1* **1998**, 1643.

[179] For the use of the descriptors *endo/exo* in ferrocenyl ligand systems, see: Togni, A.; Burckhardt, U.; Gramlich, V.; Pregosin, P. S.; Salzmann, R., *J. Am. Chem. Soc.* **1996**, *118*, 1031.

[180] Gouygou, M.; Etemadmoghadam, G.; Koenig, M. *Synthesis* **1987**, 508.

[181] Laneman, S. A.; Fronczek, F. R.; Stanley, G. G. *J. Am. Chem. Soc.* **1988**, *110*, 5585.

[182] Lamb, G. W.; Law, D.; Slawin, A. M. Z.; Clarke, M. L. *Inorg. Chim. Acta* **2009**, *362*, 4263.

[183] Gilliam, O. R.; Johnson, C. M.; Gordy, W. *Physical Review* **1950**, *78*, 140.

[184] For a detailed description see chapter 1.3.3.2 or in: A) Juge, S.; Stephan, M.; Merdes, R.; Genet, J. P.; Halutdesportes, S. *J. Chem. Soc., Chem. Commun.* **1993**, 531. B) Li, J. L.; Beak, P. *J. Am. Chem. Soc.* **1992**, *114*, 9206.

[185] Stiles, A. R.; Rust, F. F.; Vaughan, W. E. *J. Am. Chem. Soc.* **1952**, *74*, 3282.

[186] Mitchell, T. N.; Heesche, K. *J. Organomet. Chem.* **1991**, *409*, 163.

[187] Ashby, E. C.; Gurumurthy, R.; Ridlehuber, R. W. *J. Org. Chem.* **1993**, *58*, 5832.

[188] Ashby, E. C.; Sun, X.; Duff, J. L. *J. Org. Chem.* **1994**, *59*, 1270.

[189] Zhang, W. C.; Zhang, X. M. *J. Org. Chem.* **2007**, *72*, 1020.

[190] Togni, A.; Burckhardt, U.; Gramlich, V.; Pregosin, P. S.; Salzmann, R. *J. Am. Chem. Soc.* **1996**, *118*, 1031.

[191] Holz, J.; Börner, A., in *Phosphorus Ligands in Asymmetric Catalysis: Synthesis and Applications, Vol. 2* (Ed.: A. Börner), Wiley-VCH, Weinheim, **2008**, 809.

[192] Jia, G. C.; Lee, H. M.; Williams, I. D. *Organometallics* **1996**, *15*, 4235.

[193] Johnson, C. R.; Imamoto, T. *J. Org. Chem.* **1987**, *52*, 2170.

[194] Fadini, L.; Ph. D. Thesis Nr. 15593, ETH Zurich, **2004**.

[195] Fadini, L.; Togni, A. *Chimia* **2004**, *58*, 208.

[196] Pfaltz, A.; Drury, W. J. *Proc. Natl. Acad. Sci. U. S. A.* **2004**, *101*, 5723.

[197] Barbaro, P.; Bianchini, C.; Giambastiani, G.; Togni, A. *Chem. Commun.* **2002**, 2672.

[198] Barbaro, P.; Bianchini, C.; Giambastiani, G.; Masi, D.; Parisel, S. L.; Togni, A. *Synthesis* **2004**, 345.

[199] Hintermann, L.; Perseghini, M.; Barbaro, P.; Togni, A. *Eur. J. Inorg. Chem.* **2003**, 601.

[200] Barbaro, P.; Bianchini, C.; Oberhauser, W.; Togni, A. *J. Mol. Catal. A: Chem.* **1999**, *145*, 139.

[201] Fadini, L.; Togni, A. *Chem. Commun.* **2003**, 30.

[202] Fadini, L.; Togni, A. *Tetrahedron: Asymmetry* **2008**, *19*, 2555.

[203] Sadow, A. D.; Haller, I.; Fadini, L.; Togni, A. *J. Am. Chem. Soc.* **2004**, *126*, 14704.

[204] Sadow, A. D.; Togni, A. *J. Am. Chem. Soc.* **2005**, *127*, 17012.

[205] Walz, I.; Bertogg, A.; Togni, A. *Eur. J. Org. Chem.* **2007**, 2650.

[206] Walz, I.; Ph. D. Thesis Nr. 17370, ETH Zurich, **2007**.

[207] Gischig, S.; Togni, A. *Organometallics* **2004**, *23*, 2479.

[208] Drew, D.; Doyle, J. R. *Inorg. Synth.* **1990**, *28*, 346.

[209] Marion, N.; Navarro, O.; Mei, J. G.; Stevens, E. D.; Scott, N. M.; Nolan, S. P. *J. Am. Chem. Soc.* **2006**, *128*, 4101.

7 Literature

[210] Schenck, T. G.; Downes, J. M.; Milne, C. R. C.; Mackenzie, P. B.; Boucher, H.; Whelan, J.; Bosnich, B. *Inorg. Chem.* **1985**, *24*, 2334.

[211] Gottlieb, H. E.; Kotlyar, V.; Nudelman, A. *J. Org. Chem.* **1997**, *62*, 7512.

[212] *SAINT+*

[213] Sheldrick, G. M. *Acta Crystallogr., Sect. A* **1990**, *46*, 467.

[214] *SHELXL-97*, Sheldrick, G. M. (Ed). Program for Crystal Structure Refinement, Universität Göttingen, Göttingen, Germany, **1999**.

[215] Blessing, R. H. *Acta Crystallogr., Sect. B: Struct. Sci.* **1995**, *51*, 816.

[216] Flack, H. D. *Acta Crystallogr., Sect. A* **1983**, *39*, 876.

[217] Bernardinelli, G.; Flack, H. D. *Acta Crystallogr., Sect. A* **1985**, *41*, 500.

[218] *Acta Crystallogr.* **2009**, *C65*, e3.

[219] Stetter, H.; Last, W. D. *Chem. Ber.-Recl.* **1969**, *102*, 3364.

[220] Thompson, A. L. S.; Kabalka, G. W.; Akula, M. R.; Huffman, J. W. *Synthesis* **2005**, 547.

[221] Goossen, L. J.; Dezfuli, M. K. *Synlett* **2005**, 445.

[222] Kieltsch, I.; Ph. D. Thesis Nr. 17990, ETH Zurich, **2008**.

8 Appendix

8.1 Abbreviations

°	degree
°C	degree Celsius
Ac	acetyl
Ad	adamantyl
aq.	aqueous
BINAP	2,2'-bis(diphenylphosphino)-1,1'-binaphthyl
Biph.	biphenyl
Bn	benzyl
brine	saturated NaCl-solution in water
BSA	N,O-bis(trimethylsilyl)acetamide
COD	1,5-cyclooctadiene
COSY	homonuclear correlation spectroscopy
Cp	cyclopentadienyl
Cy	cyclohexyl
DBA	dibenzylidene acetone
DBU	1,8-diazabicyclo[5.4.0]undec-7-ene
DCE	1,2-dichloroethane
DMI	dimethyl itaconate
DCM	dichloromethane
DMF	dimethylformamide
dp	diastereomerically pure
Dppe	1,2-diphenylphosphino ethane
Dppf	1,1'-diphenylphosphino ferrocene
EA	elemental analysis
ee	enantiomeric excess
eq.	equivalents
ESI-MS	electrospray ionization mass spectrometry
EWG	electron-withdrawing group

8 Appendix

FC	flash column chromatography
h	hours
HMBC	heteronuclear multiple bond correlation experiment
HMQC	heteronuclear multiple-quantum coherence experiment
HPLC	high pressure liquid chromatography
HV	high vacuum
Hz	hertz
IR	infrared
L	litre
m	milli
MAA	methyl α-acetamido acrylate
MALDI-MS	matrix-assisted laser desorption/ionization mass spectrometry
Min	minutes
MS	mass spectrometry
n.d.	not determined
NMR	nuclear magnetic resonance
NOESY	nuclear Overhauser effect spectroscopy
Np	naphthyl
o-	*ortho-*
OTf	CF_3SO_3
phen	phenantroline
PHOX	phosphinooxazoline
PPFA	N,N-dimethyl-1-[2-(diphenylphosphino)ferrocenyl]ethylamine
ppm	parts per million
Py	pyridine
Pz	pyrazole
RT	room temperature
SPO	secondary phosphine oxide
T	temperature
TFA	trifluoroacetic acid
TFE	trifluoroethanol
TMS	trimethylsilyl
TLC	thin layer chromatography
TOF	turnover frequency

8.2 Crystallographic Data and Tables

(R_C,S_{Fc})-1-[Diphenylphosphino]-2-{1-[(1-adamantyl)phosphino]ethyl}ferrocene 89

Identification code	**89**
Empirical formula	$C_{34}H_{38}FeP_2$
Formula weight	564.43
Temperature	200(2) K
Wavelength	0.71073 A
Crystal system, space group	Orthorhombic, $P2_12_12_1$
Unit cell dimensions	a = 7.2177(5) Å α = 90°
	b = 12.5755(8) Å β = 90°
	c = 31.990(2) Å γ = 90°
Volume	2903.6(3) Å^3
Z, Calculated density	4, 1.291 mg/m^3
Absorption coefficient	0.651 mm^{-1}
F(000)	1192
Crystal size	0.67 x 0.26 x 0.18 mm
Data collection	Siemens SMART PLATFORM with CCD Detector Graphite monochromator
Detector distance	50 mm
Method; exposure time/frame	omega-scans; t = 5 sec
Solution by	direct methods
Refinement method	full matrix least-squares on F^2, SHELXTL
Θ range for data collection	1.27 to 27.88°
Limiting indices	$-9 \leq h \leq 9$,
	$-16 \leq k \leq 16$,
	$-41 \leq l \leq 42$
Reflections collected / unique	29290 / 6939 [R_{int} = 0.0476]
Completeness to θ = 27.88	100.0%
Absorption correction	Empirical
Max. and min. transmission	0.8907 and 0.6692
Refinement method	Full-matrix least-squares on F^2
Data / restraints / parameters	6939 / 24 / 339
Goodness-of-fit on F^2	1.151
Final R indices [I>2σ(I)]	R_1 = 0.0526, wR_2 = 0.1144
R indices (all data)	R_1 = 0.0619, wR_2 = 0.1227
Absolute structure parameter	0.047(18)
Largest diff. peak and hole	0.423 and -0.375 e Å^{-3}

8 Appendix

(S_C,R_{Fc})-1-{Bis[3,5-bis(trifluoromethyl)phenyl]phosphino}-2-{1-[(1-adamantyl)-phosphino]ethyl}ferrocene 90

Identification code	**90**
Empirical formula	$C_{38}H_{33}F_{12}FeP_2$
Formula weight	835.43
Temperature	100(2) K
Wavelength	0.71073 Å
Crystal system, space group	Monoclinic, C_2
Unit cell dimensions	a = 24.7916(18) Å, α = 90°
	b = 7.5446(5) Å, β = 117.3790(10)°
	c = 21.4478(15) Å, γ = 90°
Volume	3562.3(4) Å^3
Z, Calculated density	4, 1.558 mg/m^3
Absorption coefficient	0.605 mm^{-1}
F(000)	1700
Crystal size	0.417 x 0.183 x 0.145 mm
Data collection	Siemens SMART PLATFORM with CCD Detector Graphite monochromator
Detector distance	50 mm
Method; exposure time/frame	omega-scans; t = 2 sec
Solution by	direct methods
Refinement method	full matrix least-squares on F^2, SHELXTL
Θ range for data collection	1.66 to 27.10°
Limiting indices	-31 $\leq$ h $\leq$ 31,
	-9 $\leq$ k $\leq$ 9,
	-27 $\leq$ l $\leq$ 27
Reflections collected / unique	16755 / 7801 [R_{int} = 0.0355]
Completeness to θ = 27.10	99.9%
Absorption correction	Empirical
Max. and min. transmission	0.9174 and 0.7866
Refinement method	Full-matrix least-squares on F^2
Data / restraints / parameters	7801 / 1 / 479
Goodness-of-fit on F^2	1.027
Final R indices [I>2σ(I)]	R_1 = 0.0446, wR_2 = 0.1000
R indices (all data)	R_1 = 0.0498, wR_2 = 0.1025
Absolute structure parameter	0.035(14)
Largest diff. peak and hole	0.803 and -0.411 e Å^{-3}

8 Appendix

(R_C,S_{Fc})-1-[Diphenylphosphino]-2-{1-[(1-adamantyl)phosphinyl]ethyl}ferrocene 97

Identification code	97
Empirical formula	$C_{34}H_{38}FeOP_2$
Formula weight	580.43
Temperature	200(2) K
Wavelength	0.71073 Å
Crystal system, space group	Orthorhombic, $P2_12_12_1$
Unit cell dimensions	a = 9.5580(6) Å α = 90°
	b = 11.5112(7) Å β = 90°
	c = 26.4162(16) Å γ = 90°
Volume	2906.4(3) Å^3
Z, Calculated density	4, 1.326 mg/m^3
Absorption coefficient	0.655 mm^{-1}
F(000)	1224
Crystal size	0.58 x 0.58 x 0.58 mm
Data collection	Siemens SMART PLATFORM with CCD Detector Graphite monochromator
Detector distance	50 mm
Method; exposure time/frame	omega-scans; t(28) = 1 sec, t(55) = 4 sec
Solution by	direct methods
Refinement method	full matrix least-squares on F^2, SHELXTL
q range for data collection	1.54 to 36.39°
Limiting indices	$-15 \leq h \leq 15$,
	$-19 \leq k \leq 19$,
	$-44 \leq l \leq 44$
Reflections collected / unique	118496 / 14131 [R_{int} = 0.0574]
Completeness to θ = 36.39	99.9%
Absorption correction	Empirical
Max. and min. transmission	0.7042 and 0.7006
Refinement method	Full-matrix least-squares on F^2
Data / restraints / parameters	14131 / 18 / 343
Goodness-of-fit on F^2	1.041
Final R indices [I>2σ(I)]	R_1 = 0.0510, wR_2 = 0.1411
R indices (all data)	R_1 = 0.0583, wR_2 = 0.1476
Absolute structure parameter	0.059(11)
Largest diff. peak and hole	0.920 and -0.273 e Å^{-3}

8 Appendix

(R_C,S_P)-1-[Bis(trifluoromethyl)phosphino]-2-{1-[(1-adamantyl)phosphino]ethyl}-ferrocene 110

Identification code	**110**
Empirical formula	$C_{24}H_{27}F_6FeP_2$
Formula weight	547.25
Temperature	100(2) K
Wavelength	0.71073 Å
Crystal system, space group	Orthorhombic, $P2_12_12_1$
Unit cell dimensions	a = 6.9200(9) Å α = 90°
	b = 10.8323(14) Å β = 90°
	c = 32.179(4) Å γ = 90°
Volume	2412.1(5) Å^3
Z, Calculated density	4, 1.507 mg/m^3
Absorption coefficient	0.815 mm^{-1}
F(000)	1124
Crystal size	0.53 x 0.38 x 0.28 mm
Data collection	Siemens SMART PLATFORM with CCD Detector
	Graphite monochromator
Detector distance	50 mm
Method; exposure time/frame	omega-scans; t = 0.5 sec
Solution by	direct methods
Refinement method	full matrix least-squares on F^2, SHELXTL
Θ range for data collection	1.98 to 26.03°
Limiting indices	-8 ≤ h ≤ 8,
	-13 ≤ k ≤ 13,
	-39 ≤ l ≤ 39
Reflections collected / unique	21242 / 4750 [R_{int} = 0.0476]
Completeness to θ = 26.03	99.9%
Absorption correction	Empirical
Max. and min. transmission	0.8028 and 0.6730
Refinement method	Full-matrix least-squares on F^2
Data / restraints / parameters	4750 / 108 / 345
Goodness-of-fit on F^2	1.056
Final R indices [I>2σ(I)]	R_1 = 0.0672, wR_2 = 0.1732
R indices (all data)	R_1 = 0.0715, wR_2 = 0.1772
Absolute structure parameter	0.01(3)
Largest diff. peak and hole	2.020 and -0.676 e Å^{-3}

8 Appendix

[PdCl$_2$(89)] 112

Identification code	**112**
Empirical formula	C$_{36}$H$_{42}$Cl$_6$FeP$_2$Pd
Formula weight	911.59
Temperature	200(2) K
Wavelength	0.71073 Å
Crystal system, space group	Monoclinic, P2$_1$
Unit cell dimensions	a = 10.0760(16) Å α = 90°
	b = 14.598(2) Å β = 110.273(3)°
	c = 14.085(2) Å γ = 90°
Volume	1943.5(5) Å^3
Z, Calculated density	2, 1.558 mg/m^3
Absorption coefficient	1.356 mm^{-1}
F(000)	924
Crystal size	0.827 x 0.204 x 0.035 mm
Data collection	Siemens SMART PLATFORM with CCD Detector
	Graphite monochromator
Detector distance	50 mm
Method; exposure time/frame	omega-scans; t$_1$ = 5 sec, t$_2$ = 15 sec
Solution by	direct methods
Refinement method	full matrix least-squares on F^2, SHELXTL
Θ range for data collection	2.08 to 28.42°
Limiting indices	-13 ≤ h ≤ 13,
	-19 ≤ k ≤ 19,
	-18 ≤ l ≤ 18
Reflections collected / unique	54982 / 9704 [R$_{int}$ = 0.0596]
Completeness to θ = 28.42	99.7%
Absorption correction	Empirical
Max. and min. transmission	0.9541 and 0.5351
Refinement method	Full-matrix least-squares on F^2
Data / restraints / parameters	9704 / 2 / 419
Goodness-of-fit on F^2	1.085
Final R indices [I>2σ(I)]	R$_1$ = 0.0468, wR$_2$ = 0.1164
R indices (all data)	R$_1$ = 0.0500, wR$_2$ = 0.1182
Absolute structure parameter	0.04(2)
Largest diff. peak and hole	1.277 and -1.013 e Å^{-3}

8 Appendix

115

Identification code	**115**
Empirical formula	$C_{38}H_{44}Cl_4FeP_2Pd$
Formula weight	866.72
Temperature	200(2) K
Wavelength	0.71073 Å
Crystal system, space group	Monoclinic, $P2_1$
Unit cell dimensions	a = 10.2517(5) Å, $\alpha = 90°$
	b = 17.0419(8) Å, $\beta = 97.4060(10)°$
	c = 10.7565(5) Å, $\gamma = 90°$
Volume	1863.57(15) Å^3
Z, Calculated density	2, 1.545 mg/m^3
Absorption coefficient	1.271 mm^{-1}
F(000)	884
Crystal size	0.28 x 0.21 x 0.13 mm
Data collection	Siemens SMART PLATFORM with CCD Detector Graphite monochromator
Detector distance	50 mm
Method; exposure time/frame	omega-scans; t = 10 sec
Solution by	direct methods
Refinement method	full matrix least-squares on F^2, SHELXTL
Θ range for data collection	1.91 to 36.35°
Limiting indices	$-16 \leq h \leq 17$,
	$-28 \leq k \leq 28$,
	$-17 \leq l \leq 17$
Reflections collected / unique	62245 / 18015 [R_{int} = 0.0452]
Completeness to θ = 36.35	99.9%
Absorption correction	Empirical
Max. and min. transmission	0.8532 and 0.7204
Refinement method	Full-matrix least-squares on F^2
Data / restraints / parameters	18015 / 1 / 415
Goodness-of-fit on F^2	1.035
Final R indices [I>2σ(I)]	R_1 = 0.0511, wR_2 = 0.1334
R indices (all data)	R_1 = 0.0593, wR_2 = 0.1391
Absolute structure parameter	-0.004(16)
Largest diff. peak and hole	3.679 and -0.834 e Å^{-3}

[Rh(89)(COD)]PF$_6$ 116

Identification code	**116**
Empirical formula	C$_{46}$H$_{54}$F$_6$FeOP$_3$Rh
Formula weight	988.56
Temperature	100(2) K
Wavelength	0.71073 Å
Crystal system, space group	Monoclinic, P2$_1$
Unit cell dimensions	a = 14.6851(14) Å α = 90°
	b = 18.2672(18) Å β = 92.216(2)°
	c = 15.9074(15) Å γ = 90°
Volume	4264.1(7) Å^3
Z, Calculated density	4, 1.540 mg/m^3
Absorption coefficient	0.902 mm^{-1}
F(000)	2032
Crystal size	0.185 x 0.104 x 0.072 mm
Data collection	Siemens SMART PLATFORM with CCD Detector
	Graphite monochromator
Detector distance	50 mm
Method; exposure time/frame	omega-scans; t = 6 sec
Solution by	direct methods
Refinement method	full matrix least-squares on F^2, SHELXTL
Θ range for data collection	1.28 to 26.37°
Limiting indices	$-18 \leq h \leq 18$,
	$-22 \leq k \leq 22$,
	$-19 \leq l \leq 19$
Reflections collected / unique	38907 / 17338 [R$_{int}$ = 0.0939]
Completeness to θ = 26.37	100.0%
Absorption correction	None
Refinement method	Full-matrix least-squares on F^2
Data / restraints / parameters	17338 / 25 / 1049
Goodness-of-fit on F^2	0.683
Final R indices [I>2σ(I)]	R$_1$ = 0.0524, wR$_2$ = 0.0749
R indices (all data)	R$_1$ = 0.0849, wR$_2$ = 0.0857
Absolute structure parameter	-0.023(17)
Largest diff. peak and hole	0.776 and -0.717 e Å^{-3}

8 Appendix

[PdCl$_2$(111)] 117

Identification code	**117**
Empirical formula	C$_{24}$H$_{27}$Cl$_2$F$_6$FeOP$_2$Pd
Formula weight	740.55
Temperature	100(2) K
Wavelength	0.71073 Å
Crystal system, space group	Orthorhombic, P2$_1$2$_1$2$_1$
Unit cell dimensions	a = 8.5565(6) Å α = 90°
	b = 16.0505(11) Å β = 90°
	c = 19.3687(13) Å γ = 90°
Volume	2660.0(3) Å^3
Z, Calculated density	4, 1.849 mg/m^3
Absorption coefficient	1.603 mm^{-1}
F(000)	1476
Crystal size	0.108 x 0.092 x 0.078 mm
Data collection	Siemens SMART PLATFORM with CCD Detector
	Graphite monochromator
Detector distance	50 mm
Method; exposure time/frame	omega-scans; t = 6 sec
Solution by	direct methods
Refinement method	full matrix least-squares on F^2, SHELXTL
Θ range for data collection	2.10 to 27.49°
Limiting indices	-11 ≤ h ≤ 11,
	-20 ≤ k ≤ 20,
	-25 ≤ l ≤ 25
Reflections collected / unique	26365 / 6101 [R$_{int}$ = 0.0611]
Completeness to θ = 27.49	100.0%
Absorption correction	Empirical
Max. and min. transmission	0.8852 and 0.8459
Refinement method	Full-matrix least-squares on F^2
Data / restraints / parameters	6101 / 0 / 335
Goodness-of-fit on F^2	1.039
Final R indices [I>2σ(I)]	R$_1$ = 0.0348, wR$_2$ = 0.0707
R indices (all data)	R$_1$ = 0.0393, wR$_2$ = 0.0725
Absolute structure parameter	-0.011(19)
Largest diff. peak and hole	0.727 and -0.400 e Å^{-3}

(1S,2R,3R,S_{Fc})-2-(1-Adamantyl)-3-methyl-1-(trifluoromethyl)-2,3-dihydro-1H-ferroceno-[c] [1,2]diphosphole 129

Identification code	**129**
Empirical formula	$C_{23}H_{27}F_3FeP_2$
Formula weight	478.24
Temperature	100(2) K
Wavelength	0.71073 Å
Crystal system, space group	Monoclinic, P2$_1$
Unit cell dimensions	a = 10.743(3) Å α = 90°
	b = 8.260(3) Å β = 112.417(6)°
	c = 13.016(4) Å γ = 90°
Volume	1067.7(6) Å^3
Z, Calculated density	2, 1.488 mg/m^3
Absorption coefficient	0.889 mm^{-1}
F(000)	496
Crystal size	0.17 x 0.17 x 0.12 mm
Data collection	Siemens SMART PLATFORM with CCD Detector Graphite monochromator
Detector distance	50 mm
Method; exposure time/frame	omega-scans; t = 2 sec
Solution by	direct methods
Refinement method	full matrix least-squares on F^2, SHELXTL
Θ range for data collection	1.69 to 28.49°
Limiting indices	$-14 \leq h \leq 14$,
	$-11 \leq k \leq 11$,
	$-17 \leq l \leq 17$
Reflections collected / unique	11024 / 5342 [R$_{int}$ = 0.0626]
Completeness to θ = 28.49	99.2%
Absorption correction	None
Refinement method	Full-matrix least-squares on F^2
Data / restraints / parameters	5342 / 1 / 263
Goodness-of-fit on F^2	0.861
Final R indices [I>2σ(I)]	R$_1$ = 0.0430, wR$_2$ = 0.0721
R indices (all data)	R$_1$ = 0.0584, wR$_2$ = 0.0761
Absolute structure parameter	0.019(16)
Largest diff. peak and hole	0.494 and -0.387 e Å^{-3}

8 Appendix

(1S,2R,3R,S_{Fc})-2-Cyclohexyl-3-methyl-1-(trifluoromethyl)-2,3-dihydro-1H-ferroceno[c][1,2]diphosphole 132

Identification code	**132**
Empirical formula	$C_{19}H_{23}F_3FeP_2$
Formula weight	426.16
Temperature	0.71073 Å
Crystal system, space group	Monoclinic, P2$_1$
Unit cell dimensions	a = 12.906(2) Å α = 90°
	b = 11.692(2) Å β = 105.176(4)°
	c = 13.032(2) Å γ = 90°
Volume	1898.0(6) Å^3
Z, Calculated density	4, 1.491 mg/m^3
Absorption coefficient	0.990 mm^{-1}
F(000)	880
Crystal size	0.256 x 0.195 x 0.178 mm
Data collection	Siemens SMART PLATFORM with CCD Detector
	Graphite monochromator
Detector distance	50 mm
Method; exposure time/frame	omega-scans; t = 3 sec
Solution by	direct methods
Refinement method	full matrix least-squares on F^2, SHELXTL
Θ range for data collection	1.62 to 27.12°
Limiting indices	-16 ≤ h ≤ 16,
	-14 ≤ k ≤ 14,
	-16 ≤ l ≤ 16
Reflections collected / unique	18180 / 8312 [R$_{int}$ = 0.0450]
Completeness to θ = 27.12	100.0%
Absorption correction	Empirical
Max. and min. transmission	0.8435 and 0.7857
Refinement method	Full-matrix least-squares on F^2
Data / restraints / parameters	8312 / 1 / 453
Goodness-of-fit on F^2	0.979
Final R indices [I>2σ(I)]	R$_1$ = 0.0427, wR$_2$ = 0.0878
R indices (all data)	R$_1$ = 0.0549, wR$_2$ = 0.0933
Absolute structure parameter	-0.008(14)
Largest diff. peak and hole	0.618 and -0.282 e Å^{-3}

[PdCl$_2$(134)] 138

Identification code	**138**
Empirical formula	C$_{30}$H$_{35}$Cl$_2$F$_3$FeP$_2$Pd
Formula weight	747.67
Temperature	100(2) K
Wavelength	0.71073 Å
Crystal system, space group	Orthorhombic, P2$_1$2$_1$2$_1$
Unit cell dimensions	a = 11.2036(16) Å α = 90°
	b = 13.4658(19) Å β = 90°
	c = 19.135(3) Å γ = 90°
Volume	2886.9(7) Å^3
Z, Calculated density	4, 1.720 mg/m^3
Absorption coefficient	1.461 mm^{-1}
F(000)	1512
Crystal size	0.21 x 0.12 x 0.05 mm
Data collection	Siemens SMART PLATFORM with CCD Detector
	Graphite monochromator
Detector distance	50 mm
Method; exposure time/frame	omega-scans; t = 8 sec
Solution by	direct methods
Refinement method	full matrix least-squares on F^2, SHELXTL
Θ range for data collection	1.85 to 24.71°
Limiting indices	$-13 \leq h \leq 13$,
	$-15 \leq k \leq 15$,
	$-22 \leq l \leq 22$
Reflections collected / unique	22538 / 4937 [R$_{int}$ = 0.0912]
Completeness to θ = 24.71	100.0%
Absorption correction	Empirical
Max. and min. transmission	0.9305 and 0.7528
Refinement method	Full-matrix least-squares on F^2
Data / restraints / parameters	4937 / 240 / 352
Goodness-of-fit on F^2	1.113
Final R indices [I>2σ(I)]	R$_1$ = 0.0504, wR$_2$ = 0.1038
R indices (all data)	R$_1$ = 0.0570, wR$_2$ = 0.1064
Absolute structure parameter	0.03(3)
Largest diff. peak and hole	1.028 and -0.853 e Å^{-3}

8 Appendix

(S_P,R_C,S_{Fc})-1-[(1-Adamantyl)trifluoromethylphosphino]-2-[(1-dimethylamino)ethyl]-ferrocene 150

Identification code	**150**
Empirical formula	$C_{25}H_{33}F_3FeNP$
Formula weight	491.34
Temperature	100(2) K
Wavelength	0.71073 Å
Crystal system, space group	Monoclinic, P2$_1$
Unit cell dimensions	a = 13.862(3) Å, α = 90°
	b = 10.527(2) Å, β = 94.200(4)°
	c = 15.748(3) Å, γ = 90°
Volume	2291.8(8) Å^3
Z, Calculated density	4, 1.424 mg/m^3
Absorption coefficient	0.765 mm^{-1}
F(000)	1032
Crystal size	0.44 x 0.35 x 0.21 mm
Data collection	Siemens SMART PLATFORM with CCD Detector
	Graphite monochromator
Detector distance	50 mm
Method; exposure time/frame	omega-scans; t = 0.5 sec
Solution by	direct methods
Refinement method	full matrix least-squares on F^2, SHELXTL
Θ range for data collection	1.47 to 27.11°
Limiting indices	$-17 \leq h \leq 17$,
	$-13 \leq k \leq 13$,
	$-20 \leq l \leq 20$
Reflections collected / unique	22001 / 10021 [R$_{int}$ = 0.0465]
Completeness to θ = 27.11	99.9%
Absorption correction	Empirical
Max. and min. transmission	0.8559 and 0.7296
Refinement method	Full-matrix least-squares on F^2
Data / restraints / parameters	10021 / 13 / 684
Goodness-of-fit on F^2	0.998
Final R indices [I>2σ(I)]	R$_1$ = 0.0428, wR$_2$ = 0.0861
R indices (all data)	R$_1$ = 0.0498, wR$_2$ = 0.0895
Absolute structure parameter	0.016(11)
Largest diff. peak and hole	0.531 and -0.482 e Å^{-3}

[PtCl₂(151)] 154

Identification code	**154**
Empirical formula	$C_{36}H_{43}Cl_4F_3FeP_2Pt$
Formula weight	987.38
Temperature	100(2) K
Wavelength	0.71073 Å
Crystal system, space group	Monoclinic, $P2_1$
Unit cell dimensions	$a = 10.096(6)$ Å, $\alpha = 90°$
	$b = 12.016(7)$ Å, $\beta = 102.183(7)°$
	$c = 15.928(9)$ Å, $\gamma = 90°$
Volume	1888.7(19) Å³
Z, Calculated density	2, 1.736 mg/m³
Absorption coefficient	4.492 mm⁻¹
F(000)	976
Crystal size	0.174 x 0.173 x 0.023 mm
Data collection	Siemens SMART PLATFORM with CCD Detector
	Graphite monochromator
Detector distance	50 mm
Method; exposure time/frame	omega-scans; t = 8 sec
Solution by	direct methods
Refinement method	full matrix least-squares on F^2, SHELXTL
Θ range for data collection	1.31 to 26.03°
Limiting indices	$-12 \leq h \leq 12$,
	$-14 \leq k \leq 14$,
	$-18 \leq l \leq 19$
Reflections collected / unique	7742 / 5534 [$R_{int} = 0.0779$]
Completeness to θ = 26.03	91.8%
Absorption correction	Empirical
Max. and min. transmission	0.9037 and 0.5087
Refinement method	Full-matrix least-squares on F^2
Data / restraints / parameters	5534 / 283 / 424
Goodness-of-fit on F^2	1.015
Final R indices [I>2σ(I)]	$R_1 = 0.0852$, $wR_2 = 0.1739$
R indices (all data)	$R_1 = 0.1334$, $wR_2 = 0.1990$
Absolute structure parameter	0.073(19)
Largest diff. peak and hole	2.517 and -1.483 e Å⁻³

8 Appendix

PtCl$_2$(((S_P)-152))] 155

Identification code	**155**
Empirical formula	C$_{35}$H$_{49}$Cl$_2$F$_3$FeP$_2$Pt
Formula weight	910.52
Temperature	100(2) K
Wavelength	0.71073 Å
Crystal system, space group	Orthorhombic, P2$_1$2$_1$2$_1$
Unit cell dimensions	a = 13.5006(10) Å α = 90°
	b = 14.6436(11) Å β = 90°
	c = 17.6900(13) Å γ = 90°
Volume	3497.3(4) Å^3
Z, Calculated density	4, 1.729 mg/m^3
Absorption coefficient	4.696 mm^{-1}
F(000)	1816
Crystal size	0.22 x 0.19 x 0.13 mm
Data collection	Siemens SMART PLATFORM with CCD Detector Graphite monochromator
Detector distance	50 mm
Method; exposure time/frame	omega-scans; t = 4 sec
Solution by	direct methods
Refinement method	full matrix least-squares on F^2, SHELXTL
Θ range for data collection	1.81 to 28.33°
Limiting indices	-17 ≤ h ≤ 17,
	-19 ≤ k ≤ 19,
	-23 ≤ l ≤ 23
Reflections collected / unique	36344 / 8697 [R$_{int}$ = 0.0510]
Completeness to θ = 28.33	99.8%
Absorption correction	Empirical
Max. and min. transmission	0.5804 and 0.4248
Refinement method	Full-matrix least-squares on F^2
Data / restraints / parameters	8697 / 0 / 398
Goodness-of-fit on F^2	0.962
Final R indices [I>2σ(I)]	R$_1$ = 0.0276, wR$_2$ = 0.0544
R indices (all data)	R$_1$ = 0.0302, wR$_2$ = 0.0552
Absolute structure parameter	-0.001(4)
Largest diff. peak and hole	1.444 and -0.566 e Å^{-3}

8 Appendix

[PdCl$_2$((S_P)-142)] 156

Identification code	**156**
Empirical formula	C$_{33.04}$H$_{39}$Cl$_{5.83}$F$_3$FeP$_2$Pd
Formula weight	924.10
Temperature	100(2) K
Wavelength	0.71073 Å
Crystal system, space group	Orthorhombic, P2$_1$2$_1$2$_1$
Unit cell dimensions	a = 18.7523(15) Å α = 90°
	b = 21.7801(17) Å β = 90°
	c = 9.2700(7) Å γ = 90°
Volume	3786.1(5) Å^3
Z, Calculated density	4, 1.631 mg/m^3
Absorption coefficient	1.405 mm^{-1}
F(000)	1872
Crystal size	0.28 x 0.19 x 0.15 mm
Data collection	Siemens SMART PLATFORM with CCD Detector
	Graphite monochromator
Detector distance	50 mm
Method; exposure time/frame	omega-scans; t = 1 sec
Solution by	direct methods
Refinement method	full matrix least-squares on F^2, SHELXTL
Θ range for data collection	1.87 to 26.74°
Limiting indices	-23 ≤ h ≤ 23,
	-27 ≤ k ≤ 27,
	-11 ≤ l ≤ 11
Reflections collected / unique	35411 / 8062 [R$_{int}$ = 0.0761]
Completeness to θ = 26.74	100.0%
Absorption correction	Empirical
Max. and min. transmission	0.8191 and 0.6919
Refinement method	Full-matrix least-squares on F^2
Data / restraints / parameters	8062 / 0 / 455
Goodness-of-fit on F^2	1.037
Final R indices [I>2σ(I)]	R$_1$ = 0.0517, wR$_2$ = 0.1126
R indices (all data)	R$_1$ = 0.0608, wR$_2$ = 0.1174
Absolute structure parameter	0.00(3)
Largest diff. peak and hole	1.185 and -0.732 e Å^{-3}

8 Appendix

[PdCl$_2$((R_P)-142)] 157

Identification code	**157**
Empirical formula	C$_{64}$H$_{82}$Cl$_8$F$_6$Fe$_2$P$_4$Pd$_2$
Formula weight	1697.28
Temperature	100(2) K
Wavelength	0.71073 Å
Crystal system, space group	Monoclinic, P2$_1$
Unit cell dimensions	a = 9.7782(3) Å α = 90°
	b = 31.8585(9) Å β = 101.9290(10)°
	c = 11.2392(3) Å γ = 90°
Volume	3425.61(17) Å^3
Z, Calculated density	2, 1.645 mg/m^3
Absorption coefficient	1.393 mm^{-1}
F(000	1720
Crystal size	0.331 x 0.24 x 0.185 mm
Data collection	Siemens SMART PLATFORM with CCD Detector
	Graphite monochromator
Detector distance	50 mm
Method; exposure time/frame	omega-scans; t$_1$ = 3 sec, t$_2$ = 6 sec
Solution by	direct methods
Refinement method	full matrix least-squares on F^2, SHELXTL
Θ range for data collection	1.85 to 36.34°
Limiting indices	-16 ≤ h ≤ 16,
	-52 ≤ k ≤ 52,
	-18 ≤ l ≤ 18
Reflections collected / unique	139022 / 31698 [R$_{int}$ = 0.0603]
Completeness to θ = 36.34	99.5%
Absorption correction	Empirical
Max. and min. transmission	0.7826 and 0.6556
Refinement method	Full-matrix least-squares on F^2
Data / restraints / parameters	31698 / 1 / 777
Goodness-of-fit on F^2	1.146
Final R indices [I>2σ(I)]	R$_1$ = 0.0646, wR$_2$ = 0.1463
R indices (all data)	R$_1$ = 0.0681, wR$_2$ = 0.1479
Absolute structure parameter	0.068(14)
Largest diff. peak and hole	4.449 and -2.543 e Å^{-3}

8 Appendix

[Rh((S$_P$)-142)COD)]PF$_6$ 158

Identification code	**158**
Empirical formula	C$_{40}$H$_{53}$Cl$_2$F$_9$FeP$_3$Rh
Formula weight	1027.39
Temperature	100(2) K
Wavelength	0.71073 Å
Crystal system, space group	Monoclinic, P2$_1$
Unit cell dimensions	a = 12.394(2) Å α = 90°
	b = 10.5591(18) Å β = 93.505(4)°
	c = 15.967(3) Å γ = 90°
Volume	2085.7(6) Å^3
Z, Calculated density	2, 1.636 mg/m^3
Absorption coefficient	1.057 mm^{-1}
F(000)	1048
Crystal size	0.152 x 0.073 x 0.019 mm
Data collection	Siemens SMART PLATFORM with CCD Detector
	Graphite monochromator
Detector distance	50 mm
Method; exposure time/frame	omega-scans; t = 12 sec
Solution by	direct methods
Refinement method	full matrix least-squares on F^2, SHELXTL
Θ range for data collection	1.28 to 26.38°
Limiting indices	-15 ≤ h ≤ 15,
	-13 ≤ k ≤ 13,
	-19 ≤ l ≤ 19
Reflections collected / unique	18949 / 8514 [R$_{int}$ = 0.0908]
Completeness to θ = 26.38	100.0%
Absorption correction	Empirical
Max. and min. transmission	0.9802 and 0.8559
Refinement method	Full-matrix least-squares on F^2
Data / restraints / parameters	8514 / 7 / 506
Goodness-of-fit on F^2	0.998
Final R indices [I>2σ(I)]	R$_1$ = 0.0663, wR$_2$ = 0.1020
R indices (all data)	R$_1$ = 0.0919, wR$_2$ = 0.1116
Absolute structure parameter	0.03(3)
Largest diff. peak and hole	1.034 and -0.789 e Å^{-3}

8 Appendix

[Rh(Cl)((S$_P$)-142)(CO)] 160

Identification code	**160**
Empirical formula	$C_{32}H_{39}ClF_3FeOP_2Rh$
Formula weight	752.78
Temperature	100(2) K
Wavelength	0.71073 Å
Crystal system, space group	Orthorhombic, $P2_12_12_1$
Unit cell dimensions	a = 11.2236(15) Å $\alpha = 90°$
	b = 15.063(2) Å $\beta = 90°$
	c = 18.745(2) Å $\gamma = 90°$
Volume	3168.9(7) Å^3
Z, Calculated density	4, 1.578 mg/m^3
Absorption coefficient	1.206 mm^{-1}
F(000)	1536
Crystal size	0.549 x 0.297 x 0.245 mm
Data collection	Siemens SMART PLATFORM with CCD Detector
	Graphite monochromator
Detector distance	50 mm
Method; exposure time/frame	omega-scans; t = 1 sec
Solution by	direct methods
Refinement method	full matrix least-squares on F^2, SHELXTL
Θ range for data collection	1.73 to 28.35°
Limiting indices	$-14 \leq h \leq 14$,
	$-20 \leq k \leq 20$,
	$-24 \leq l \leq 24$
Reflections collected / unique	32984 / 7848 [R_{int} = 0.0364]
Completeness to θ = 28.35	99.5%
Absorption correction	Empirical
Max. and min. transmission	0.7565 and 0.5572
Refinement method	Full-matrix least-squares on F^2
Data / restraints / parameters	7848 / 0 / 371
Goodness-of-fit on F^2	1.039
Final R indices [I>2σ(I)]	R_1 = 0.0262, wR_2 = 0.0624
R indices (all data)	R_1 = 0.0277, wR_2 = 0.0631
Absolute structure parameter	-0.001(12)
Largest diff. peak and hole	0.817 and -0.351 e Å^{-3}

8 Appendix

(S_P)-{(R_C)-1-[(S_{Fc})-2-(Bis(trifluoromethyl)phosphino)ferrocenyl]ethyl}-{(R_C)-1-[(S_{Fc})-2-(diphenylphosphino)ferrocenyl]ethyl}adamantylphosphine 179

Identification code	**179**
Empirical formula	$C_{48}H_{44}F_6Fe_2P_3$
Formula weight	939.44
Temperature	100(2) K
Wavelength	0.71073 Å
Crystal system, space group	Orthorhombic, $P2_12_12_1$
Unit cell dimensions	a = 10.9110(7) Å $\alpha = 90°$
	b = 16.4225(11) Å $\beta = 90°$
	c = 23.5898(16) Å $\gamma = 90°$
Volume	4227.0(5) Å^3
Z, Calculated density	4, 1.476 mg/m^3
Absorption coefficient	0.861 mm^{-1}
F(000)	1932
Crystal size	0.51 x 0.23 x 0.20 mm
Data collection	Siemens SMART PLATFORM with CCD Detector
	Graphite monochromator
Detector distance	50 mm
Method; exposure time/frame	omega-scans; t_1 = 1 sec, t_2 = 3 sec
Solution by	direct methods
Refinement method	full matrix least-squares on F^2, SHELXTL
Θ range for data collection	1.51 to 28.36°
Limiting indices	$-14 \leq h \leq 14$,
	$-21 \leq k \leq 21$,
	$-31 \leq l \leq 31$
Reflections collected / unique	113274 / 10534 [R_{int} = 0.0691]
Completeness to θ = 28.36	99.7%
Absorption correction	Empirical
Max. and min. transmission	0.8433 and 0.6669
Refinement method	Full-matrix least-squares on F^2
Data / restraints / parameters	10534 / 0 / 578
Goodness-of-fit on F^2	1.071
Final R indices [I>2σ(I)]	R_1 = 0.0291, wR_2 = 0.0703
R indices (all data)	R_1 = 0.0309, wR_2 = 0.0713
Absolute structure parameter	0.005(8)
Largest diff. peak and hole	0.539 and -0.321 e Å^{-3}

8 Appendix

(S_P)-{(R_C)-1-[(S_{Fc})-2-(Bis(trifluoromethyl)phosphino)ferrocenyl]ethyl}-{(S_C)-1-[(R_{Fc})-2-(diphenylphosphino)ferrocenyl]ethyl}adamantylphosphine 182

Identification code	**182**
Empirical formula	$C_{49}H_{51}Cl_2F_6Fe_2P_3$
Formula weight	1029.41
Temperature	200(2) K
Wavelength	0.71073 Å
Crystal system, space group	Orthorhombic, $P2_12_12_1$
Unit cell dimensions	a = 11.0430(10) Å α = 90°
	b = 12.3799(11) Å β = 90°
	c = 34.240(3) Å γ = 90°
Volume	4680.9(7) Å^3
Z, Calculated density	4, 1.461 mg/m^3
Absorption coefficient	0.895 mm^{-1}
F(000)	2120
Crystal size	0.41 x 0.21 x 0.19 mm
Data collection	Siemens SMART PLATFORM with CCD Detector
	Graphite monochromator
Detector distance	50 mm
Method; exposure time/frame	omega-scans; t = 2 sec
Solution by	direct methods
Refinement method	full matrix least-squares on F^2, SHELXTL
Θ range for data collection	1.19 to 28.32°
Limiting indices	$-14 \leq h \leq 14$,
	$-16 \leq k \leq 16$,
	$-45 \leq l \leq 44$
Reflections collected / unique	49000 / 11639 [R_{int} = 0.0977]
Completeness to θ = 28.32	99.9%
Absorption correction	None
Max. and min. transmission	0.8470 and 0.7133
Refinement method	Full-matrix least-squares on F^2
Data / restraints / parameters	11639 / 50 / 607
Goodness-of-fit on F^2	0.826
Final R indices [I>2σ(I)]	R1 = 0.0517, wR^2 = 0.0870
R indices (all data)	R^1 = 0.0958, wR^2 = 0.0969
Absolute structure parameter	0.002(15)
Largest diff. peak and hole	0.565 and -0.561 e Å^{-3}

[PdCl(179)]PF$_6$ 183

Identification code	**183**
Empirical formula	H$_{51}$Cl$_5$F$_{12}$Fe$_2$P$_4$Pd
Formula weight	1399.14
Temperature	100(2) K
Wavelength	0.71073 Å
Crystal system, space group	Triclinic, P$_1$
Unit cell dimensions	a = 12.2973(11) Å α = 80.834(2)°
	b = 13.4782(12) Å β = 82.089(2)°
	c = 18.4298(16) Å γ = 64.800(2)°
Volume	2720.3(4) Å^3
Z, Calculated density	2, 1.708 mg/m^3
Absorption coefficient	1.294 mm^{-1}
F(000)	1404
Crystal size	0.38 x 0.30 x 0.26 mm
Data collection	Siemens SMART PLATFORM with CCD Detector
	Graphite monochromator
Detector distance	50 mm
Method; exposure time/frame	omega-scans; t = 1 sec
Solution by	direct methods
Refinement method	full matrix least-squares on F^2, SHELXTL
Θ range for data collection	1.68 to 25.68°
Limiting indices	-14 ≤ h ≤ 14,
	-16 ≤ k ≤ 16,
	-22 ≤ l ≤ 22
Reflections collected / unique	23266 / 19715 [R$_{int}$ = 0.0243]
Completeness to θ = 25.68	99.7%
Absorption correction	Empirical
Max. and min. transmission	0.7264 and 0.6426
Refinement method	Full-matrix least-squares on F^2
Data / restraints / parameters	19715 / 3 / 1352
Goodness-of-fit on F^2	1.036
Final R indices [I>2σ(I)]	R$_1$ = 0.0527, wR$_2$ = 0.1338
R indices (all data)	R$_1$ = 0.0580, wR$_2$ = 0.1387
Absolute structure parameter	0.012(16)
Largest diff. peak and hole	1.362 and -1.232 e Å^{-3}

198

Identification code	**198**
Empirical formula	$C_{35}H_{52}Cl_2FeP_2Pt$
Formula weight	856.55
Temperature	100(2) K
Wavelength	0.71073 Å
Crystal system, space group	Orthorhombic, $P2_12_12_1$
Unit cell dimensions	a = 12.8506(15) Å $\alpha = 90°$
	b = 14.6385(17) Å $\beta = 90°$
	c = 18.060(2) Å $\gamma = 90°$
Volume	3397.3(7) Å^3
Z, Calculated density	4, 1.675 mg/m^3
Absorption coefficient	4.816 mm^{-1}
F(000)	1720
Crystal size	0.33 x 0.20 x 0.16 mm
Data collection	Siemens SMART PLATFORM with CCD Detector Graphite monochromator
Detector distance	50 mm
Method; exposure time/frame	omega-scans; t = 1 sec
Solution by	direct methods
Refinement method	full matrix least-squares on F^2, SHELXTL
Θ range for data collection	1.94 to 28.44°
Limiting indices	$-17 \leq h \leq 16$,
	$-19 \leq k \leq 19$,
	$-24 \leq l \leq 24$
Reflections collected / unique	35148 / 8515 [R_{int} = 0.0813]
Completeness to θ = 28.44	99.5%
Absorption correction	Empirical
Max. and min. transmission	0.5204 and 0.3036
Refinement method	Full-matrix least-squares on F^2
Data / restraints / parameters	8515 / 0 / 370
Goodness-of-fit on F^2	0.996
Final R indices [I>2σ(I)]	R_1 = 0.0461, wR_2 = 0.0967
R indices (all data)	R_1 = 0.0552, wR_2 = 0.1003
Absolute structure parameter	-0.008(7)
Largest diff. peak and hole	3.737 and -2.693 e Å^{-3}

8 Appendix

8.3 Fit Results of 113

By R. Verel, NMR Facility LAC, ETH Zürich

8.3.1 Experimental Data

File	JB_074_2_31P 1 1
Acq.	Date 21 Feb 2008 / 17:30:51
Solvent	CD_2Cl_2
Obs. Nucl.	^{31}P
Apod	0.5 Hz / EM
SF 121.49 MHz	
Resolution	0.557 Hz/pnt

8.3.2 Fit Parameters

Fit Program	GammaSpecFit.g
Date	21 Jan 2011
Input File	JB_074_2_exp.dat
Initial Pars.	JB_074_AAMM.sys
Output File	JB_074_2_fit.dat
Command line	./GammaSpecFit.g JB_074_2_exp.dat p JB_074_AAMM.sys 22 16000 18000 JB_074_2_fit.dat
Spin Simulation	Gamma Package (ver. 4.1.2)
Minimization	Minuit2 (ver. 5.27.02)

8.3.3 Summary Results

Isotropic Spin System JB_074_AAMM
(Static Bo Field Of 7.046161 T)

Spin Index	0	1	2	3
Isotope	^{31}P	^{31}P	^{31}P	^{31}P
Momentum	1/2	1/2	1/2	1/2
Shifts	-9.14 KHz	-9.14 KHz	-0.00 Hz	-0.00 Hz
	-75.26 ppm	-75.24 ppm	-4.38 ppm	-4.38 ppm
Js Spin 0		-249.88 Hz	39.52 Hz	392.04 Hz
Js Spin 1			395.13 Hz	41.60 Hz
Js Spin 2				2.43 Hz
Omega	121.44 MHz	121.44 MHz	121.44 MHz	121.44 MHz
Linewidth	22.2 Hz			

8 Appendix

8.3.4 Detailed Results

minimum:
Minuit did successfully converge.
# of function calls: 734
minimum function Value: 1.222480570841

# ext.	Name	type	Value	Error +/-
0	CSiso_0	free	-25.06397162095	0.0460700591
1	CSiso_1	free	-25.05235423426	0.0462151214
2	CSiso_2	free	45.81347164923	0.0362366477
3	CSiso_3	free	45.81589894170	0.0374908759
4	J_01	limited	-249.88407645240	7.7020122881
5	J_02	limited	39.52407660889	26.3371414762
6	J_03	limited	392.03856100480	26.1055497249
7	J_12	limited	395.12516486530	23.4746598608
8	J_13	limited	41.59938254051	23.6818324907
9	J_23	limited	2.43261715292	7.9811569995
10	lb	limited	22.19282961540	27.1169365870
11	inten	limited	0.93287169419	0.39758785214

8.3.5 Spectrum 1

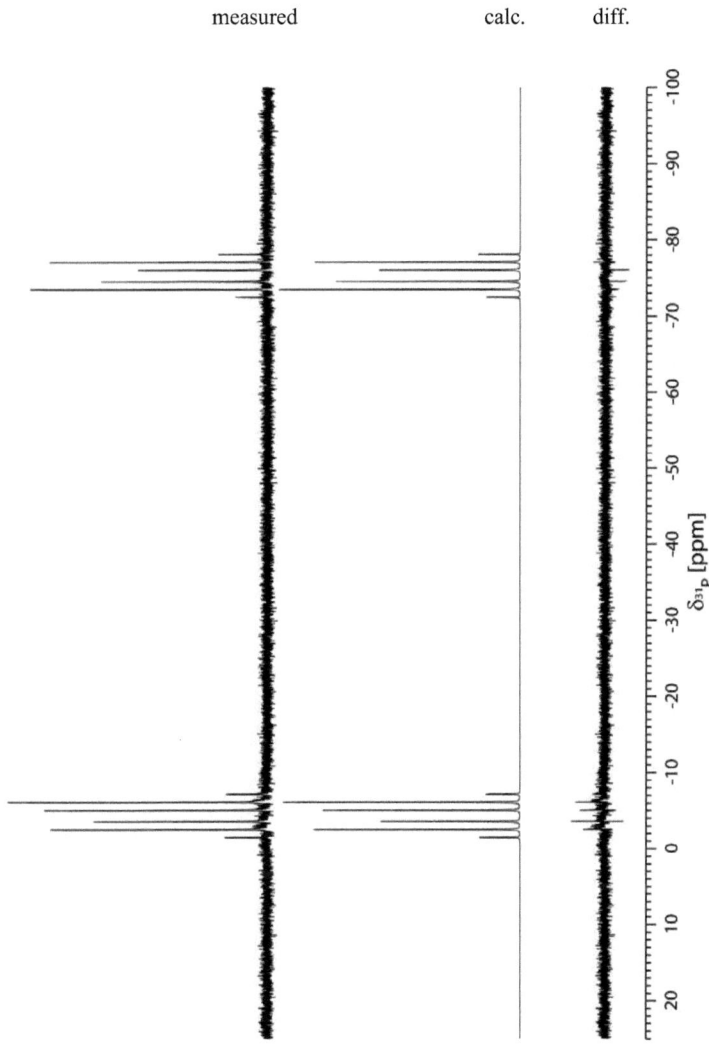

8 Appendix

8.3.6 Spectrum 2 (enlarged)

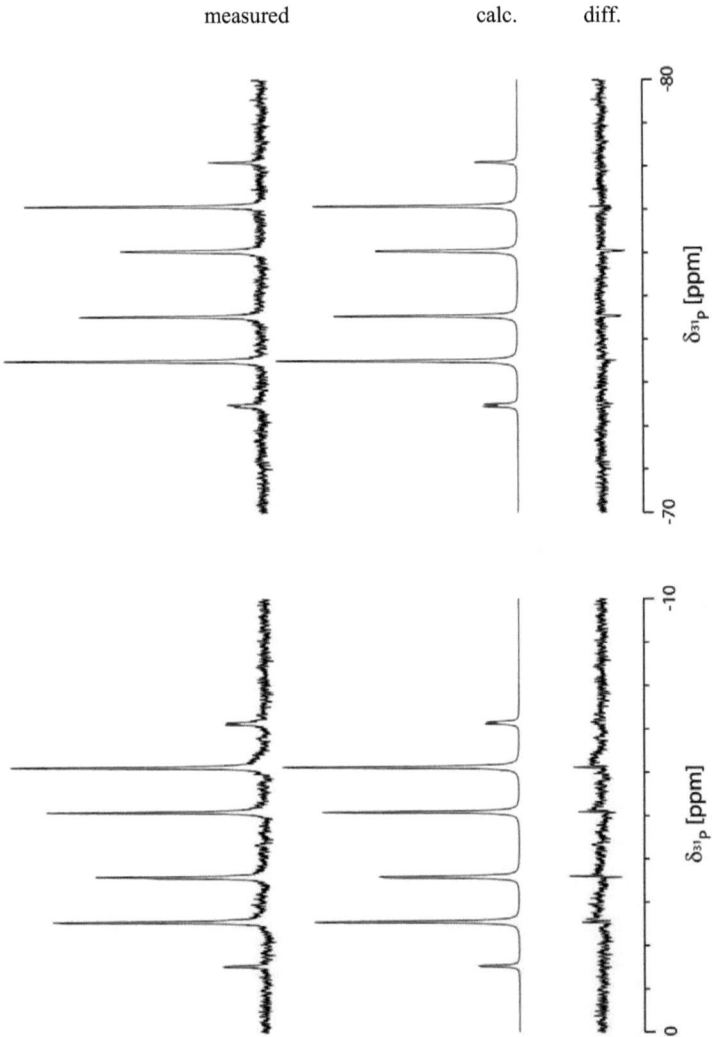

8.4 Danksagung

Während der letzten Jahre wurde ich von zahlreichen Leuten unterstützt. Ihnen möchte ich ganz herzlich danken:

Prof. Dr. Antonio Togni, meinem Doktorvater danke ich für die Möglichkeit, an einem interessanten und herausfordernden Projekt zu arbeiten. Das mir entgegengebrachte Vertrauen sowie die enorme Freiheit bei der Projektplanung und Durchführung der Arbeit habe ich sehr geschätzt. Für die Unterstützung und Motivation während der letzten Jahre möchte ich mich ganz herzlich bedanken.

Prof. Dr. Paul S. Pregosin danke ich für die freundliche und unkomplizierte Übernahme des Korreferats.

René Verel wird gedankt für die Simulation der Phosphorspektren. Weiterer Dank gehört *Aitor Moreno* und seinem Vorgänger *Dr. Heinz Rüegger* für die Hilfe beim Messen und Auswerten zahlreicher NMR-Spektren. *Pietro Butti*, *Katrin Niedermann* und *Raphael Aardoom* wird gedankt für das Messen und Lösen einer beträchtlichen Anzahl Kristallstrukturen, die sich während dieser Arbeit angehäuft haben und die wohl nicht immer ganz einfach zu lösen waren. Ein grosses Dankeschön gebührt auch „unserem" Laboranten *Philip Battaglia* für die umfangreichen synthetischen Beiträge sowie *Cyril Bachmann* und *Carla Rigling* für die enthusiastische und fleissige Durchführung ihrer Semesterprojekte.

Dr. Jan Welch, *Katrin Niedermann*, *Rino Schwenk* und *Elisabeth Männel* danke ich herzlich für das sorgfältige Korrekturlesen des Manuskripts. Speziell *Jan* gebührt grosser Dank für die mühselige Arbeit aus meiner Aneinanderreihung englischer Sätze eine gut und flüssig lesbare Arbeit zu formen.

Dr. Benoît Pugin und der Firma *Solvias AG* danke ich für die grosszügige Bereitstellung von Ugi-Amin, sowie für die Hilfe bei den Hydrierungsreaktionen und der Bestimmung der Aktivitäten der Liganden. Der Förderagentur des Bundes „Komission für Innovation und Technologie (*KTI*)" und der *ETH Zürich* danke ich für die finanzielle und materielle Untersützung während des Doktorats.

Nicht vergessen gehen darf natürlich ein grosses Dankeschön an die ehemaligen und aktuellen Mitglieder der Gruppen *Togni/Mezzetti* für die gute Arbeitsatmosphäre und die vielen interessanten Gespräche verschiedenster Art. Neben den intensiven und hilfreichen chemischen Diskussionen waren auch die erfrischenden Gespräche über die etwas profaneren

8 Appendix

Dinge des Lebens und die traditionellen Dienstag- und Freitagabend-Runden eine grosse Bereicherung zum (Chemie)-Alltag. Speziell zu Erwähnen sind die ehemaligen und aktuellen Mitglieder des Labors H222; *Irene Walz, Iris Kieltsch, Raffael Koller, Philip Battaglia, Katrin Niedermann, Natalja Früh* und *Jolanda Winkler* für ihre praktische und mentale (jede Sch... isch ä Chance) Unterstützung und das gute (Arbeits)klima im Labor. Den beiden erstgenannten danke ich für die hilfreichen Tipps bei meinen ersten Schritten als Doktorand und die intensiven, (feucht-)fröhlichen Aufenthalte in Oxford, London, Hawai'i oder am Oktoberfest in München, die zwar nicht immer unbedingt erholsam waren, aber dennoch neue Energie für die Forschungsarbeit lieferten. *Raffi* danke ich für die grossartige Zeit wärend des Studiums und des Doktorats, die interessanten chemischen Diskussionen, die harten Töggelimatches als Partner (Sieg im grössten Finale) oder Gegner, die gemeinsame Erforschung der CF_3-phosphine und des Nachtlebens in allen möglichen kleinen (Oxford, Bern, Basel) und grossen Städten (Zürich, London, München), sowie die After-Work-Bierdegustationen an den verschiedensten möglichen und unmöglichen Orten. *Katrin* möchte ich danken für die gemeinsame Zeit in Hawai'i, die langen, lustigen Abende bei Lachs, Wein und anderen Getränken, die mehr oder weniger anstrengenden sportlichen Aktivitäten, sowie die rasanten, nächtlichen Fahrradausflüge durch den Norden Zürichs.

Nicht vergessen gehen darf natürlich ein Dankeschön an alle Mitglieder der traditionellen Bistro/Alumni-Lounge-Runde, *Katrin, Michelle, Elli, Nico, Jan, Barbara, Rino, Peter,* für die entspannenden und erheiternden Abende nach Laborschluss.

Meinem Mitbewohner *Sandro* danke ich für die tolle WG-Zeit, die sportlichen und weniger sportlichen Aktivitäten sowie zusammen mit *Michelle* die unvergessliche Zeit auf der anderen Seite der Welt. *Ebi* möchte ich danken für die temporäre Aufnahme in den DCC, die entspannenden Jogging-Touren während des Schreibens, sowie die vielen intensiven Gespräche über Chemie, Sport, Politik und andere Aspekte des Lebens. Weiterer Dank gehört den ehemaligen Mitstudenten und Angehörigen der freitäglichen Mittagstischrunde, *Gisela, Wulli, Johannes, Paolo, Martin, Roger* und *Matthias* für die fröhlichen Stunden, die wir neben der harten Zeit während Studium und Doktorat zusammen geniessen durften.

Meinen Eltern *Armin* und *Helen* danke ich für ihre Hilfe in allen Belangen während des Studiums wie auch des Doktorats und und meiner Schwester *Simone* für das Vorspuren des Weges zum naturwissenschaftlichen Forscher. Ohne die Unterstützung meiner Familie wäre diese Dissertation so nicht möglich gewesen.

Die VDM Verlagsservicegesellschaft sucht für wissenschaftliche Verlage abgeschlossene und herausragende

Dissertationen, Habilitationen, Diplomarbeiten, Master Theses, Magisterarbeiten usw.

für die kostenlose Publikation als Fachbuch.

Sie verfügen über eine Arbeit, die hohen inhaltlichen und formalen Ansprüchen genügt, und haben Interesse an einer honorarvergüteten Publikation?

Dann senden Sie bitte erste Informationen über sich und Ihre Arbeit per Email an *info@vdm-vsg.de*.

Sie erhalten kurzfristig unser Feedback!

VDM Verlagsservicegesellschaft mbH
Dudweiler Landstr. 99   Telefon +49 681 3720 174
D - 66123 Saarbrücken   Fax   +49 681 3720 1749
www.vdm-vsg.de

Die VDM Verlagsservicegesellschaft mbH vertritt

Printed by Books on Demand GmbH, Norderstedt / Germany